Analytical Methods for Food Additives

Analytical Methods for Food Additives

Contributors
_____

Christian W. Huck et al.

AURIS
Reference

www.aurisreference.com

Analytical Methods for Food Additives

Contributors: Christian W. Huck et al.

Published by Auris Reference Limited

www.aurisreference.com

United Kingdom

Analytical Methods for Food Additives

ISBN: 978-1-78154-879-0

British Library Cataloguing in Publication Data
A CIP record for this book is available from the British Library

Printed in the United Kingdom

Exclusively distributed by CBS Publishers & Distributors Pvt. Ltd.

Sales & Distribution Rights only for India, Pakistan, Bangladesh, Sri Lanka, Nepal and Bhutan.This book is not to be sold outside these territories.

Contents

List of Abbreviations

ADI	Acceptable Daily Intake
AP	Active packaging
AGE	advanced glycation end products
ATR-IRMS	Attenuated Total Reflectance Infrared Microspectroscopy
BBD	Box-Behnken design
CA	Controlled atmosphere
CE	Capillary electrophoresis
CFU	Colony forming units
CFAA	China Food Additives Association
COS	Chitosan oligosaccharides
CID	Collision induced dissociation
CKD	Chronic kidney disease
CL	Cryptococcus laurentii
DADs	Diode array detectors
DMSO	Dimethylsulphoxide
CDC	Disease Control and Prevention
EPA	Environmental Protection Agency
EOF	Electroosmotic flow
EO	Essential oils
EC	European Community
EFSA	European Food Safety Authority
FDA	Food and Drug Administration
GFR	Glomerular filtration rate
GFDA	Ghana Food and Drugs Authority
HEC	Hydroxyethyl cellulose
HHP	High hydrostatic pressure
HPLC	High performance liquid chromatography
HTPs	Hevein-like peptides
HV	Hydroxyl value
IEC	Ion-exchange chromatography
INS	International Numbering System
JECFA	Joint FAO/WHO Expert Committee on Food Additives
KDIGO	Kidney Disease—Improving Global Outcomes
LODs	Limits of detection
LOQ	Limit of quantification
LMMC	Low molecular mass chitosans
MAP	Modified atmosphere packaging
MIC	Minimal inhibitory concentration
MRSA	Methicillin-resistant Staphylococcus aureus
OSFG	Oxygen scavenging fish gelatin
OSS	Oxygen scavenging system

PE	Polyethylene
PEP	Phosphat-Einheiten-Programm
PET	Polyethylene terephthalate
PGPR	Polyglycerol polyricinoleate
PI	Prebiotic index
PCA	Principal Component analysis
PEF	Pulsed electric fields
RPP	Refined potato protein
RSD	Relative standard deviation
SHIME	Simulators of the human intestinal microbial ecosystem
SIMCA	Soft independent modeling of class analogy
SOS	Soybean oligosaccharides
TBHQ	Tertiary butylhydroquinone
USFDA	United States Food and Drug Administration
VLDL	Very low density lipoproteins
TSAYE	Yeast Extract added

List of Contributors

Christian W. Huck
Institute of Analytical Chemistry and Radiochemistry, CCB – Center for Chemistry and Biomedicine, Innsbruck, Austria

G.Z. Qin
S.P. Tian, Key Laboratory of Photosynthesis and Environmental Molecular Physiology, Institute of Botany, Chinese Academy of Sciences, Xiangshan Nanxincun 20, Haidian District, Beijing 100093, China.

S.P. Tian
S.P. Tian, Key Laboratory of Photosynthesis and Environmental Molecular Physiology, Institute of Botany, Chinese Academy of Sciences, Xiangshan Nanxincun 20, Haidian District, Beijing 100093, China.

Y. Xu,
S.P. Tian, Key Laboratory of Photosynthesis and Environmental Molecular Physiology, Institute of Botany, Chinese Academy of Sciences, Xiangshan Nanxincun 20, Haidian District, Beijing 100093, China.

Z.L Chan,
S.P. Tian, Key Laboratory of Photosynthesis and Environmental Molecular Physiology, Institute of Botany, Chinese Academy of Sciences, Xiangshan Nanxincun 20, Haidian District, Beijing 100093, China.

B.Q. Li
S.P. Tian, Key Laboratory of Photosynthesis and Environmental Molecular Physiology, Institute of Botany, Chinese Academy of Sciences, Xiangshan Nanxincun 20, Haidian District, Beijing 100093, China.

Marco Iammarino,
Istituto Zooprofilattico Sperimentale della Puglia e della Basilicata, Foggia, Italy.

Aurelia Di Taranto
Istituto Zooprofilattico Sperimentale della Puglia e della Basilicata, Foggia, Italy.

Marlus Chorilli
[1]Departamento de Fármacos e Medicamentos, Faculdade de Ciências Farmacêuticas de Araraquara, Universidade Estadual Paulista (UNESP), Araraquara, Brazil

Hérida Regina Nunes Salgado
Departamento de Fármacos e Medicamentos, Faculdade de Ciências Farmacêuticas de Araraquara, Universidade Estadual Paulista (UNESP), Araraquara, Brazil

Fabiana De Santana Santos
Faculdade de Americana, Americana, Brazil

Lucélia Magalhães Da Silva
Programa de Pós-Graduação em Ciências Farmacêuticas, Faculdade de Ciências Farmacêuticas de Araraquara, Universidade Estadual Paulista (UNESP), Araraquara, Brazil

Renato Souza Cruz
Technology Department, State University of de Feira de Santana, Feira de Santana, BA, Brazil

Geany Peruch Camilloto
Food Tecnhology Department, Federal University of Viçosa, Viçosa, MG, Brazil

Ana Clarissa dos Santos Pires
Food Tecnhology Department, Federal University of Viçosa, Viçosa, MG, Brazil

Bürge Aşçı, Şule Dinç Zor
Department of Chemistry, Faculty of Science and Arts, Yildiz Technical University, Davutpasa, 34220 Istanbul, Turkey

Özlem Aksu Dönmez
Department of Chemistry, Faculty of Science and Arts, Yildiz Technical University, Davutpasa, 34220 Istanbul, Turkey

Eberhard Ritz
Nierenzentrum Heidelberg

Kai Hahn[2],
[2]Nephrologische Gemeinschaftspraxis/Dialyse, Dortmund

Markus Ketteler
Nephrologische Klinik, Klinikum Coburg GmbH, Coburg

Martin K. Kuhlmann
Vivantes Klinikum im Friedrichshain, Berlin

Johannes Mann
Klinik für Nieren-, Hochdruck- und Rheumakrankheiten, Städtisches Klinikum München Schwabing

Emad I. Hussein
Department of Biological Sciences, Yarmouk University, Irbid, Jordan

Ghassan J. M. Kanan
[1]Department of Biological Sciences, Yarmouk University, Irbid, Jordan

Khalid M. Al- Batayneh
Department of Biological Sciences, Yarmouk University, Irbid, Jordan

Khalaf Alhussaen
Department of Plant Production and Protection, Faculty of Agriculture, Jerash, University, Jerash, Jordan

Wesam Al Khateeb
Department of Biological Sciences, Yarmouk University, Irbid, Jordan

Janti Qar
[1]Department of Biological Sciences, Yarmouk University, Irbid, Jordan

Jacob H. Jacob
Department of Biological Sciences, Al al-Bayt University, Mafraq, Jordan

Riyadh Muhaidat
Department of Biological Sciences, Yarmouk University, Irbid, Jordan

Mohamed I. Hegazy
Department of Microbiology, Faculty of Agriculture, Zagazig University, Egypt

Courage Kosi Setsoafia Saba
Department of Biotechnology, Faculty of Agriculture, University for Development Studies, Tamale Ghana

Hélène Barreteau
Laboratoire des Enveloppes Bactériennes et Antibiotiques, IBBMC, UMR 8619 CNRS, Bâtiment 430, Université de Paris-Sud, FR-91405 Orsay, France

Cédric Delattre
Vellore Institut of Technology – Deemed University (VIT), Vellore 632014, Tamilnadu, India

Philippe Michaud
Laboratoire de Génie Chimique et Biochimique, Université Blaise Pascal – CUST, 24 avenue des Landais, BP206, FR-63174 Aubière cedex, France

Josefa Bastida-Rodríguez
Department of Chemical Engineering, University of Murcia, Campus de Espinardo, 30100 Murcia, Spain

Laura Espina
Departamento de Produccio´n Animal y Ciencia de los Alimentos, Facultad de Veterinaria, Universidad de Zaragoza, Zaragoza, Spain,

Tilahun K. Gelaw
Departament d'Enginyeria Quı́mica, Universitat Rovira i Virgili, Avinguda Paı̈ssos
Catalans, Tarragona, Spain

Sı́lvia de Lamo-Castellvı
Departament d'Enginyeria Quı́mica, Universitat Rovira i Virgili, Avinguda Paı̈ssos
Catalans, Tarragona, Spain

Rafael Pagaʹn
Departamento de Produccioʹn Animal y Ciencia de los Alimentos, Facultad de Veterinaria, Universidad de Zaragoza, Zaragoza, Spain,

Diego Garcı́a-Gonzalo
Departamento de Produccioʹn Animal y Ciencia de los Alimentos, Facultad de Veterinaria, Universidad de Zaragoza, Zaragoza, Spain,

Tana Hintz
Department of Food Science, Rutgers, The State University of New Jersey, New
Brunswick, NJ 08901, USA

Karl K. Matthews
Department of Food Science, Rutgers, The State University of New Jersey, New
Brunswick, NJ 08901, USA

Rong Di
Department of Plant Biology, Rutgers, The State University of New Jersey, New
Brunswick, NJ 08901, USA

S. Roller
Biotechnology Unit, Leatherhead Food Research Association, Leatherhead, Surrey,
UK

Preface

The accurate measurement of additives in food is essential in meeting both regulatory requirements and the need of consumers for accurate information about the products they eat. Analytical methods for food additives is a standard work for the food industry in ensuring the accurate measurement of additives in foods. The text *Analytical Methods for Food Additives* covers a wide range of food additives and reviews current techniques, their respective strengths and weaknesses. A novel analytical tools for quality control in food science have been presented in first chapter. The aim of second chapter is to evaluate beneficial effect of two food additives, ammonium molybdate (NH_4-Mo) and sodium bicarbonate (NaBi), on antagonistic yeasts for control of brown rot caused by *Monilinia fructicola* in sweet cherry fruit under various storage conditions. In third chapter, the entire procedure of extraction, purification, chromatographic separations and quali/quantitative determination of seven food additives (sorbic acid, benzoic acid, lactic acid, acetic acid, nitrites, nitrates and phosphates) has been developed and applied for the analysis of different types of cheese (mozzarella, cheese spread, semi-hard and hard cheeses). Fourth chapter focuses on the biotechnological development of new food preservatives. Oxygen scavengers, an approach on food preservation has been proposed in fifth chapter. In sixth chapter, a new RP-HPLC method has been developed, using experimental design, for simultaneous determination of five synthetic food additives in soft drinks, including three synthetic colorants (carmoisine, allura red, and ponceau 4R), two preservatives (potassium sorbate and sodium benzoate), and caffeine. Seventh chapter discusses phosphate additives in food. The aim of eighth chapter is to evaluate the antifungal activity of some plant extracts, cinnamon bark and sticky fleabane leaves, food preservatives and their mixtures, i.e., plant extracts and food preservatives against *P. digitatum* and *P. italicum*. The objective of ninth chapter is to survey labels of processed and packaged foods in Ghana and document all the potentially harmful food additives in processed and packaged foods in some Ghanaian food products. Tenth chapter focuses on the knowledge in the area of food bioactive oligosaccharides and catalogues the processes employed to generate them. Structure, applications, and production methods of food additive polyglycerol polyricinoleate (E-476) have been focused in eleventh chapter. Last chapter explores the bactericidal effect of (+)-limonene, the major constituent of citrus fruits' essential oils, against *E. coli*.

Chapter 1

NOVEL ANALYTICAL TOOLS FOR QUALITY CONTROL IN FOOD SCIENCE

Christian W. Huck

Institute of Analytical Chemistry and Radiochemistry, CCB – Center for Chemistry and Biomedicine, Innsbruck, Austria

INTRODUCTION

Due to the fast technological and data treatment advancements new insights into food can be considered. The application of these novel analytical techniques belongs to the responsibility of food chemists and analysts. Thereby, an increase in efficiency is based on an improved lower limit of detection (LOD), selectivity to separate analytes of interest and speed of analysis.

High-performance liquid chromatography (HPLC) belongs to the traditional separation techniques applied to a broad range of hydrophilic and hydrophobic ingredients in both the reversed-phase (RP) [1] as well as normal-phase (NP) [2] mode. In a conventional HPLC system the inner diameter of the separation column, which is the core of the separation unit, is 4.6 mm. During the last decade miniaturization down to 20 µm allowed to increase on one side the sensitivity and on the other side speed of analysis could be enhanced dramatically. Therefore, novel stationary phases mainly based on polymers have been designed and brought to the market to enable both the separation of low and high-molecular weight analytes [3]. As an alternative separation technique capillary electrophoresis (CE), which separates analytes due to their different ion mobility based on charge and molecular weight in an electric field within a fused silica capillary having an inner diameter of approximately 200 µm can be applied [4]. Thereby, the appearance of the electroosmotic flow (EOF) can influence the separation efficiency by either speeding up the separation process or by improving the resolution. Capillary electrochromatography (CEC) is a hybrid technique of both HPLC and CE in which both pressure and an electrical field are applied and enables extreme high resolution. The drawback of this separation method is the fact, that real samples can hardly be analysed due to the disturbance by the matrix [5]. In many cases the analyte of interest is only available in very low concentrations. Therefore, selective enrichment and purification steps are the method of choice, which can be accomplished by solid-phase extraction (SPE). Therefore, a material designed for a special analytical question is filled into a cartridge or pipette tip and the sample of interest is put onto the material in liquid form [6]. In the following, analytes of interest can interact with the functional groups of the stationary phase and compounds being not of interest can simply be washed away. In the final elution step, only some micro liters of liquid are required to elute the analytes of interest from the stationary phase being available in relatively high concentrations for the following analytical steps. The following analytical procedure can be either a separation or spectroscopic method. Spectroscopic methods at this stage of the analytical procedure either include mass spectrometry (MS) and/or vibrational spectroscopy, respectively. In MS most of the samples are analysed applying electrospray ionization (ESI) as an interface with different types of mass detectors including e.g. time of flight (TOF), ion trap, ion cyclotron and quadrupoles. As an alternative, matrix assisted laser desorption ionization time of flight mass spectrometry (MALDI-TOF/MS) can be applied for the determination of high molecular weight compounds including proteins, peptides and lipids. For the analysis of low molecular ingredients < 1000 Da the so called matrix-free laser desorption ionization (mf-LDI) MS technique must be applied [7]. Vibrational spectroscopy in the field of food analysis is mainly applied in the mid ($400 - 4000$ cm^{-1}) as well as in the near infrared ($4000 - 12000$ cm^{-1}) of the electromagnetic spectrum. In combination with chemo metrical algorithms these methods can be used for the authentication of the material on one hand, on the other hand quantitative analysis allows to control selected quality parameters [8].

In the following a systematic analytical approach is introduced, which allows combining the different analytical techniques in a synergistic manner to get deeper insights into the composition and origin of food samples.

SYSTEMATIC ANALYTICAL APPROACH

The key technologies described in the above chapter can be combined according to the scheme depicted in Figure 1. In this approach extraction of the material for the further analytical steps and individual procedures can be linked to sample enrichment/purification, separation, vibrational spectroscopy and mass spectroscopy followed by database analysis. The different parts are described in the following sub-chapters.

Sample Enrichment/Purification

In many cases interesting analytes are only available in extremely low concentrations and/or in very complex matrices, respectively. Therefore, pre-concentration steps based on solid-phase extraction (SPE) can be very helpful. Nano-materials such as nanotubes, fullerenes, diamond offer excellent physiochemical properties due to a high ratio of surface to size, which results in a high capacity and allows analyte detection with high sensitivity down to the femtomole range in the case when mass spectrometry is applied for detection. Especially carbon nano materials can be easily further derivatised with a number of different functional groups including reversed-phase (RP), normal-phase (NP), ion exchange (IEX), immobilized affinity (IMAC) and so on depending on the specific demand. As an alternative they can be incorporated into a polymer matrix for highly selective extraction by certain compound characteristics. For the practical handling pipette tips have been tested to be most suitable and this special type of SPE is called "hollow monolithic incorporated tip" as it has an open flow channel in the middle enabling an easy pipetting procedure. For the highly efficient pre-concentration of phosphopeptides nano particular TiO_2, ZrO_2 and mixtures thereof are incorporated into a polymer matrix as depicted in Figure 2 [9]. By this technique hundreds of microliters can be flushed over the system and finally elution of the desired compounds to be analysed is carried out with only a few microliters causing a dramatic increase in concentration from which further analytical investigations can benefit due to the easier handling of the systematic investigation.

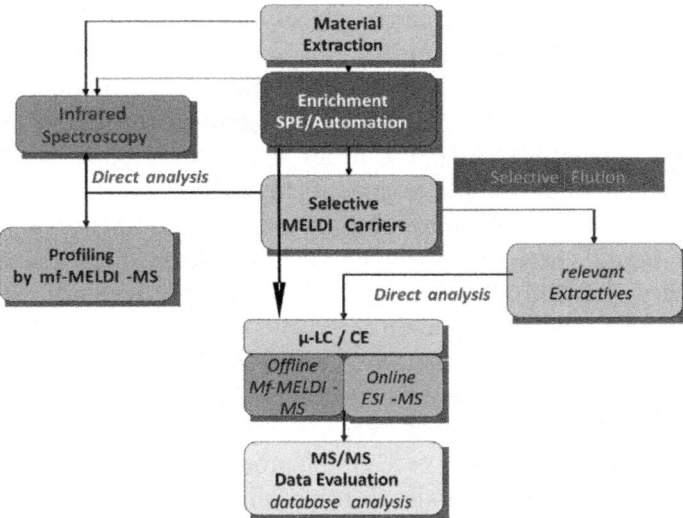

Figure 1: Multidimensional analytical approach

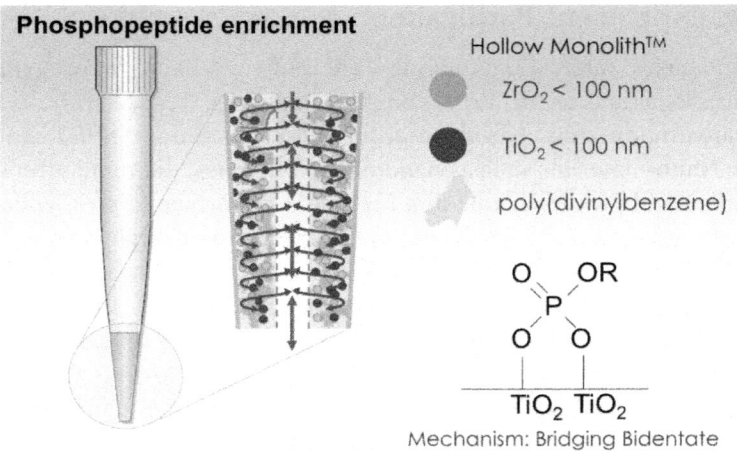

Figure 2: TiO_2, ZrO_2 incorporated into a polymer matrix for phosphopeptide enrichment

Immobilisation of such polymers into pipette tips with trypsin can be used for fast digestion of peptides and proteins within only a few minutes ensuring high capacity and sequence coverage (Figure 3) even in the high-throughput mode using robotic pipetting systems [10]. In comparison to this quite young approach the conventional digestion procedure lasts approximately 24 hours and doesn´t show in any case better results by higher sequence coverages. For this reason this approach is of high interest for the routine analysis and/or diagnostics, respectively. As a carrier glycidylmethacrylate-co-divinylbenzene (GMA/DVB) polymerized in pipette tips was chosen. The major advantages of in-tip digestion are easy handling and small sample amount required for analysis. Microwave-assisted digestion was applied for highly efficient and time saving proteolysis. Adaption to an automated robotic system allowed fast and reproducible sample treatment. Investigations with matrix-assisted laser desorption/ionization time-of-flight mass spectrometry (MALDI-TOF/MS) and liquid chromatography coupled to electrospray-ionization mass spectrometry (LC-ESI/MS) attested high sequence coverages (SCs) for the three standard proteins, myoglobin (Myo, 89%), bovine serum albumin (BSA, 78%) and alpha-casein (α-Cas, 83%). Compared to commercially available trypsin tips clear predominance concerning the digestion performance was achieved. Storability was tested over a period of several weeks and results showed only less decrease (<5%) of protein sequence coverages. The application of microwave-assisted in-tip digestion (2 minutes) with full automation by a robotic system allows high-throughput analysis (96 samples within 80 minutes) and highly effective proteolysis.

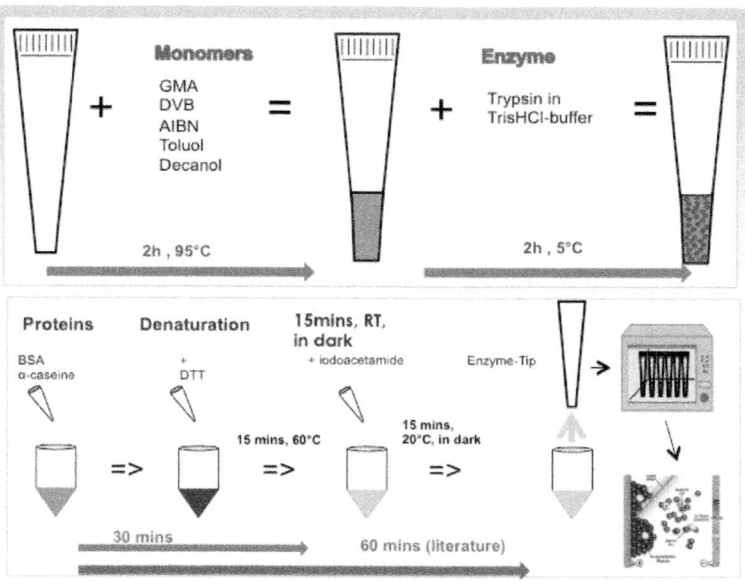

Figure 3: Trypsin immobilized pipette tips for high-throughput analysis of peptides

Selective Meldi-Carriers

Material enhanced laser desorption ionisation (MELDI) is a method, which is based on the conventional matrix assisted laser desorption ionisation time of flight mass spectrometric (MALDI-TOF/MS) detection with the significant difference that before LDI MS step a selective enrichment procedure is carried out for the distinct analysis of a certain compound class. Compared to other similar techniques in this field, this approach benefits from the physical properties of the material itself (pore size, surface area, capacity, etc.) and its chemical derivatisation/functionalisation. In the past this technique was proven to be highly efficient for the analysis of biomarkers following an optimised strategy (Figure 4). In the first step a selected material including e.g., nanotubes, fullerenes, nano-crystalline diamond, polymers, cellulose, etc., which are derivatised with functional groups (C18, IMAC (immobilised metal affinity chromatography), IEX and others) is activated and the serum sample of interest is incubated. During this step, selective binding of molecules according to their functional group is achieved and finally undesired components can be washed away applying an optimised protocol. In the next step the incubated material is put onto a conventional steel target used in MALDI-TOF/MS, a matrix substance is added (e.g., sinapinic acid) and finally the mass spectrum is generated by the laser desorption ionisation process. The result is a mass spectrum being characteristic for a patient and/or the nutrition profile. Multivariate analysis (MVA) can be applied for further data analysis and interpretation, a clustering into certain stages of an illness can be achieved, respectively. From the mass spectrum potent biomarker molecules can be selected and identified by further analytical steps. The biomarker itself and/or the profile of the corresponding mass spectrum can be used for the screening of certain diseases, stages therefrom, allergies, nutrition effects and so on [7].

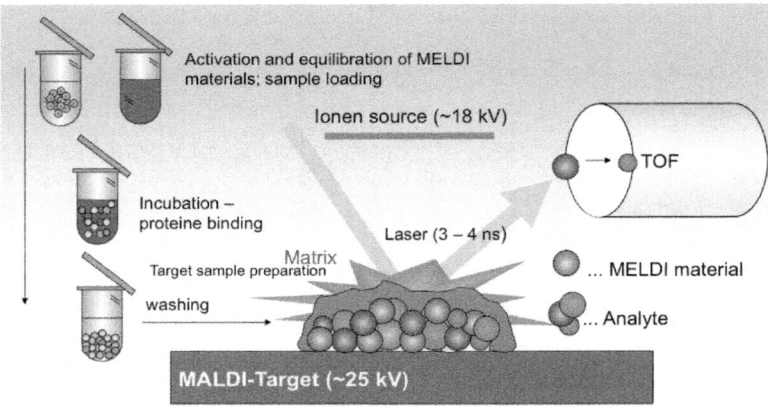

Figure 4: Principle of material enhanced laser desorption ionization (MELDI)

For the analysis of low-molecular weight compounds (MW < 1000 Da) the conventional MELDI approach is replaced by the matrix-free (mf) MELDI approach for which the addition of a matrix substance is not required so that no disturbing peaks appear. In this approach a conventional steel target with a 50 nm thick titanium oxide layer can be applied fulfilling all requirements for a successful laser desorption ionization process [11].

As an alternative the incubated analytes of interest can be selectively eluted from the functinalised carrier material and further analysed by liquid chromatography (LC) or capillary electrophoresis (CE).

Liquid Chromatography, Capillary Electrophoresis and Electro-chromatography

Novel materials used in miniaturised liquid chromatography (μ-LC) are mainly polymer based, e.g. poly(1,2-bis(p-vinylphenyl)ethane). These polymers possess the huge advantage that chemical (composition of the polymer) and physical parameters including mainly porosity can be adjusted [12]. Extensive investigations on polymerisation time and temperature have been carried out enabling a tailored design of micro-, meso- and macro-pore distribution [13,

14]. This results in the applicability of such capillaries with an inner diameter between 20 and 200 μm for even the separation of high- and low-molecular weight compounds. These capillaries can be highly successfully applied analysing peptides, proteins, oligonucleotides, DNA fragments as well as "small molecules" such as phenols, flavonoids, catechins, acids etc. Figure 5 shows as an example the separation of olive oil ingredients. This separation is characterised by a very high ratio of flow to back pressure, which is of high interest to perform extremely rapid Coupling to mass spectrometry enables a highly efficient analysis even of crude samplesoffering all the possibilities of collision induced dissociation (CID) and database search [15].

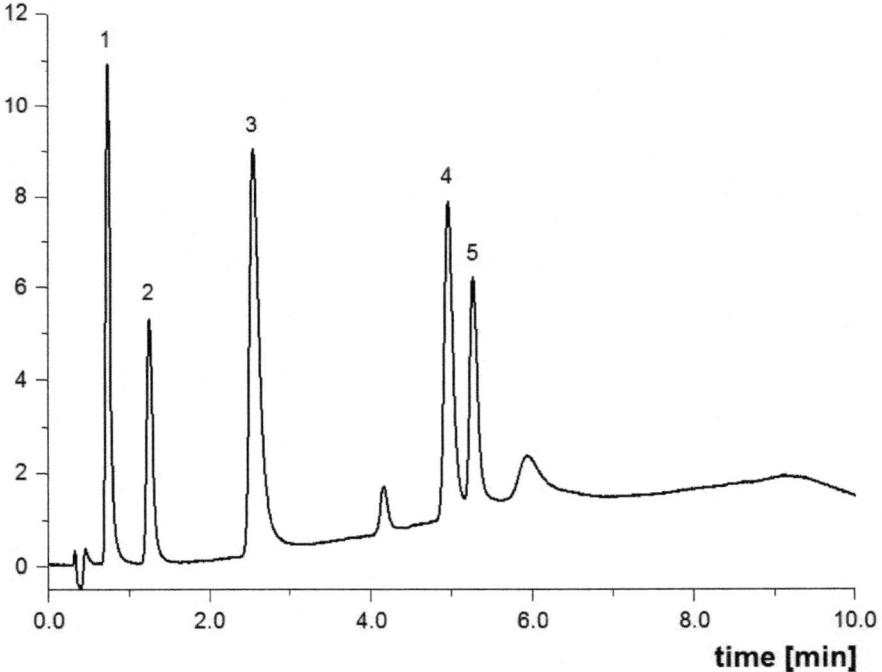

Column, PA/PDA (Kap. 155, polymerised for 10 min), 80 x 0.2 mm; chromatographic conditions: mobile phase, A: 0.1 % TFA, B: ACN, 0.1 % TFA; gradient, 5-45 % B in 10 min; flow rate, 8 μl/min; temperature, RT; detection, UV 210 nm;
Peak identification: (1) Hydroxytyrosol, (2) Tyrosol, (3) Caffeic acid, (4) Vanillin and (5) Oleuropein.

Figure 5: Separation of olive oil ingredients using a monolithic capillary column. Conditions: capillary 80 x 0.2 mm; mobile phase, A: 0.1% TFA; B: CAN; gradient, 5-45%B in 10 min; Flow rate 8 μl/min; temperature, RT; detection, UV 210 nm. Peak assignment, (1) hydroxytyrosol, (2) tyrosol, (3) caffeic acid, (4) vanillin, (5) oleuropein.

As an alternative separation method capillary electrophoresis (CE) and / or electrochromatography (CEC) can be applied. In CE separation of analytes is achieved due to their different ion mobility based on charge and molecular weight in an electric field within a fused silica capillary having an inner diameter of approximately 200 μm [4]. As has already been remarked the electroosmotic flow (EOF) has a main influence on the separation and can be used for speeding up. In CEC both an electrical field and high pressure are applied resulting in high resolution. This technique can be applied to check the identification and purity of standards compounds with very high efficiency. For the reproducible separation and analysis of food ingredients such as phenols, acids, peptides, lipids, coating of the capillary's inner wall was shown being advantageous as irreversible analyte adsorption by free hydroxyl-groups from the silanole of the fused silica capillary can be avoided. Latex-diol and fullerene coated capillaries were successfully introduced and as a detection system on-line hyphenation to MALDI-TOF/MS was shown to be highly efficient not only for the investigation of flavonoids but also for peptides, especially phosphorylated (Figure 6) [16, 17]. This system can be used for the investigation of the casein profile in milk offering the advantage over all other more classical analysis tools that in this case also higher phosphorylated species can be separated and detected. From the ratio of different phosphorylation degrees several interpretations concerning the quality but also the origin of the milk can be carried out.

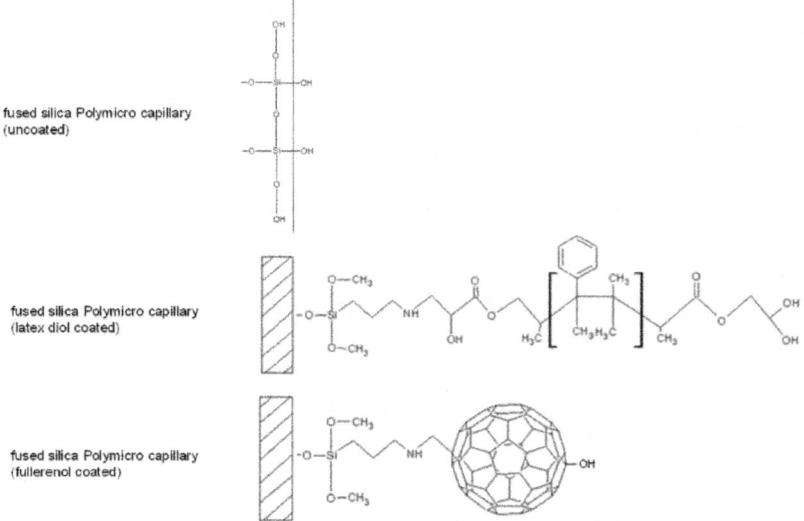

Figure 6: Inner capillary wall coatings applied in CE and CEC

Vibrational Spectroscopy

For quality control both mid- (MIR, $400 - 4000$ cm^{-1}) and near-infrared (NIR, $4000 - 12000$ cm^{-1}) can be conducted. In MIR fundamental stretching and bending vibrations occur, in NIR the corresponding overtones and combination vibrations are detected. This means that NIR-spectra can contain a lot of more vibrational information, which is an advantage for the analysis of highly complex samples. Therefore, during the last decade several applications in the field of food analysis were developed in the NIR region. Samples can be analysed either in transmission, reflectance and interactance mode (Figure 7) so that liquid as well as solid samples can be investigated. Due to the quite broad bands compared to MIR, chemometrical spectra treatment is required for establishing adequate calibration models and to analyse data. These are mainly multivariate (MVA) methods allowing to correct baseline, atmospheric noise etc. For qualitative analysis in most cased principal component analysis (PCA), for quantitative partial least square regression (PLSR) are applied [18].

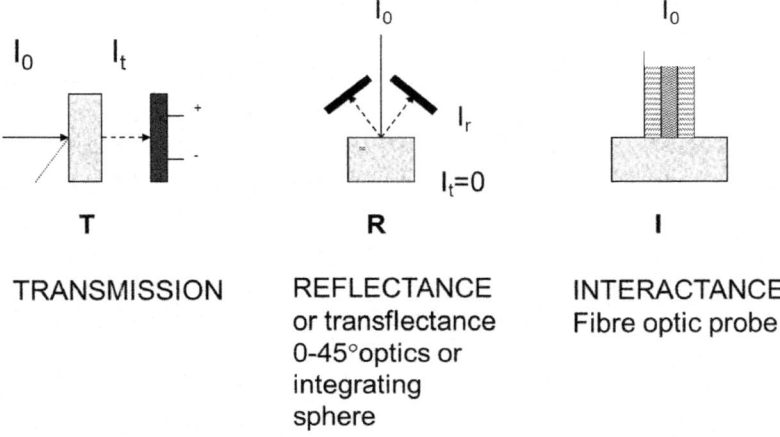

| TRANSMISSION | REFLECTANCE or transflectance 0-45°optics or integrating sphere | INTERACTANCE Fibre optic probe |

Figure 7: Sample measurement modes in NIR

An impressive example for the successful implementation in the food related production is the quality control of wine. It has been shown that NIR can be used to identify grapes, vines, age by qualitative (Figure 8) and its ingredients (acids, carbohydrates, pH etc.) simultaneously, non-invasively within a few seconds by quantitative analysis [19]. Another big advantage of this method can be found by the fact that the sample is not destroyed and can therefore be used for further purposes including following analytical steps.

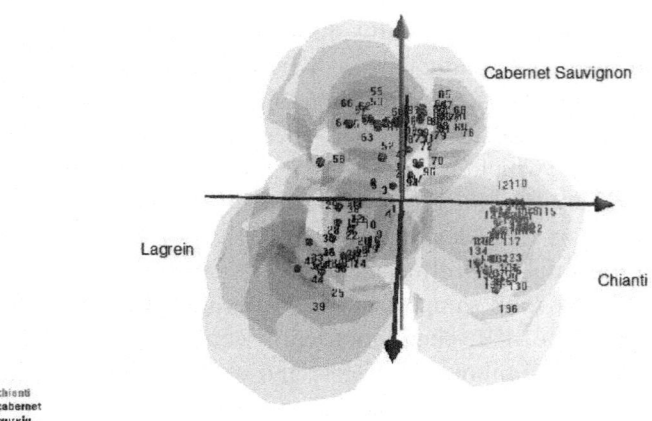

Figure 8: Factor plot of 141 specta of different wines (Lagrein, Chianti, Cabernet Sauvignon). Conditions: Normalisation, 1. derivative; wavenumber range, 4500 - 10000 cm^{-1}; thickness 3 mm; scans, 10; temperature, 23°C.

Quantitative NIRS methods, which allow determining the carbohydrate, total acid, tartaric acid, malic acid, pH in grape variety and the polyphenol content in grapes were established [19]. The method can control the quality already at a very early stage during the wine production and allows improvement of its quality by this. Grapes of 12 different vines (*Weißburgunder, Chardonnay, Ruländer, Silvaner, Müller Thurgau, Gewürztraminer, Sauvignon, Lagrein, Grossvernatsch, Blauburgunder, Cabernet, Merlot*) were harvested in autumn 2000 and squeezed. The obtained grape variety was thermo stated at 23°C and analyzed quantitatively by NIRS in the transflection mode using an optical thin layer thickness of 1 mm. In order to establish a calibration model 252 spectra of samples with lower and upper concentration as a reference were recorded. 76 % of all spectra were randomly used for calibration, 24% for validation. Data preparation was carried out in order to minimize technical influences, which mainly cause a drift in baseline. Quantitative analysis was carried out by partial least square regression (PLSR).

Carbohydrates. Data preparation comprised normalization between 0 and 1 and following calculation of the first derivative using a wavenumber range from 4500 - 7548 cm^{-1}. The PRESS function showed that 3 factors were needed for the calculation of the model. Calculation with 3 factors resulted in a good conformity between SEE and SEP. Linear regression between true and predicted values resulted in a value for the correlation coefficient of R^2=0.99 for calibration and R^2=0.99 for validation. Results for SEE and SEP: 0.13°

KMW and 0.11 °KMW, the BIAS value is 2.30×10^{-15}. *Total acids.* PLSR in a concentration range between 5 and 11 g/l included normalization between 0 and 1, full multiplicative scatter correction (MSC) and calculation of the 1st derivative (Taylor 3 points) between 4500 and 7548 cm⁻¹. 3 factors were necessary to obtain a minimum for PRESS and an agreement between SEE (0.60 g/l) and SEP (0.61 g/l) of nearly 100%.The highly linear model allows determining the total acid content with a prediction error of 0.61 g/l.

Tartaric acid. After normalization, performing of the 1st derivative over between 4500 and 7308 cm⁻¹, four factors were used for creation of the highly linear model depicted in Figure 6c with $R^2 = 0.91$ for calibration and $R^2 = 0.87$ for validation). Despite the small concentration range between 3.1 and 6.7 g/l used for calibration this system allows to determine the tartaric acid content in grape variety with an absolute error of estimation of 0.40 g/l and prediction of 0.54 g/l.

Malic acid. Malic acid often shows 2-5 times higher values compared to tartaric acid. Calibration between 2.9 and 7.0 g/l after normalization between 0 and 1 and calculation of a second smoothened derivative was carried out using three factors, SEE and SEP showing acceptable agreement.Absolute values for SEE, and BIAS were 0.43 g/l and $-4.25 \times 10^{-15.}$ Straight line for calibration showed a linearity of $R^2 = 0.89$ and allowed a prediction of the malic acid content with an absolute error of 0.55 g/l.

pH. Normalization and calculation of the smoothed 2nd derivative between 4500 and 7308 cm⁻¹ showed an optimum for BIAS at five factors. Despite the narrow calibration range of pH 3.09 - 3.74 the calibration equation shows a R^2 of 0.82.

In order to enable the determination of these parameters with only one single measurement, simultaneous analysis of the carbohydrate, total acid, tartaric acid, malic acid content and pH was achieved by performing normalization (between 0 and 1) and calculating its 2nd derivative (Taylor 3 points). Four factors over a wavenumber range from 4500 to 7308 cm⁻¹ showed 73-100% agreement between SEE and SEP. Linear regression showed high linearity for each investigated parameter with slightly lower values for R^2. Compared to the above-described single analysis this method allows a quantitative analysis of all parameters at once within a few seconds. Values for SEP are slightly increased (Table 1).

Polyphenols mainly influence taste, sensory properties and color of a wine. Therefore, a rapid method to analyze its quantity is important. The method according to Folin - Ciocalteu was used as a reference method (see Materials and Methods). Gallic acid-1-hydrate was used as reference standard in a concentration range from 0 to 4.93 µg/ml with equidistant steps. 24 gallic acid-1-hydrate solutions in a concentration range between 0.442 and 7.08 mg/ml were measured in the transmission mode threefold and in random order by NIRS. Evaluation using PLSR was achieved by dividing 72-recorded spectra randomly into a calibration (54 spectra) and validation (18 spectra) set. Data pretreatment comprised normalization between 0 and 1 and calculation of the 1st derivative (Savitzky-Golay) between 4008-7512 cm^{-1}. Using three factors, the PRESS function showed a minimum and a good agreement between SEE (0.45 mg/ml) and SEP (0.46 mg/ml). Linear regression between predicted and true values allowed to predict the gallic acid-1-hydrate concentration between 0 and 7 mg/ml with R^2=0.98.

In order to determine the total polyphenol concentration 30 must samples were measured in the transmission mode threefold and in random order. 90 spectra were divided into 72 calibration and 18 validation spectra. Normalization and performing of the 1st derivative allowed minimizing shifts in the baseline. 4 factors were necessary to obtain a minimum for the PRESS function and to get a maximum agreement of SEE and SEP. Linear regression allowed correlating true and predicted values with a R^{2}of 0.97. Compared to the traditionally used Folin - Ciocalteu method in a winery, which is very time-consuming and expensive due to the usage of different chemicals, the NIRS method is very simple, precise and incomparably fast.

Table 1: Prediction results for the determination of the carbohydrate, total acid, tartaric acid, malic acid content and pH

Note. a Single analysis; b Simultaneous analysis

Parameter	Unit	SEE		SEP		BIAS	
		a	b	a	b	a	b
Carbohydrates	KMW	0.13	0.21	0.11	0.19	2.30 × 10^{-15}	3.33 × 10^{-16}
Total acids	g/l	0.60	0.43	0.61	0.53	7.17 × 10^{-15}	-1.08 × 10^{-14}
Tartaric acid	g/l	0.40	0.41	0.54	0.55	-1.08 × 10^{-14}	-3.43 × 10^{-15}
Malic acid	g/l	0.43	0.49	0.55	0.65	-4.25 × 10^{-15}	-2.44 × 10^{-15}
pH		0.07	0.09	0.06	0.09	-1.26 × 10^{-15}	-7.15 × 10^{-15}

Quality control of coffee ingredients including caffeine, theobromine and theophylline [20] and of food additives deriving from the highly interesting field of Traditional Chinese Medicine (TCM) [21] can be carried out in a similar way. Thereby, emphasis must be put onto the calibration method for which the above mentioned techniques can be applied as a reference. A new analytical method based on near infrared spectroscopy (NIRS) for the quantitation of the three main alkaloids caffeine (Caf), theobromine (Tbr) and theophylline (Tph) in roasted coffee after discrimination of the rough green beans into Arabic and Robusta was established. This validated method was compared to the most commonly used liquid chromatography (LC) connected to UV and mass spectrometric (MS) detection. As analysis time plays an important role in choosing a reference method for the calibration of the NIR-spectrometer, the non-porous silica-C18 phase offers a very fast method. Coupling of the optimised LC method to a mass spectrometer (MS) via an electrospray ionisation (ESI) interface not only allowed to identify Caf, Tbr and Tph by their characteristic fragmentation pattern using collisionally induced dissociation (CID), but also to quantitate the content of the three analytes, which was found to be 6% higher compared to UV-detection. The validated LC–UV method was chosen as a reference method for the calibration of the NIRS system. Analysis of 83 liquid coffee extracts in random order resulted for Caf and Tbr in values for S.E.E. (standard error of estimation) of 0.34, 0.40 g/100 g, S.E.P. (standard error of prediction) of 0.07 and 0.10 g/100 g with correlation coefficients of 0.86 and 0.85 in a concentration range between 0.10 and 4.13 g/100 g. Compared to LC the lower limit of detection (LOD) of the NIRS-method is found at 0.05 g/100 g compared to 0.244–0.60 ng/100 g in LC, which makes it impossible to analyse Tph by NIRS.

The possibility to hyphenate a MIR/NIR spectrometer to a microscope unit allows determining the distribution of active ingredients within a tissue sample down to a resolution of 1.2 μm [22]. A "hyperspectral cube" is recorded with the dimensions of the sample on the x- and y-axis and the absorbance on the z-axis from which the image can be extracted (Figure 9).

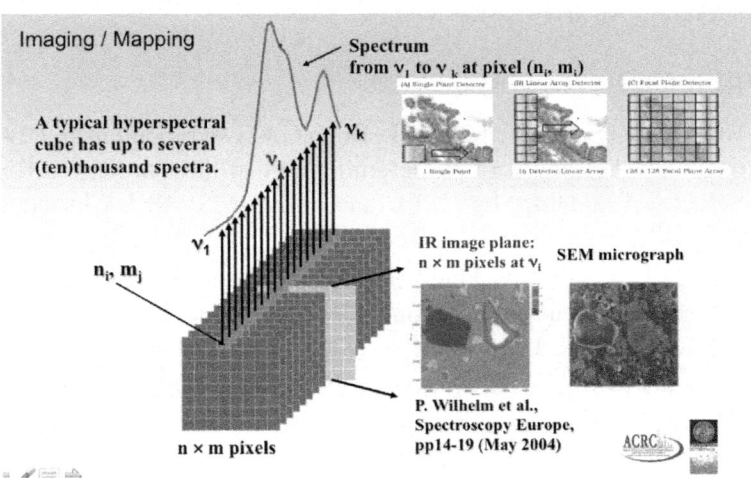

Figure 9: Principle of hyperspectral cube

Fourier Transform Infrared (FTIR) spectroscopic imaging and mapping techniques have become essential tools for the detection and characterization of the molecular components of biological tissues and the modern analytical techniques enabling molecular imaging of complex samples. These techniques are based on the absorption of IR radiations by vibrational transitions in covalent bonds and their major advantage is the acquisition of local molecular expression profiles, while maintaining the topographic integrity of the tissue by avoiding time-consuming extraction, purification and separation steps. These new techniques enable global analysis of biological samples with high spatial resolution and provide unique chemical-morphological information about the tissue status. With these non-destructive examination methods it is possible to get qualitative and quantitative information of heterogeneous samples.

Additionally, MALDI-TOF/MS imaging can be applied from the same sample of interest to get knowledge concerning the molecular weight distribution. This method is also suitable for studying the effect of nutrition onto different kinds of diseases, e.g. prostate cancer.

CONCLUSIONS

The techniques described can be applied according to the scheme depicted in Figure 1. This systematic analytical strategy allows getting multifacial knowledge and insights into food and samples derived therefrom.

ACKNOWLEDGEMENTS

This work was financially supported by the Efre-project "Originalp" (EU) and Agrarmarketing Austria (AMA, Vienna, Austria)

REFERENCES

1.   Huck C.W., Popp M., Scherz H., Bonn G.K.Development and Evaluation of a New Method for the Determination of the Carotenoid Content in Selected Vegetables by HPLC and HPLC-MS/MS.J. Chromatogr. Sci. 2000 38 441

2.   W. M. Stöggl, C. W. Huck, H. Scherz, G. K. Bonn, of. Analysis, E. Vitamin, Food. in, Preparations. Phytopharmaceutical, H. P. L. C. by, H. P. L. C. -A, P. C. I. -M, S. M. S. , Chromatographia 2001 54 179

3.   JakschitzT., HuckC.W., LubbadS., Bonn G.K.Monolithic Poly(TMSM-DMSBMS) Stationary Phases for the fast Separation of Oligonucleotides and Proteins.J. Chromatogr. A 2007 1147 53

4.   Huck C.W., Bakry R., Bonn G.K.Progress in capillary electrophoresis

of biomarkers and metabolites between 2002 and 2005.Electrophoresis 2006 27 111

5. Stöggl W.M., Huck C.W., Stecher G., Bonn G.K., Capillary Electrochromatography of Biologically Relevant Flavonoids. Electrophoresis 2006 27 787

6. Huck C.W., Bonn G.K.Review: Polymer based sorbents for solid-phase extraction.J. Chromatogr. A 2000 885 51

7. Feuerstein I., Najam-ul-Haq M., Rainer M., Trojer L., Bakry R., Hidayat Aprilita N., Stecher G., Huck C.W., Klocker H., Bartsch G., Guttman A., BonnG.K.Material Enhanced Laser Desorption/Ionization (MELDI)- a new protein profiling tool utilizing specific carrier materials for TOF-MS Analysis.J. Am. Soc. Mass Spectrom. 2006 17 1203

8. Petter C.H., Heigl N., Bachmann S., Huck-Pezzei V.A.C., Najam-ul-Haq M., Bakry R., Bernkop-Schnürch A., Bonn G.K., Huck*C.W.Near Infrared Spectroscopy Compared to Liquid Chromatography Coupled to Mass Spectrometry and Capillary Electrophoresis as a Detection Tool for Peptide Reaction Monitoring.Amino Acids 2008

9. M. Rainer, H. Sonderegger, R. Bakry, C. W. Huck, S. Morandell, L. A. Huber, D. T. Gjerde, G. K. Bonn, Analysis of protein phosphorylation by monolithic extraction columns based on poly(divinylbenzene) containing embedded titanium dioxide and zirconium dioxide nano-powders. Proteomics 2008 8 21 4593 4602

10. H. Hahn, M. Rainer, T. Ringer, D. Gjerde, C. W. Huck, G. K. Bonn, Ultra-fast Microwave-Assisted In-Tip Digestion of Proteins. J. Proteom. Res. 2009 8 9 4225 4230

11. G. K. Bonn, R. Bakry, C. W. Huck, R. Vallant, Z. Szabo, Analysis of low molecular weight molecules by maldi-ms. Eur. Pat. Appl. (2008pp. CODEN: EPXXDW EP 1973142 A1 20080924 AN 2008:1151584

12. Greiderer A., Clark Ligon S.Jr., Huck C.W., Bonn G.K.Monolithic poly(1 2 -bis(p-vinylphenyl)ethane capillary columns for simultaneous separation of low- and high-molecular weight compounds.J.Sep. Sci. 2009

13. Greiderer A., Trojer L., Huck C.W., Bonn G.K.Influence of the Polymerisation time on the Porous and Chromatographic Properties of Monolithic Poly(1,2-bis(p-vinylphenyl)ethane Capillary Columns.J.Chromatogr. A 2009 1216 45 7747 7754

14. N. Heigl, A. Greiderer, C. H. Petter, H. W. Siesler, G. K. Bonn, C. W. Huck, Simultaneous Determination of Physical and Chemical

Parameters of Monolithic Porous Polymers with a Combined Use of Fourier-Transform Near Infrared Diffuse Reflection Spectroscopy and Multivariate Techniques. Anal. Chem. 2008 80 22 8493 8500

15.   L. Trojer, A. Greiderer, C. P. Bisjak, W. Wieder, N. Heigl, C. W. Huck, G. K. Bonn, Handbook. in, H. P. L. C. of, Edition. Second, by. edited, D. Corradini, Press, 2010 3 45Print 978-1-57444-554-1eBook ISBN: 978-1-4200-1694-9

16.   Bachmann S., Vallant R., Bakry R., Huck C.W., Corradini D., Bonn G.K.Capillary electrophoresis coupled to matrix assisted laser desorption ionization with novel covalently coated capillaries.Electrophoresis, 2010 31 618 629

17.   Stöggl W.M., Huck C.W., Stecher G., Bonn G.K.Capillary Electrochromatography of Biologically Relevant Flavonoids. Electrophoresis, 2006 27 787

18.   Petter C.H., Heigl N., Bachmann S., Huck-Pezzei V.A.C., Najam-ul-Haq M., Bakry R., Bernkop-Schnürch A., Bonn G.K., Huck C.W.Near Infrared Spectroscopy Compared to Liquid Chromatography Coupled to Mass Spectrometry and Capillary Electrophoresis as a Detection Tool for Peptide Reaction Monitoring.Amino Acids, 2008 34 4 605 616

19.   W. Guggenbichler, C. W. Huck, A. Kobler, G. K. Bonn, Infrared. Near, Spectroscopy. Reflectance, A. Tool, Quality. for, In. Control, Production. Wine, J. Food Agric. Environm., 2006 4 2 98 106

20.   Huck C.W., Guggenbichler W., Bonn G.K., Analysis of Caffeine, Theobromine and Theophylline in Coffee by Near Infrared Reflectance Spectroscopy (NIRS) Compared to High Performance Liquid Chromatography (HPLC) Coupled to Mass Spectrometry.Anal. Chim. Acta, 2005 538 195

21.   C. Mattle, N. Heigl, G. K. Bonn, C. W. Huck, Near Infrared Diffuse Reflection Spectroscopy and Multivariate Calibration Hyphenated to Thin Layer Chromatography for Quality Control and Simultaneous Quantification of Methoxylated Flavones in a Phytomedicine. J. Planar Chromatogr., 2010 23 5 348 352

22.   C. Pezzei, J. D. Pallua, G. Schaefer, C. Seifarth, V. Huck-Pezzei, L. K. Bittner, H. Klocker, G. Bartsch, G. K. Bonn, C. W. Huck, cancer. Prostate, by. Characterization, Transform. Fourier, microspectroscopy. Infrared, Mol. Biosys., 2010 6 2287

Chapter 2

COMBINATION OF ANTAGONISTIC YEASTS WITH TWO FOOD ADDITIVES FOR CONTROL OF BROWN ROT CAUSED BY MONILINIA FRUCTICOLA ON SWEET CHERRY FRUIT

G.Z. Qin, S.P. Tian, Y. Xu, Z.L. Chan, B.Q. Li

S.P. Tian, Key Laboratory of Photosynthesis and Environmental Molecular Physiology, Institute of Botany, Chinese Academy of Sciences, Xiangshan Nanxincun 20, Haidian District, Beijing 100093, China.

ABSTRACT

Aims: To evaluate beneficial effect of two food additives, ammonium molybdate (NH_4-Mo) and sodium bicarbonate (NaBi), on antagonistic yeasts for control of brown rot caused by *Monilinia fructicola* in sweet cherry fruit under various storage conditions. The mechanisms of action by which food additives enhance the efficacy of antagonistic yeasts were also evaluated.

Methods and Results: Biocontrol activity of *Pichia membranefaciens* and *Cryptococcus laurentii* against brown rot in sweet cherry fruit was improved by addition of 5 mmol l^{-1} NH_4-Mo or 2% NaBi when stored in air at 20 and 0°C, and in controlled atmosphere (CA) storage with 10% O_2 + 10% CO_2 at 0°C. Population dynamics of *P. membranefaciens* in the wounds of fruit were inhibited by NH_4-Mo at 20°C after 1 day of incubation and growth of *C. laurentii* was inhibited by NH_4-Mo at 0°C in CA storage after 60 days. In contrast, NaBi did not significantly influence growth of the two yeasts in fruit wounds under various storage conditions except that the growth of *P. membranefaciens* was stimulated after storage for 45 days at 0°C in CA storage. When used alone, the two additives showed effective control of brown rot in sweet cherry fruit and the efficacy was closely correlated with the concentrations used. The result of *in vitro* indicated that growth of *M. fructicola* was significantly inhibited by NH_4-Mo and NaBi.

Conclusion: Application of additives improved biocontrol of brown rot on sweet cherry fruit under various storage conditions. It is postulated that the enhancement of disease control is directly because of the inhibitory effects of additives on pathogen growth, and indirectly because of the relatively little influence of additives on the growth of antagonistic yeasts.

Significance and Impact of the Study: The results obtained in this study suggest that an integration of NH_4-Mo or NaBi with biocontrol agents has great potential in commercial management of postharvest diseases of fruit.

INTRODUCTION

Biological control using microbial antagonists has shown potential as an alternative for natural control of plant pathogens instead of synthetic chemical fungicides (McLaughlin *et al.* 1992; Droby *et al.* 1998; Fan and Tian 2000; Qin *et al.* 2004). Many yeasts and bacteria have been reported to effectively inhibit postharvest decay (Fan and Tian 2000; Tian *et al.* 2002b). By comparison with antagonistic bacteria, yeasts have been pursued actively in recent years, as production of antibiotics or other toxic secondary metabolites was not involved in their activities against postharvest pathogens. The mode of action of antagonistic yeasts may be competition for space and nutrients (Droby *et al.* 1989; Piano *et al.* 1997; Janisiewicz *et al.* 2000), production of cell-wall lytic enzymes (Wisniewski *et al.* 1991; El-Ghaouth *et al.* 1998), and induction of host resistance (Arras 1996; Droby *et al.* 2002). However, application of antagonistic micro-organisms alone did not provide commercially acceptable control of postharvest diseases. In large-scale tests, the use of biological control often needed to be combined with low doses of synthetic fungicides to obtain a level of decay control equivalent to synthetic fungicides (Droby *et al.* 2003). In order to completely eliminate the use of synthetic fungicides, more environmentally friendly and harmless compound(s) should be explored to improve the activity of the antagonist. Selected chemicals such as calcium chloride (Wisniewski *et al.* 1995; Tian *et al.* 2002a), chitosan (El-Ghaouth *et al.* 2000a), 2-deoxy-D-glucose (Janisiewicz 1994; El-Ghaouth *et al.* 2000b) and salicylic acid (Qin *et al.* 2003) in combination with biological control agents have been demonstrated to give beneficial effects on control of fruit decay.

Ammonium molybdate and sodium bicarbonate have been reported to play an important role in the inhibition of postharvest decay when used alone (Nunes *et al.* 2001; Gamagae *et al.* 2003; Karabulut *et al.* 2003,2004). Previous research has indicated that ammonium molybdate and sodium bicarbonate could enhance the efficacy of biological control (Nunes *et al.* 2002a,b; Porat *et al.* 2002; Droby *et al.* 2003; Hamid *et al.* 2003; Obagwu and Korsten 2003; Gamagae *et al.* 2004; Yao *et al.* 2004;Zhang *et al.* 2004). Nunes *et al.* (2002a)) reported that the performance of ammonium molybdate combined with the yeast *Candida sake* exhibited better control of blue and grey mould caused by *Penicillium expansum* and *Botrytis cinerea*, respectively, on pears than that of individual treatments. Droby *et al.* (2003) observed that biocontrol activity by *Candida oleophila* against *P. expansum* and *B. cinerea* in apples and *Monilinia fructicola* and *Rhizopus stolonifer* in peaches was enhanced by the addition of sodium bicarbonate. Although ammonium molybdate and sodium bicarbonate were demonstrated to improve the activity of biological control of fruit decay, little information is given about integrating biocontrol agents with the two materials under controlled atmosphere (CA) condition. CA storage in reduced O_2 and elevated CO_2 environments is a common commercial practice, which extends the storage life of many commodities by slowing fruit respiration rates and maintaining important quality attributes (Volz *et al.* 1998). It is likely that the successful commercial use of biological control will depend on the compatibility of antagonists with postharvest practices such as low-temperature and CA storage (Tian *et al.* 2002b).

The main objective of this study was to investigate the performance of ammonium molybdate and sodium bicarbonate, which are widely used in the food industry, in combination with *Pichia membranefaciens* Hansen or *Cryptococcus laurentii* (Kufferath) Skinner for control of brown rot caused by *M. fructicola* (Wint.) Honey on sweet cherry (*Prunus avivum* L. cv. Hongdeng) fruit under various storage conditions. In addition, the mechanisms by which ammonium molybdate and sodium bicarbonate enhanced the efficacy of biocontrol agents were evaluated.

MATERIALS AND METHODS

Plant material

Sweet cherry fruit were harvested at commercial maturity and sorted based on size and the absence of physical injuries or infections. The firmness of fruit was 13·7 N as determined by a penetrometer (FT-327, UC Fruit Firmness Tester, Milan, Italy), and soluble solids content was 18·7%. Prior to use, fruit were disinfected with 2% (v/v) sodium hypochlorite for 2 min, washed with tap water and dried in air.

Yeast antagonist and pathogen

Pichia membranefaciens and *C. laurentii* were isolated from the surfaces of apple fruits with the method of **Wilson and Chalutz (1989)** and identified by CABI Bioscience Identification Services (International Mycological Institute,

UK). The yeasts were cultured in nutrient yeast dextrose broth (NYDB: 8 g of nutrient broth, 5 g of yeast extract and 10 g of dextrose in 1000 ml water) for 48 h at 25°C with a shaker (HZQ-C, Dong Ming Co., China) at 200 rev min^{-1}. Yeast cells were collected by centrifugation at 6000g for 10 min. The concentration of the yeast was determined with a hemacytometer and adjusted to 1 × 10^7 cells ml^{-1} with sterile distilled water.

Monilinia fructicola was isolated from infected sweet cherry fruit and maintained on potato dextrose agar (PDA) for 14 days at 25°C. The spores were removed from the surface of the cultures, suspended in 5 ml of sterile distilled water containing 0·05% (v/v) Tween 80, and filtered through four layers of sterile cheesecloth in order to remove adhering mycelia. Spore concentrations of *M. fructicola* were determined with a hemacytometer and adjusted to the required concentrations with sterile distilled water.

Chemical products

The ammonium molybdate [(NH$_4$)$_6$Mo$_7$O$_{24}$·4H$_2$O] (hereafter NH$_4$-Mo) and sodium bicarbonate (NaHCO$_3$) (hereafter NaBi) used in this study was purchased from Sigma Chemical Co. (St. Louis, MO, USA).

Enhancement of biocontrol activity with NH$_4$-Mo or NaBi at various storage conditions

In a preliminary study, we found that NH$_4$-Mo at 5 mmol l^{-1} and NaBi at 2% caused no surface injury to the fruit and showed the best effect in improving the biocontrol activity of antagonists (data not shown). In this experiment, sweet cherry fruit were wounded (3 mm deep × 3 mm wide) with a sterile nail at the equator of each fruit. Each wound was added with 20 μl of the treatment suspensions as follows: *P. membranefaciens* (1 × 10^7 cells ml^{-1}),*P. membranefaciens* (1 × 10^7 cells ml^{-1}) + NH$_4$-Mo (5 mmol l^{-1}), *P. membranefaciens*(1 × 10^7 cells ml^{-1}) + NaBi (2%), *C. laurentii* (1 × 10^7 cells ml^{-1}), *C. laurentii* (1 × 10^7 cells ml^{-1}) + NH$_4$-Mo (5 mmol l^{-1}), *C. laurentii* (1 × 10^7 cells ml^{-1}) + NaBi (2%), NH$_4$-Mo (5 mmol l^{-1}) and NaBi (2%). Fruit treated with sterile distilled water served as control. When the wounds were air-dried, sweet cherry fruit were challenge-inoculated with 15 μl of a conidial suspension of *M. fructicola* at 1 × 10^4 spores ml^{-1}. Treated fruit were put in 200 × 130 × 50 mm plastic boxes with sterile water to maintain high humidity (RH about 95%) and stored in air at 20 and 0°C. Fruit stored in CA storage were placed in the plastic boxes as described above and transferred to a CA cabinet (Fruit s.r.l. Control, type FC-701, Milan, Italy) with 10% O$_2$ + 10% CO$_2$ at 0°C (RH about 95%). Disease incidences of fruit stored at 20°C were measured after 3 and 4 days. Fruit stored at 0°C in air and in CA storage were observed after 40 and 60 days, and then fruit were moved to 20°C in air for 3 days to allow symptom development. Each treatment contained three replicates with 13 fruits per replicate and the entire experiment was repeated.

Influence of NH_4-Mo or NaBi on growth of biocontrol agents in wounds of sweet cherry fruit

Fruit were wounded and 20 μl of a cell suspension of *P. membranefaciens* or *C. laurentii* at 1×10^7 cells ml^{-1}, alone or containing 5 mmol l^{-1} NH_4-Mo or 2% NaBi, was pipetted into each wound and stored in air at 20 and 0°C, or in CA storage with 10% O_2 + 10% CO_2 at 0°C. Fruit samples were taken as described by **Janisiewicz et al. (1992)** at different times after treatment by removing the wound tissue with a cork borer. The resulting cylinders of excised tissue (10 mm deep × 10 mm wide) from five fruit were placed in a mortar with 10 ml of sterile distilled water and ground with a pestle. Then, 100 μl of the serial tenfold dilutions were plated on nutrient yeast dextrose agar (NYDA) medium. Colonies were counted after incubated at 20°C for 72 h and expressed as log$_{10}$ colony forming units (CFU) wound^{-1}. Each treatment contained three replicates and the entire experiment was repeated.

Effect of NH_4-Mo or NaBi with various concentrations for control of fruit decay

Sweet cherry fruit were wounded then, 20 μl of NH_4-Mo at 0, 1, 5, 10 and 15 mmol l^{-1} or NaBi at 0·5, 1, 2 and 4% was added to each wound. When the wounds were air dried, fruit were inoculated with 15 μl of a conidial suspension of *M. fructicola* at 1×10^4 spores ml^{-1}. The treated fruit were put in 200 mm × 130 mm × 50 mm plastic boxes with sterile water to maintain high humidity (about 95%) and stored at 20°C. Disease incidences were measured after 3 days. Each treatment contained three replicates of 13 fruits each and the entire experiment was repeated.

Inhibition of growth of *M. fructicola* by NH_4-Mo or NaBi *in vitro*

The inhibition of spore germination and germ tube elongation of *M. fructicola* by NH_4-Mo or NaBi was assayed in potato dextrose broth (PDB) using the method of **Tian et al. (2002a)**. Spore suspensions were prepared as described. Aliquots of 100 μl of pathogen suspensions were transferred to a glass tube (180 mm × 16 mm) containing 5 ml of PDB to obtain a final concentration of 5 × 10^5 spores ml^{-1}. The PDB contained various concentration of NH_4-Mo (0, 1, 5, 10 and 15 mmol l^{-1}) or NaBi (0·5%, 1%, 2% and 4%). All tubes were put on a rotary shaker at 100 rev min^{-1} at 25°C and incubated for 15 h. Approximately

200 spores of *M. fructicola* were measured for germination and germ tube length per treatment within each replicate. Each treatment was replicated three times and the experiment was repeated.

Statistical analysis

Data for yeast populations (CFU wound^{-1}) were transformed to logarithms to improve the homogeneity of variances. The incidence of decay, population dynamics and growth of pathogen in PDB were analysed by analysis of variance with SPSS (SPSS Inc., Chicago, IL, USA). When the treatment effects were statistically significantly ($P < 0.05$), the least significant difference (LSD) test was used for means separation. Data analysis indicated that treatment effects were similar among the experimental repeats. Results presented were pooled across repeated experiments.

RESULTS

Enhancement of biocontrol activity by NH$_4$-Mo or NaBi at 20°C

A single application of biocontrol agent *P. membranefaciens* or *C. laurentii* was effective at controlling brown rot caused by *M. fructicola*, but better control was achieved when NH$_4$-Mo or NaBi was added to the suspensions of the antagonists (Fig. 1).

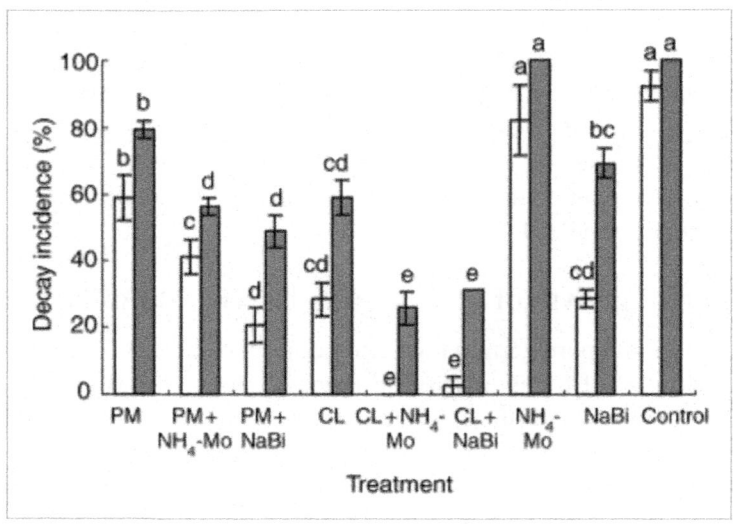

Figure 1: The effect of ammonium molybdate (NH$_4$-Mo) or sodium bicarbonate (NaBi) on*Pichia membranefaciens* (PM) or *Cryptococcus laurentii* (CL) against brown rot

caused by *Monilinia fructicola* on sweet cherry fruit at 20°C. Disease incidences were measured after 3 (☐) and 4 (■) days. Bars represent standard errors of the treatment means of pooled data. Values followed by the same letter are not statistically different by the least significant difference (LSD) test ($P < 0.05$).

After incubation for 3 days at 20°C, fruit treated with *P. membranefaciens* in combination with NH_4-Mo or NaBi showed 18% and 38·5% less decay of brown rot, respectively, than *P. membranefaciens* treatment alone. Decay in fruit treated with *C. laurentii* + NH_4-Mo was completely controlled. In contrast, the addition of NaBi to the suspension of *C. laurentii* reduced the decay incidence by 25·6% compared with *C. laurentii* treatment alone. A similar effect was observed as storage time increased.

Enhancement of bicontrol activity by NH_4-Mo or NaBi at 0°C in air and in CA storage

NH_4-Mo or NaBi enhanced biocontrol activities of both *P. membranefaciens* and *C. laurentii* against brown rot of sweet cherry fruit after 40 and 60 days of storage at 0°C in air (Fig. 2a). Decay developed rapidly as fruit were transferred to 20°C for 3 days. CA storage was effective in controlling brown rot of sweet cherry fruit (Fig. 2b). Decay was visible only after 60 days of storage under CA storage, whereas fruit stored at 0°C in air had reached an advanced stage of decay at that time. The addition of NH_4-Mo or NaBi to the suspension of *P. membranefaciens* or *C. laurentii* resulted in a significant improvement of the biocontrol activities of both antagonists against brown rot after 60 days under CA storage and 3 days of shelf life at 20°C (Fig. 2b) ($P < 0.05$).

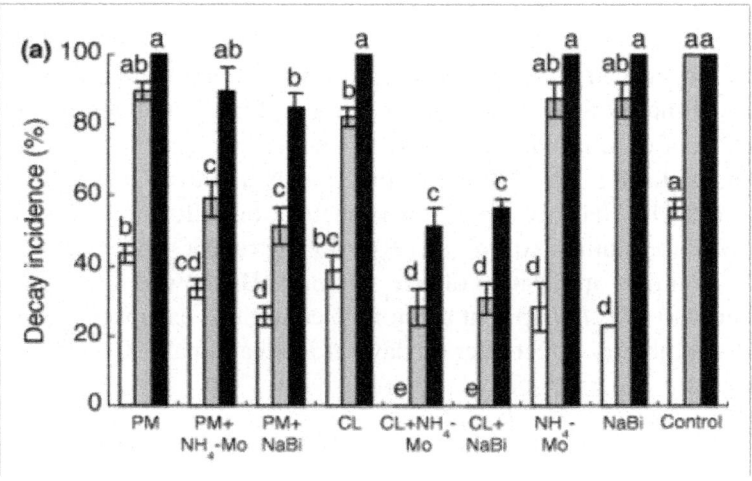

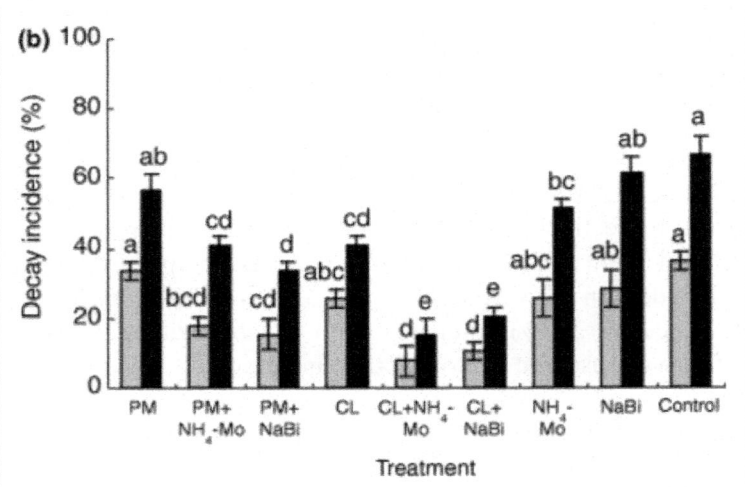

Figure 2: The effect of ammonium molybdate (NH_4-Mo) or sodium bicarbonate (NaBi) on *Pichia membranefaciens* (PM) or *Cryptococcus laurentii* (CL) against brown rot caused by *Monilinia fructicola* on sweet cherry fruit at 0°C in (a) air and in (b) control atmosphere storage with 10% O_2 + 10% CO_2. Disease incidences were measured after 40 (□) and 60 (▧) days at 0°C, following by 3 (■) days of shelf life at 20°C in air. Bars represent standard errors of the treatment means of pooled data. Values followed by the same letter are not statistically different by the least significant difference (LSD) test ($P < 0.05$).

Growth of antagonistic yeasts in wounds of sweet cherry fruit

Pichia membranefaciens and *C. laurentii* multiplied quickly in the wounds of sweet cherry fruit at 20°C (Table 1). Low temperature in air and in CA storage suppressed the growth of *P.membranefaciens*, but did not significantly affect the growth of *C. laurentii*, regardless of whether the yeasts were used alone or combined with NH_4-Mo or NaBi. Population dynamics of *P. membranefaciens* was inhibited by NH_4-Mo after 1 day of incubation at 20°C ($P < 0.05$), whereas the growth was stimulated by NaBi after 45 days at 0°C under CA storage ($P < 0.05$). In contrast, application of NH_4-Mo or NaBi showed no statistical effect on the growth of *C.laurentii* at all storage conditions except that NH_4-Mo had a slightly inhibitory effect after 60 days of incubation at 0°C in CA storage.

Table 1: Population dynamics of *Pichia membranefaciens* (PM) and *Cryptococcus laurentii*(CL), alone or combined with either ammonium molybdate (NH_4-Mo) or sodium bicarbonate (NaBi) in the wounds of sweet cherry fruit stored in air at 20 and 0°C, or in control atmosphere (CA) storage with 10% O_2 + 10% CO_2 at 0°C

Storage conditions and period (days)	Population dynamics [$\log_{10}$ colony forming units (CFU) wound^{-1}]					
	PM	PM + NH_4-Mo	PM + NaBi	CL	CL + NH_4-Mo	CL + NaBi
1.	*Fruit were wounded and a cell suspension of *P. membranefaciens* or *C. laurentii*, alone or containing 5 mmol l^{-1} NH_4-Mo or 2% NaBi, was pipetted into each wound and stored in air at 20 and 0°C, or in CA storage with 10% O_2 + 10% CO_2 at 0°C. Fruit samples were taken at different times after treatment.					
2.	Colonies were expressed as $\log_{10}$ colony forming units (CFU) wound^{-1}.					
3.	Data are treatment means of pooled data ± standard errors.					
20°C						
0	4·50 ± 0·06*	4·50 ± 0·06	4·50 ± 0·06	4·66 ± 0·03	4·66 ± 0·03	4·66 ± 0·03
1	6·06 ± 0·03	6·07 ± 0·03	6·01 ± 0·03	6·28 ± 0·03	6·15 ± 0·03	6·20 ± 0·06
2	6·34 ± 0·03	6·05 ± 0·03	6·38 ± 0·03	6·53 ± 0·03	6·48 ± 0·06	6·76 ± 0·03
3	6·41 ± 0·10	6·20 ± 0·13	6·45 ± 0·20	6·68 ± 0·07	6·44 ± 0·15	6·82 ± 0·06
4	6·44 ± 0·07	6·17 ± 0·09	6·45 ± 0·06	6·59 ± 0·06	6·53 ± 0·07	6·61 ± 0·06
0°C						
15	4·24 ± 0·06	4·20 ± 0·07	4·24 ± 0·06	5·78 ± 0·07	5·52 ± 0·03	5·78 ± 0·09
30	4·15 ± 0·09	4·19 ± 0·07	4·24 ± 0·17	5·59 ± 0·06	5·47 ± 0·07	5·89 ± 0·03
45	3·82 ± 0·12	3·73 ± 0·07	3·87 ± 0·12	6·06 ± 0·07	6·00 ± 0·1	6·27 ± 0·03
60	3·73 ± 0·10	3·73 ± 0·07	3·78 ± 0·09	6·44 ± 0·03	6·40 ± 0·06	6·52 ± 0·03
CA storage						
15	4·59 ± 0·09	4·62 ± 0·1	4·71 ± 0·06	5·28 ± 0·03	5·12 ± 0·10	4·93 ± 0·12
30	4·29 ± 0·03	4·34 ± 0·09	4·46 ± 0·07	5·62 ± 0·12	5·64 ± 0·03	5·67 ± 0·03
45	4·00 ± 0·06	4·10 ± 0·12	4·49 ± 0·06	6·31 ± 0·06	6·16 ± 0·03	6·06 ± 0·03
60	3·94 ± 0·09	3·90 ± 0·24	4·43 ± 0·07	6·33 ± 0·03	5·97 ± 0·09	6·16 ± 0·03

Efficacy of various concentrations of NH_4-Mo or NaBi

The results demonstrated that the effects of NH_4-Mo or NaBi against brown rot cause by *M. fructicola* in sweet cherry fruit was positively correlated with the concentrations of the two materials. As concentrations of NH_4-Mo and NaBi increased, decay incidence of brown rot decreased. NH_4-Mo significantly ($P < 0.05$) reduced decay incidence of brown rot only at the concentration of 15 mmol l^{-1} (Fig. 3a), whereas NaBi showed marked effects of decay control at all concentrations (Fig. 3b).

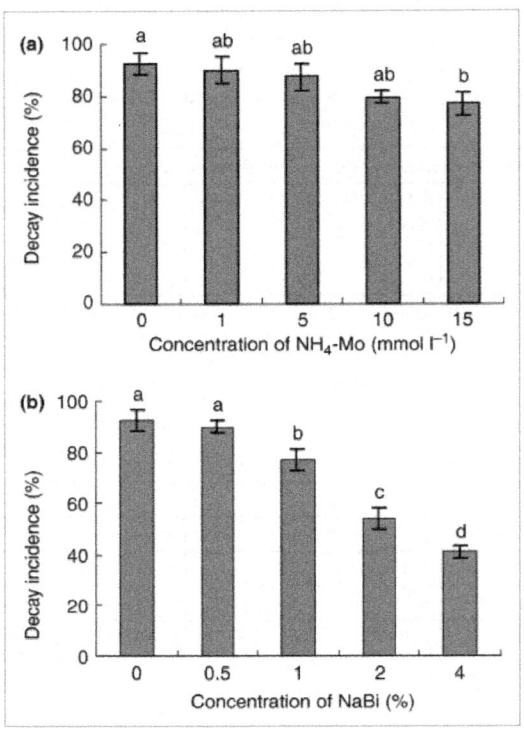

Figure 3: Activity of (a) ammonium molybdate (NH$_4$-Mo) and (b) sodium bicarbonate (NaBi) with various concentrations against brown rot caused by *Monilinia fructicola* on sweet cherry fruits at 20°C. Disease incidences were measured after 3 days. Bars represent standard errors of the treatment means of pooled data. Values followed by the same letter are not statistically different by the least significant difference (LSD) test ($P < 0.05$).

Inhibitory effect of NH$_4$-Mo or NaBi on pathogen growth *in vitro*

The inhibitory effect of NH$_4$-Mo or NaBi with various concentrations on spore germination and germ tube elongation of *M. fructicola in vitro* is shown in Table 2. The results indicated that the inhibition was positively correlated with the concentration of the two materials. Spore germination of *M. fructicola* was significantly ($P < 0.05$) inhibited by NH$_4$-Mo when the concentration reached 5 mmol l^{-1}. In contrast, the inhibitory effect of NaBi was significant at all concentrations ($P < 0.05$). NaBi at 2% (w/v) completely inhibited the growth of *M. fructicola*. Spore germination was less sensitive to NH$_4$-Mo or NaBi than germ tube elongation.

Table 2: Effect of ammonium molybdate (NH_4-Mo) and sodium bicarbonate (NaBi) on spore germination and germ tube elongation of *Monilinia frutcicolain vitro*

Treatment	Spore germination (%)	Germ tube length (mm)
1. *Germination and germ tube length were measured microscopically after 15 h of incubation at 25°C in potato dextrose broth.		
2. Data are treatment means of pooled data ± standard errors. Values of each column followed by the same letter are not significantly different according to least significant difference (LSD) test (P < 0·05).		
Water-treated control	100·0 ± 0·0 a*	338·7 ± 7·0 a
NH_4-Mo (1 mmol l^{-1})	94·4 ± 2·7 a	307·5 ± 8·7 b
NH_4-Mo (5 mmol l^{-1})	49·8 ± 1·6 b	117·8 ± 2·9 c
NH_4-Mo (10 mmol l^{-1})	38·8 ± 3·1 c	54·7 ± 1·2 e
NH_4-Mo (15 mmol l^{-1})	5·8 ± 1·1 f	28·2 ± 1·2 f
NaBi (0·5%)	31·2 ± 3·0 d	90·4 ± 2·7 d
NaBi (1%)	15·2 ± 2·5 e	61·8 ± 2·2 e
NaBi (2%)	0·0 ± 0·0 f	0·0 ± 0·0 g
NaBi (4%)	0·0 ± 0·0 f	0·0 ± 0·0 g

DISCUSSION

In recent years, as the use of synthetic fungicide is being curtailed, alternative methods, particularly biological control, have been pursued actively for the control of postharvest decay of fruit (Janisiewicz and Korsten 2002). However, biological control alone is often less effective than many of the commercial fungicides currently in use (Spadaro and Gullino 2004). Increased interest has been shown in combining biocontrol agents with other approaches such as heat treatment (Obagwu and Korsten 2003; Wszelaki and Mitcham 2003), application of chemical components generally regarded as safe (Nunes *et al.* 2002a,b; Droby *et al.* 2003), and CA storage (Usall *et al.* 2000) to increase their activity against postharvest pathogens. In this study, the antagonistic yeasts *P. membranefaciens* or *C. laurentii* were applied along with NH_4-Mo or NaBi to control brown rot caused by *M. fructicola* on sweet cherry fruit. The results indicated that fruit treated with*P. membranefaciens* or *C. laurentii* combined with NH_4-Mo or NaBi showed less decay incidence of brown rot than fruit treated with antagonists or chemicals alone at 20°C. Similar results were obtained by Nunes *et al.* (2002a) testing *C. sake* combined with NH_4-Mo against *P. expansum* and *B. cinerea* in pears and by Droby *et al.* (2003) using

*C. oleophila*combined with NaBi against *P. expansum* and *B. cinerea* in apples and *M. fructicola* and *R. stolonifer* in peaches.

The ultimate potential of biocontrol agent depends on its effectiveness and compatibility with routine postharvest operations such as low temperatures and controlled or modified atmosphere during storage (Usall *et al.* 2000; Qin and Tian 2004). CA storage with reduced O_2 and elevated CO_2 is a useful practice to reduce diseases and extend the shelf life of postharvest fruit (Sams and Conway 1987;Rogiers and Knowles 2000; Tian *et al.* 2002c). We have recently demonstrated that *P. membranefaciens* and *C.laurentii* were effective against *Alternaria alternata* and *P. expansum* in sweet cherry fruit at 0°C in air and under CA condition with 10% O_2 + 10% CO_2 (Qin *et al.* 2004). This work further demonstrated that combination of the two antagonists with NH_4-Mo or NaBi under low temperature in air and in CA storage provided better control of brown rot than the antagonists or the chemicals alone on sweet cherry fruit. To our knowledge, this is the first report in which a biocontrol agent was integrated with NH_4-Mo or NaBi under high CO_2environment to get beneficial effects against postharvest diseases of fruit.

The modes of action through which NH_4-Mo and NaBi resulted in an increase in the level of biological control are uncertain. It has been suggested that yeast antagonists exert biological control activity primarily through competition for nutrients and spaces (**Roberts 1990, Janisiewicz *et al.* 2000**). Our previous work indicated that effective colonization and high population size of *P. membranefaciens* and *C. laurentii* are important factors in the successful control of fruit diseases (**Fan and Tian 2000; Qin *et al.* 2004**). The influence of NH_4-Mo and NaBi on the growth of the two antagonists was investigated in this experiment.*Cryptococcus laurentii* multiplied quickly in the wounds of sweet cherry fruit under various storage conditions, regardless of whether the yeast were used alone or combined with NH_4-Mo or NaBi. Application of the additives did not inhibit the growth of *C. laurentii* except that NH_4-Mo showed a slightly inhibitory effect after 60 days of incubation at 0°C in CA storage. In contrast, NH_4-Mo reduced the population of this yeast at 20°C, whereas NaBi stimulated the growth of *P. membranefaciens* after 45 days of storage under CA condition. However, the reduction in the population did not affect the enhancement of the biocontrol activity. This result was in accordance with previous reports of **Nunes *et al.* (2002b)** and **Wan *et al.* (2003)**, who found that NH_4-Mo improved the ability of biocontrol agents even if the population of antagonists was decreased by the addition of chemical. The enhancement of biocontrol activity may be because of a synergistic effect (**Nunes *et al.* 2002b**). In this study, brown rot caused by *M. fructicola* in sweet cherry fruit was controlled significantly by NH_4-Mo or NaBi when used alone. The two chemicals exhibited direct inhibition of spore germination and germ tube elongation of *M. fructicola* in vitro. These results agree with those of **Droby et al. (2003)**, who reported that NaBi could inhibit the growth of postharvest pathogens *in vitro*.

In summary, NH_4-Mo or NaBi improved the efficacy of the antagonistic yeasts to control brown rot of sweet cherry fruit under various storage conditions. The enhancement of disease control may be directly because of the inhibitory effects of additives on pathogen growth, and indirectly because of the relatively little influence of additives on the growth of antagonistic yeasts. These results suggest that NH_4-Mo or NaBi may be utilized as environmentally friendly additive to enhance the performance of antagonistic yeast against postharvest decays of fruit.

ACKNOWLEDGEMENTS

This work was supported by the National Science Fund for Distinguished Young Scholars of China (30225030) and the National Natural Science Foundation of China (30430480; 30500351).

REFERENCES

1.   Arras, G. (1996) Mode of action of an isolate of *Candida famata* in biological control of *Penicillium digitatum* in orange fruits. *Postharvest Biol Technol* **8**, 191–198.

2.   Droby, S., Chalutz, E., Wilson, C.L. and Wisniewski, M. (1989) Characterization of the biocontrol activity of*Debaryomyces hansenii* in the control of *Penicillium digitatum* of grapefruit. *Can J Microbiol* **35**, 794–800.

3.   Droby, S., Cohen, L., Daus, A., Weiss, B., Horev, B., Chalutz, E., Katz, H., Keren-Tzur, M. *et al.* (1998)Commercial testing of Aspire: a yeast preparation for the biological control of postharvest decay of citrus. *Biol Control* **12**, 97–101.

4.   Droby, S., Vinokur, V., Weiss, B., Cohen, L., Daus, A., Goldschmidt, E.E. and Porat, R. (2002) Induction of resistance to *Penicillium digitatum* in grapefruit by the yeast biocontrol agent *Candida oleophila*. *Phytopathology* **92**, 393–399.

5.   Droby, S., Wisniewski, M., El-Ghaouth, A. and Wilson, C. (2003) Influence of food additives on the control of postharvest rots of apple and peach and efficacy of the yeast-based biocontrol product Aspire. *Postharvest Biol Technol* **27**, 127–135.

6. El-Ghaouth, A., Smilanick, J.L., Brown, G.E., Ippolito, A., Wisniewski, M. and Wilson, C.L. (2000a)Applications of *Candida saitoana* and glycolchitosan for the control of postharvest diseases of apple and citrus fruit under semi-commercial conditions. *Plant Dis* **84**, 243–248.

7. El-Ghaouth, A., Smilanick, J.L., Wisniewski, M. and Wilson, C.L. (2000b) Improved control of apple and citrus fruit decay with a combination of *Candida saitoana* and 2-deoxy-d-glucose. *Plant Dis* **84**, 249–253.

8. El-Ghaouth, A., Wilson, C.L. and Wisniewski, M. (1998) Ultrastructural and cytochemical aspects of the biological control of *Botrytis cinerea* by *Candida saitoana* in apple fruit. *Phytopathology* **88**, 282–291.

9. Fan, Q. and Tian, S.P. (2000) Postharvest biological control of *Rhizopus* rot of nectarine fruits by *Pichia membranefaciens*. *Plant Dis* **84**, 1212–1216.

10. Gamagae, S.U., Sivakumar, D., Wijeratnam, R.S.W. and Wijesundera, R.L.C. (2003) Use of sodium bicarbonate and *Candida oleophila* to control anthracnose in papaya during storage. *Crop Prot* **22**,775–779.

11. Gamagae, S.U., Sivakumar, D. and Wijeratnam, R.L.C. (2004) Evaluation of post-harvest application of sodium bicarbonate-incorporated wax formulation and *Candida oleophila* for the control of anthracnose of papaya. *Crop Prot* **23**, 575–579.

12. Hamid, M., Siddiqui, I. and Shaukat, S. (2003) Improvement of *Pseudomonas fluorescens* CHA0 biocontrol activity against root-knot nematode by the addition of ammonium molybdate. *Lett Appl Microbiol* **36**, 239–244.

13. Janisiewicz, W.J. (1994) Enhancement of biocontrol of blue mold with the nutrient analog 2-deoxy-d-glucose on apples and pears. *Appl Environ Microbiol* **60**, 2671–2676.

14. Janisiewicz, W.J. and Korsten, L. (2002) Biological control of postharvest diseases of fruits. *Annu Rev Phytopathol* **40**, 411–441.

15. Janisiewicz, W.J., Usall, J. and Bors, B. (1992) Nutritional enhancement of biocontrol of blue mold on apples. *Phytopathology* **82**, 1364–1370.

16. Janisiewicz, W.J., Tworkoski, T.J. and Sharer, C. (2000) Characterizing the mechanism of biological control of postharvest diseases on fruits with a simple method to study competition for nutrients.*Phytopathology* **90**, 1196–1200.

17. Karabulut, O.A., Smilanick, J.L., Gabler, F.M., Mansour, M. and Droby, S. (2003) Near-harvest applications of *Metschnikowia fructicola*, ethanol, and sodium bicarbonate to control postharvest diseases of grape in central

California. *Plant Dis* **87**, 1384–1389.

18. Karabulut, O.A., Arslan, U. and Kuruoglu, G. (2004) Control of postharvest diseases of organically grown strawberry with preharvest applications of some food additives and postharvest hot water dips. *J Phytopathol* **152**, 224–228.

19. McLaughlin, R.J., Wilson, C.L., Droby, S., Ben-Arie, R. and Chalutz, E. (1992) Biological control of postharvest diseases of grape, peach, and apple with the yeasts *Kloeckera apiculata* and *Candida guilliermondii*. *Plant Dis* **76**, 470–473.

20. Nunes, C., Usall, J., Teixidó, N., Eribe X.O.D. and Viñas, I. (2001) Control of post-harvest decay of apples by pre-harvest and post-harvest application of ammonium molybdate. *Pest Manage Sci* **57**, 1093–1099.

21. Nunes, C., Usall, J., Teixidó, N., Abadias, M. and Viñas, I. (2002a) Improved control of postharvest decay of pears by the combination of *Candida sake* (CPA-1) and ammonium molybdate. *Phytopatholgy* **92**,281–287.

22. Nunes, C., Usall, J., Teixidó, N. and Viñas, I. (2002b) Improvement of *Candida sake* biocontrol activity against post-harvest decay by the addition of ammonium molybdate. *J Appl Microbiol* **92**, 927–935.

23. Obagwu, J. and Korsten, L. (2003) Integrated control of citrus green and blue molds using *Bacillus subtilis* in combination with sodium bicarbonate or hot water. *Postharvest Biol Technol* **28**, 187–194.

24. Piano, S., Neyrotti, V., Migheli, Q. and Gullino, M.L. (1997) Biocontrol capability of *Metschnikowia pulcherrima* against *Botrytis* postharvest rot of apple. *Postharvest Biol Technol* **11**, 131–140.

25. Porat, R., Daus, A., Weiss, B., Cohen, L. and Droby, S. (2002) Effects of combining hot water, sodium bicarbonate and biocontrol on postharvest decay of citrus fruit. *J Hortic Sci Biotechnol* **77**, 441–445.

26. Qin, G.Z. and Tian, S.P. (2004) Biocontrol of postharvest diseases of jujube fruit by *Cryptococcus laurentii* combined with a low dosage of fungicides under different storage conditions. *Plant Dis* **88**,497–501.

27. Qin, G.Z., Tian, S.P., Xu, Y. and Wan, Y.K. (2003) Enhancement of biocontrol efficacy of antagonistic yeasts by salicylic acid in sweet cherry fruit. *Physiol Mol Plant Pathol* **62**, 147–154.

28. Qin, G.Z., Tian, S.P. and Xu, Y. (2004) Biocontrol of postharvest diseases on sweet cherries by four antagonistic yeasts in different storage conditions. *Postharvest Biol Technol* **31**, 51–58.

29. Roberts, R.G. (1990) Postharvest biological control of gray mold of apple

by *Cryptococcus laurentii.Phytopathology* **80**, 526–530.

30.   Rogiers, S.Y. and Knowles, N.R. (2000) Efficacy of low O_2 and high CO_2 atmospheres in maintaining the postharvest quality of saskatoon fruit (*Amelanchier alnifolia* Nutt.). *Can J Plant Sci* **80**, 623–630.

31.   Sams, C.E. and Conway, W.S. (1987)Additive effects of controlled-atmospheres storage and calcium chloride on decay, firmness retention, and ethylene production in apples. *Plant Dis* **71**, 1003–1005.

32.   Spadaro, D. and Gullino, M.L. (2004) State of the art and future prospects of the biological control of postharvest fruit diseases. *Int J Food Microbiol* **91**, 185–194.

33.   Tian, S.P., Fan, Q., Xu, Y. and Jian, A.L. (2002a) Effects of calcium on biocontrol activity of yeast antagonists against the postharvest fungal pathogen *Rhizopus stolonifer*. *Plant Pathol* **51**, 352–358.

34.   Tian, S.P., Fan, Q., Xu, Y. and Liu, H.B. (2002b) Biocontrol efficacy of antagonist yeasts to gray mold and blue mold on apples and pears in controlled atmospheres. *Plant Dis* **86**, 848–853.

35.   Tian, S.P., Xu, Y., Jiang, A.L. and Gong, Q.Q. (2002c) Physiological and quality responses of longan fruit to high O_2 or high CO_2 atmospheres in storage. *Postharvest Biol Technol* **24**, 335–340.

36.   Usall, J., Teixidó, N., Fons, E. and Viñas, I. (2000) Biological control of blue mould on apple by a strain of*Candida sake* under several controlled atmosphere conditions. *Int J Food Microbiol* **58**, 83–92.

37.   Volz, R.K., Biasi, W.V., Grant, J.A. and Mitcham, E.J. (1998) Prediction of controlled atmosphere-induced flesh browning in 'Fuji' apple. *Postharvest Biol Technol* **13**, 97–107.

38.   Wan, Y.K., Tian, S.P. and Qin, G.Z. (2003) Enhancement of biocontrol activity of yeasts by adding sodium bicarbonate or ammonium molybdate to control postharvest disease of jujube fruits. *Lett Appl Microbiol***37**, 249–253.

39.   Wilson, C.L. and Chalutz, E. (1989) Postharvest biological control of *Penicillium* rots of citrus with antagonistic yeasts and bacteria. *Sci Hortic* **40**, 105–112.

40.   Wisniewski, M., Biles, C., Droby, S., McLaughlin, R., Wilson, C. and Chalutz, E. (1991) Mode of action of the postharvest biocontrol yeast, *Pichia guilliermondii*. I: characterization of attachment to *Botrytis cinerea*. *Physiol Mol Plant Pathol* **39**, 245–258.

41.   Wisniewski, M., Droby, S., Chalutz, E. and Eilam, Y. (1995) Effects of Ca^{2+} and Mg^{2+} on *Botrytis cinerea*and *Penicillium expansum in vitro* and

on the biocontrol activity of *Candida oleophila*. *Plant Pathol* **44**,1016–1024.

42. Wszelaki, A.L. and Mitcham, E.J. (2003) Effect of combinations of hot water dips, biological control and controlled atmospheres for control of gray mold on harvested strawberries. *Postharvest Biol Technol* **27**,255–264.

43. Yao, H.J., Tian, S.P. and Wang, Y.S. (2004) Sodium bicarbonate enhances biocontrol efficacy of yeasts on fungal spoilage of pears. *Int J Food Microbiol* **93**, 297–304.

44. Zhang, H.Y., Fu, C.X., Zheng, X.D., He, D., Shan, L.J. and Zhan, X. (2004) Effect of s*Cryptococcus laurentii*(Kufferath) Skinner in combination with sodium bicarbonate on biocontrol of postharvest green mold decay of citrus fruit. *Bot Bull Acad Sin* **45**, 159–164.

Chapter 3

DEVELOPMENT AND VALIDATION OF AN ION CHROMATOGRAPHY METHOD FOR THE SIMULTANEOUS DETERMINATION OF SEVEN FOOD ADDITIVES IN CHEESES

Marco Iammarino, Aurelia Di Taranto

Istituto Zooprofilattico Sperimentale della Puglia e della Basilicata, Foggia, Italy.

ABSTRACT

Cheeses are characterized by several chemical-physical properties that make it difficult for the microorganism's growth, consequently. The actual European legislation allows the addition of few food additives in this type of food products. In this work, the entire procedure of extraction, purification, chromatographic separations and quali/quantitative determination of seven food additives (sorbic acid, benzoic acid, lactic acid, acetic acid, nitrites, nitrates and phosphates) was developed and applied for the analysis of different types of cheese (mozzarella, cheese spread, semi-hard and hard cheeses). Through validation procedure it was possible to evaluate the most important validation parameters. Extended calibration curves ($r > 0.990$) were obtained for all the analyzed compounds. Recovery values ranged from 72.8% to 98.4% and a good repeatability was obtained, with precision levels in the range of 0.03% - 0.11% ($n = 6$). The potential and feasibility of the method were tested by analysing real samples, such as mozzarella, cheese spread, semi-hard and hard cheeses, confirming that the method is well suited to satisfy the demands for accurate confirmation analyses of seven food additives in cheeses, which is especially valuable in official check analyses and in monitoring schemes.

INTRODUCTION

Dairy products are characterized by several chemicalphysical properties that make it difficult for the microorganism's growth. These characteristics are low water activity, low pH, high concentration of sodium chloride, lacking of fermentescible carbohydrates and the possible presence of batteriocins produced by microbial starters and/or of anaerobic condition [1]. These conditions do not make it necessary to the addition of particular food additives, in particular preservatives, in this type of food products; consequently, the actual legislation allows the addition of few food additives in dairy products (**Table 1**). Among these additives, the most used ones are two preservatives (with some restrictions, see **Table 1**): sorbic acid and nitrates, and three acidity regulators: lactic acid, acetic acid and phosphates. Another two important food preservatives, benzoic acid and nitrites, are not admitted [2], but they are considered in this study due to their toxicity and large use in many foodstuffs. In several foodstuffs the contrast action towards different microorganisms, yeast and moulds is exercised through the addition of a mixture of sorbic acid and benzoic acid [3]. The first one cannot be considered as harmful for humans since it can be metabolized like caproic acid [4,5]. For this reason, some limits of use related to sorbic acid are particularly high (up to 2000 mg·kg^{-1} in processed cheeses). As it concerns benzoic acid, it may be considered harmless for humans, at employment doses, because it is not accumulated, but completely eliminated by urine as hippuric acid [6-8]. Nevertheless, high doses of benzoic acid (until to 1000 mg·day^{-1}) may cause oesophagus burning, nausea and headache [9]; moreover, it causes occasional allergic reactions (hives, dermatitis, rhinitis, etc.) in sensitive persons [10].

Table 1: Food additives analysed and relating legal limits

Food additive	E-number	Legal limit (mg·kg^{-1})	Restrictions
Sorbic acid—sorbates	E200, 202, 203	1000	Only cheese, prepacked, sliced and cut; layered cheese and cheese with added foods
		quantum satis	Only ripened products surface treatment
		2000	Processed cheese
Acetic acid	E260	quantum satis	Only mozzarella and whey cheese
Benzoic acid—benzoates	E210, 211, 212, 213	Not admitted	-
Lactic acid	E270	quantum satis	Only mozzarella and whey cheese
Nitrites	E249, 250	Not admitted	-
Nitrates	E251, 252	150	Only hard, semi-hard and semi-soft cheese
Phosphoric acid—phosphates di-tri and polyphosphates	E338, 339, 340, 341, 450, 451, 452	2000	Only unripened products except mozzarella
		20,000	Processed cheese

For these reasons an acceptable daily intake (ADI) for benzoic acid, equal to 0 - 5 mg·kg bw, was established [11] and this food preservative was deleted from the list of authorized food additives in dairy products, since it was considered unnecessary. It is important to underline that the actual legislation indicates that a quantifiable amount of this substance may be detected in certain fermented products resulting from the fermentation process following good manufacturing practice. Consequently, recent studies have deepened the dairy products characteristic and a maximum admissible limit for benzoic acid in cheeses was proposed [12]. Among the most common food preservatives, nitrites and nitrates are well-known due to their large use in meat curing. Notoriously, these food additives are harmful for humans. They may cause methemoglobinemia, which is the pathology due to reaction between nitrates, haemoglobin and myoglobin resulting in the synthesis of methaemoglobin and metmyoglobin [13-15]. Moreover, the reaction between secondary amines and nitrous acid, deriving from nitrates, brings about the synthesis of N-nitroso compounds that are notorious carcinogenic agents [16]. In the actual European Legislation, only nitrates are admitted in dairy products, limited to hard, semi-hard and semi-soft cheese; the limit of use is equal to 150 mg·kg^{-1}. For lactic acid, acetic acid and phosphates, there are no legal limits established by the actual Normative (lactic acid and phosphates are naturally present in dairy products). This is due to not-toxicity of these compounds at the employment doses. However it is compulsory to indicate the addition of these food additives on the product label [17], moreover, high intakes of phosphates may cause significant decreases in some oligoelements absorption [18] and they have been correlated with the bile ducts stones pathogenesis [19]. Actually, an analytical confirmatory method for the simultaneous determination of these food additives in cheeses is not available, consequently, in this work, the entire procedure of extraction, purification, chromatographic separations and quali/ determination of seven food additives (sorbic acid, benzoic acid, lactic acid, acetic acid, nitrites, nitrates and phosphates) (**Figure 1**) was developed and applied for the most common types of cheeses (mozzarella cheese, cheese spread, semi-hard and hard cheese). The analytical method was fully validated following an in-house validation model according to European Regulations and then evaluating the most important validation parameters.

MATERIALS AND METHODS

Chemicals

Lactic acid solution ($\geq$85%), acetic acid ($\geq$99.7%), potassium phosphate tribasic ($\geq$98%), nitrite ion standard solution (1000 mg·L^{-1}), nitrate ion standard solution (1000 mg·L^{-1}), sorbic acid ($\geq$99%) and benzoic acid ($\geq$99.5%) were purchased from Sigma-Aldrich (Stenheim, Germany). Sodium carbonate anhydrous ($\geq$99.5%) and sodium hydroxide (50% w/w) were supplied by J. T. Baker (Deventer, Netherlands). All solutions used for ion-exchange chromatography were prepared with ultrapure water (minimal resistance 18.2 MΩ-cm), supplied by Milli-Q RG unit from Millipore (Bedford, MA, USA). Sodium carbonate solutions used as eluents were prepared by dilution in ultrapure water, degassing with nitrogen. The samples used for validation procedure were collected in local stores and they were characterized by the absence of food additives declared on the label. The sample fortified for accuracy tests (precision and recovery) was a semi-hard cheese produced with raw milk of cow origin, characterized by a ripening time of 8 months. The samples used for ruggedness evaluation were 4 mozzarella cheeses (2 from cow milk and 2 from buffalo milk), 4 different brands of cheese spreads and 4 hard cheeses (2 produced with raw milk of sheep origin, 1 with pasteurized milk of cow origin and 1 with raw milk of goat origin). They were characterized by ripening times ranging from 16 to 18 months.

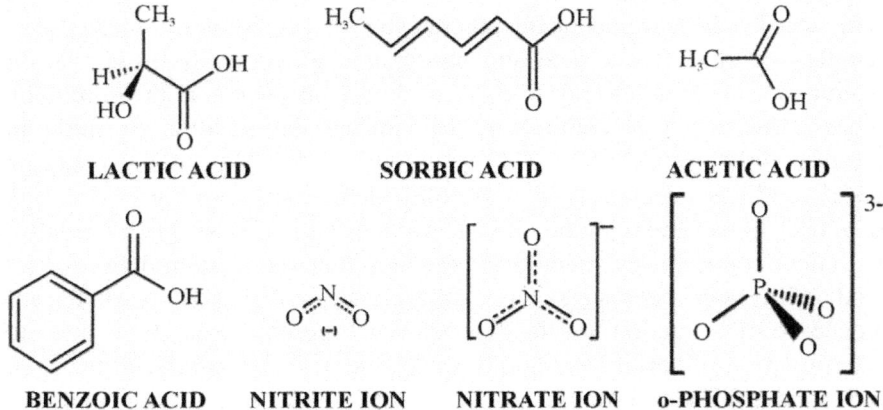

Figure 1: Structures of seven food additives analysed.

Instrumentation

All chromatographic determinations were performed on a Dionex HPLC system DX500 (Dionex Corporation, Sunnyvale, CA) equipped with an electrochemical detector (model ED40) fit to conductivity mode, a temperature compensated conductivity cell, a GP50 quaternary gradient pump, an injection loop (25 μL) and a Rheodyne injection valve (model RH9125, Cotati, CA, USA). A Dionex anion self-regenerating suppressor (ASRS II, 4 mm) was used for electrochemical suppression at an operating current of 50 mA. The chromatographic separations were accomplished by using an anion-exchange column IonPac® AS9-HC (250 mm × 4 mm i.d., particle size: 9 μm, Dionex Corporation, Sunnyvale, CA) eluted in gradient mode at a flow rate of 1.0 mL·min^{-1}. The mobile phase consisted of 0.9 mM Na_2CO_3 (A) and 28.5 mM Na_2CO_3 (B). The experimental separation conditions involved a multilinear gradient operating at room temperature and consisting of a linear gradient from 0.9 mM to 3.7 mM in 5 minutes, from 3.7 mM to 9.2 mM in 1 min, an isocratic step for 19 minutes, then a linear gradient from 9.2 mM to 28.5 mM in 1 minute and 4 minutes at this eluent concentration. The system was then re-equilibrated for 10 min at the initial Na_2CO_3 concentration. The mobile phase reservoir bottles (DX500 2 L bottles, Dionex) were closed and pressurized with pure nitrogen to 0.8 MPa. The system was interfaced via software (PeakNet™, Dionex Corporation, Sunnyvale, CA) to a personal computer for instrumentation control, data acquisition and processing.

Sample Preparation

A 4-g portion of sample, homogenized by blade ho-mogenizer, was placed in a Falcon tube and mixed with 40 mL of a NaOH 8.5×10^{-3} M solution. This particular solution allows an optimal analytes extraction from matrix, moreover it is essential for the preparation of standard solutions because it brings about an increment of method sensibility (by increasing acids dissociation) and it makes more neutral the solution pH, stabilising the nitrite ion.

The extraction was obtained by placing the tube in ultrasonic bath (Transsonic Digitals, Elma Instruments, Singen, Germany. Ultrasound power: 80%; Heating: 40°C) for 10 minutes and then vortexing (Digital vortex mixer, VWR Int., Milan, Italy) for 1 minute. The sample purification was carried out through a centrifugation (Jouan BR4i centrifuge, Thermo Fisher, Milan, Italy. Speed: $1500 \times g$, 10 minutes at room temperature) and a filtration of ~5 mL of supernatant with Minisart®GF syringe filters (0.2 μm, Sartorius AG, Goettingen, Germany). A final removing of excess chlorides is necessary. This was obtained through a purification of ~1 mL of filtrate with OnGuard II Ag chromatography filters (Dionex Corporation, Sunnyvale, CA) previously activated with 1 mL of ultrapure water, prior to chromatographic analysis (injection volume: 25 μL).

Method Validation

The optimized analytical method was submitted to a validation procedure, following an in-house validation model developed according to the Regulation No. 882/ 2004/EC [20] and Decision No. 657/2002/EC [21]. These Regulations describe the analytical parameters to appraise in order to assure the method reliability. These parameters are linearity, detection and quantification limits (LOD and LOQ), selectivity, accuracy (precision and recovery), ruggedness and measurement uncertainty.

Method linearity was verified by injecting 5 standard solutions, obtained by diluting the stock solution (100.0 mg·L⁻¹ lactic acid, acetic acid, nitrites, nitrates and phosphates, 1000.0 mg·L⁻¹Benzoic Acid and 2000.0 mg·L⁻¹

sorbic acid) with NaOH 8.5×10^{-3} M. The limit of determination (LOD) and quantification (LOQ) were elaborated according to Miller & Miller [22]: LOD = 3.3 s_a/b and LOQ = 10 s_a/b, where s_a is the standard deviation of the intercept and b is the slope of the regression line obtained from the calibration curve.

The selectivity is the method capacity to distinguish the analytes from other matrix components. This parameter was investigates through the analyses of twenty samples of cheeses (five mozzarella cheeses, five cheese spreads, five semi-hard cheeses and five hard cheeses) and verifying the absence of interfering peaks in the retention time-window of interest (±2.5% of the retention time of each analyte). It is important to underline that lactic acid and phosphate are substances naturally present in dairy products, consequently, for these additives, the selectivity was evaluated indirectly through accuracy tests.

Method accuracy (precision and recovery) was evaluated by analysing 6 cheese samples spiked with stock solution to obtain final concentrations equal to 20.0 $mg \cdot kg^{-1}$ in lactic acid, 20.0 $mg \cdot kg^{-1}$ in acetic acid, 400.0 $mg \cdot kg^{-1}$ in sorbic acid, 20.0 $mg \cdot kg^{-1}$ in nitrites, 200.0 $mg \cdot kg^{-1}$ in benzoic acid, 20.0 $mg \cdot kg^{-1}$ in nitrates and 20.0 $mg \cdot kg^{-1}$ in phosphates. For lactic acid and phosphates, that are naturally present in dairy products, the concentrations obtained by analysing the "blank" sample (not spiked) were subtracted from those obtained from spiked samples.

Method ruggedness was ascertained under major changes conditions toward different types of cheeses (mozzarella, cheese spread, semi-hard and hard cheeses) by using the Youden factorial experimental design [23].

Finally, the measurement uncertainty for each food additive was calculated by using the bottom-up method together with validation data obtained from each step of the analytical procedure [24]. The final values were obtained by the following equation:

$$\bar{u} = \sqrt{\left(\bar{u}(C)\right)^2 + \left(\bar{u}(V_f)\right)^2 + \left(\bar{u}(w)\right)^2}$$

where $\bar{u}$ indicates the relative uncertainty, V_f is the volume of final extract, and w is the sample weight. Four sources of uncertainty were considered for the determination of $\bar{u}(C)$: (a) standards preparation; (b) repeatability; (c) recovery; (d) calibration curve.

RESULTS AND DISCUSSION

Method Oprimization

The ion chromatography method for the simultaneous determination of seven food additives in cheeses was optimized starting from a chromatographic separation of organic acids proposed by Dionex Corporation [25]. The optimized gradient elution guaranteed a good resolution of analytes towards endogenous interfering peaks and a good retention time repeatability for each analyte (<0.5%, n = 6).

Different solutions, to use both for standards dilution and for samples extraction, were tested. The most important analytical problem to solve was the nitrites oxidation to nitrates due to the contemporary presence of different acids in the mixture [26]. It is not possible to use buffer solutions, since they cause an increase of interfering signals that compromise method selectivity. In order to basify the solution, NaOH solutions at different concentrations were tested. The best analytes stability (one week) was obtained by using NaOH 8.5 $\times$ 10^{-3} M compared to NaOH 8.5 $\times$ 10^{-2} M, NaOH 8.5 $\times$ 10^{-4} M and ultrapure water. Another analytical problem was the high chlorides concentration of this type of food product that represents an important chromatographic interference for sorbic acid. This inconvenience was solved by using OnGuard II Ag chromatography filters (Dionex Corporation, Sunnyvale, CA) that remove about 90% of total chlorides and make possible the identification of sorbic acid.

Method Validation

The most important validation parameters appraised through validation procedure are reported in**Table 2**.

All calibration curves showed a determination coefficient (r^2) higher than 0.990. This parameter together with evaluation of intercept (not significantly different from 0), of slope (t_s/b < 0.22) and of Mandel's test (linear regression preferable to quadratic) confirmed method linearity for the food additives considered [27]. In **Figure 2** a chromatogram related to an injection of a standard solution is shown.

As it concerns limits of detection, the lower values were obtained for lactic acid (LOD and LOQ equal to 0.6 and 1.9 mg·kg^{-1} respectively) whereas the highest values were registered for sorbic acid (LOD and LOQ equal to 59.4 and 196.0 mg·kg^{-1} respectively). These high values related to sorbic acid do not represent a weakness of this method, since the established legal limits for this food additive in dairy products are relatively high (up to 2000 mg·kg^{-1}); subsequently, there is no necessity to reach high method sensibility.

No interfering peaks were observed by analysing twenty samples of cheeses (mozzarella cheeses, cheese spreads, semi-hard cheeses and hard cheeses), so, method selectivity was verified for acetic acid, sorbic acid, benzoic acid, nitrites and nitrates. For lactic acid and phosphates, method selectivity was assured through accuracy tests.

Method accuracy (precision and recovery) was evaluated by injecting 6 spiked samples. Precision was evaluated as CV%, obtaining values in the range 3.0% - 10.9%. Recovery percentages were in the range 72.8% - 98.4%. Since there are no established reference values for the evaluation of accuracy of an analytical method for the determination of food additives in food products, the obtained values were compared with those reported in

Table 2: Analytical method validation parameters.

Food additive	Linearity (r^2)	Specificity	LOD[a] mg·kg^{-1} in the sample	LOQ[b] mg·kg^{-1} in the sample	Recovery (%)	CV%[c] (n = 6)
Lactic acid	0.992	-	0.6	1.9	92.3	9.4
Acetic acid	0.998	No interferences	3.6	11.9	79.6	3.0
Sorbic acid	0.990	No interferences	59.4	196.0	72.8	10.2
Nitrite	0.994	No interferences	2.8	9.3	78.6	10.3
Benzoic acid	0.999	No interferences	16.5	54.5	95.0	7.0
Nitrate	0.997	No interferences	2.1	6.8	88.1	10.9
Phosphate	0.997	-	2.1	7.0	98.4	8.5

[a]Limit of Detection. [b]Limit of Quantification. [c]Coefficient of Variation.

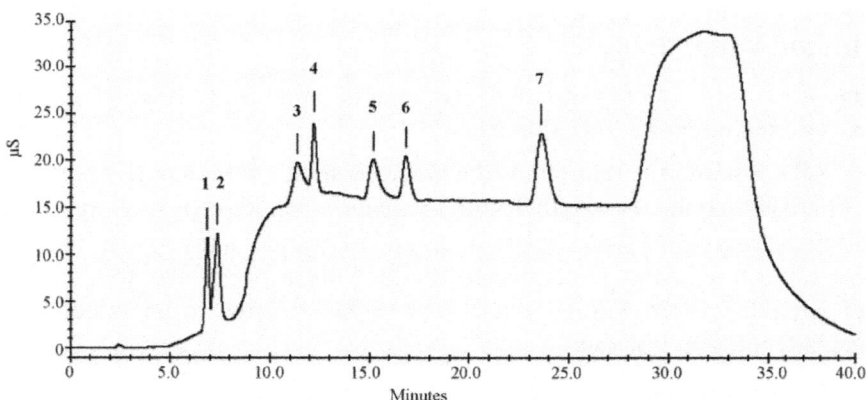

Figure 2: Chromatogram of a standard solution. (1): lactic acid (5.0 mg·L^{-1}), (2): acetic acid (5.0 mg·L^{-1}), (3): sorbic acid (100.0 mg·L^{-1}), (4): nitrites (5.0 mg·L^{-1}), (5): benzoic acid (50.0 mg·L^{-1}), (6): nitrates (5.0 mg·L^{-1}), (7): phosphates (5.0 mg·L^{-1}).

Decision No. 657/2002/EC, resulting substantially conform. In Figures 3 and 4 comparisons between cheese samples spiked with a standard solution and not-spiked are shown.

Method ruggedness under major changes conditions was evaluated by using Youden factorial experimental design. Different sets of real samples: 4

mozzarella cheeses, 4 cheese spreads, 4 semi-hard cheeses and 4 hard cheeses, spiked at the same fortification level used for accuracy tests. The seven factors chosen as variables for Youden test were the matrix and six fictitious factors. The Youden experimental design requires twelve independent experiments: four with validation matrix (semihard cheese) and four with each alternative matrices. The results obtained for each alternative matrix, expressed as standard deviation of difference S_{Di} of each food additive, were lower than the estimated method precision (evaluated as twice the repeatability standard deviation). These results confirmed that the matrix variation does not compromise the analytical performances and, consequently, the method is applicable to mozzarella cheese, cheese spread, semi-hard and hard cheese samples.

Measurement uncertainty was calculated for each analyte. The obtained values were in the range 8.6% - 12.7%, confirming method reliability.

Real Samples Analyses

In order to confirm the reliability of described analytical method, 20 different types of cheese samples (8 mozzarella, 6 cheese spreads, 3 semi-hard cheeses and 3 hard cheeses) were collected in local stores and analysed. No food additives were declared on the label of these samples (except for two mozzarella samples in which citric acid was added). Quantifiable amounts of nitrites, sorbic acid, acetic acid and benzoic acid were not registered. Nitrates quantifiable amounts were detected in two samples: a semi-hard cheese (23.6 mg·kg⁻¹) and a hard cheese sample (11.5 mg·kg⁻¹). The presence of quantifiable residues of nitrates in cheeses was already reported by several authors [28-32]. Lactic acid and phosphates were quantified in the ranges 40.3 - 162.1 mg·kg⁻¹ and 17.3 - 71.4 mg·kg⁻¹respectively.

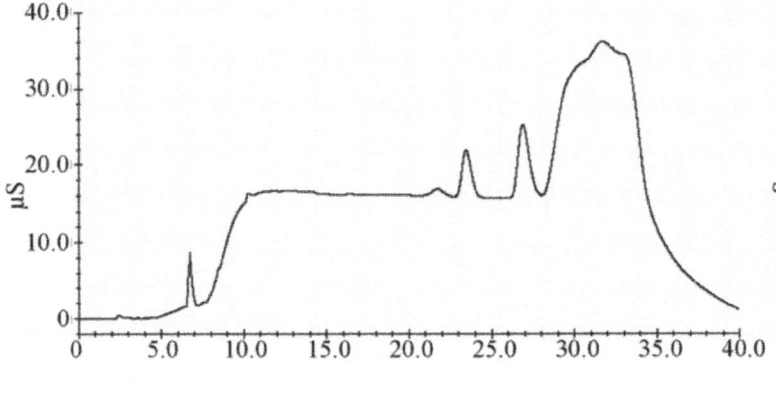

(a)

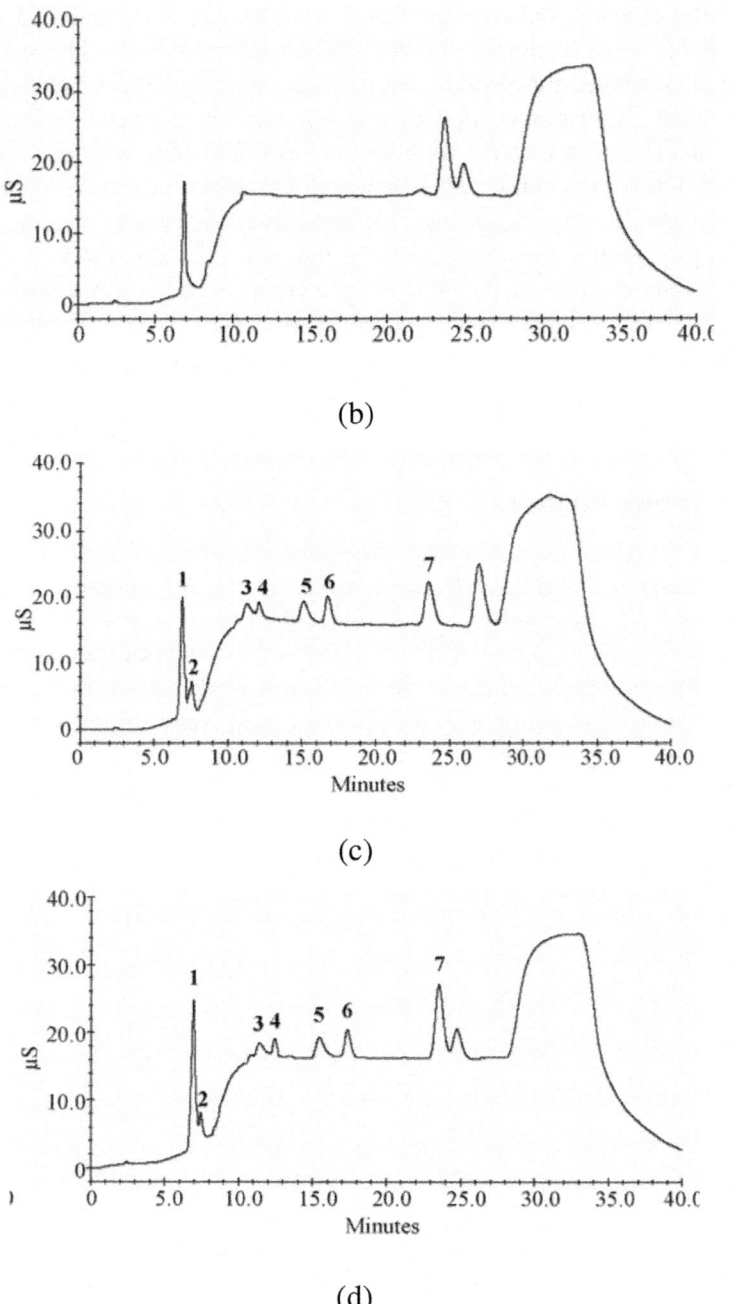

(b)

(c)

(d)

Figure 3. Chromatograms comparison between spiked samples (lactic acid 20.0 mg·kg^{-1}, acetic acid 20.0 mg·kg^{-1}, sorbic acid 400.0 mg·kg^{-1}, nitrites 20.0 mg·kg^{-1}, benzoic acid 200 mg·kg^{-1}, nitrates 20.0 mg·kg^{-1}, phosphates 20.0 mg·kg^{-1}) and not-spiked: Not-spiked Mozzarella cheese (a); Spiked Mozzarella cheese (b); Not-spiked cheese spread (c); Spiked cheese spread (d).

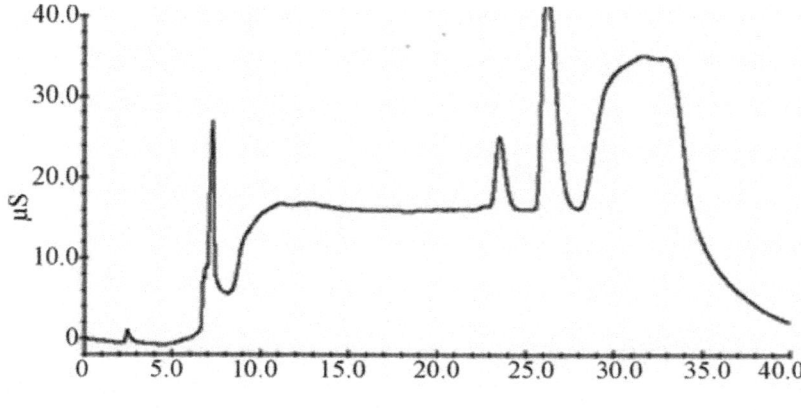

(a)

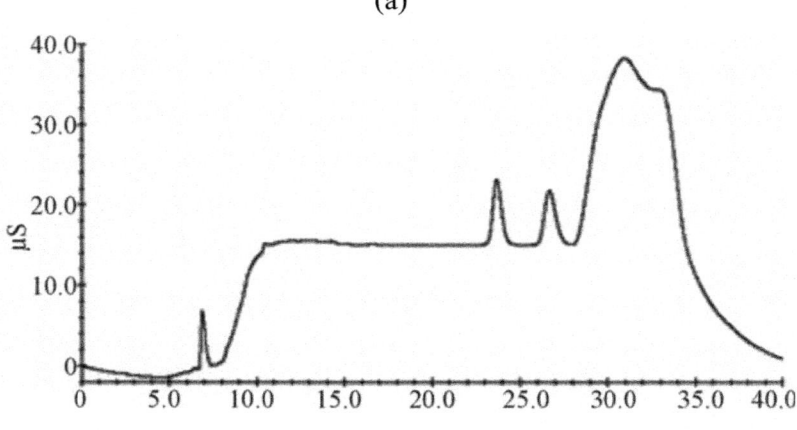

(b)

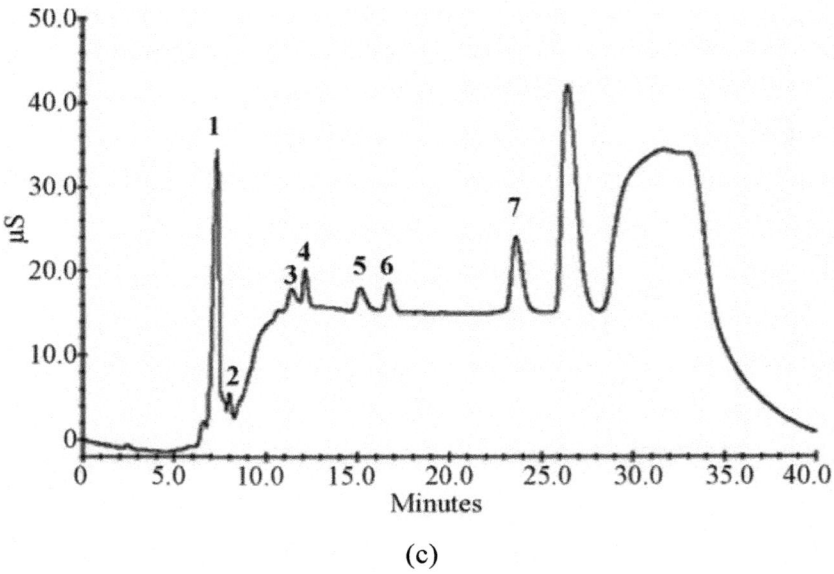

(c)

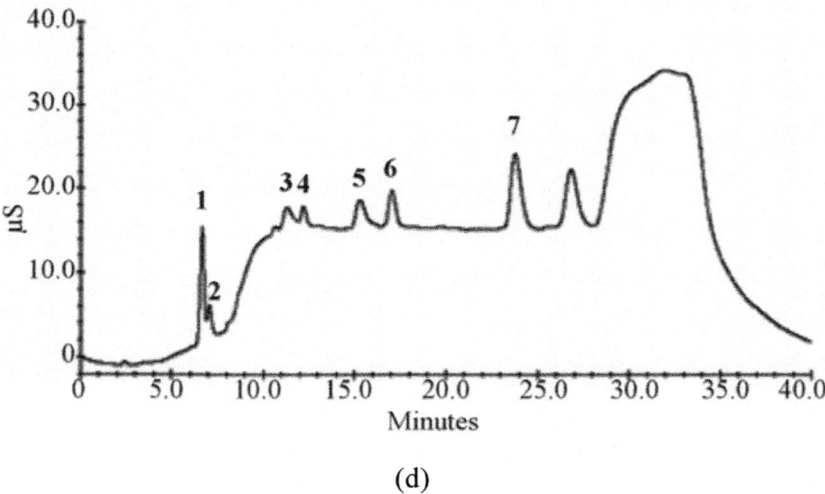

(d)

Figure 4: Chromatograms comparison between spiked samples (lactic acid 20.0 mg·kg⁻¹, acetic acid 20.0 mg·kg⁻¹, sorbic acid 400.0 mg·kg⁻¹, nitrites 20.0 mg·kg⁻¹, benzoic acid 200 mg·kg⁻¹, nitrates 20.0 mg·kg⁻¹, phosphates 20.0 mg·kg⁻¹) and not-spiked: Not-spiked semi-hard cheese (a); Spiked semi-hard cheese (b); Not-spiked hard cheese (c); Spiked hard cheese (d).

CONCLUSIONS

In this work, an analytical method for the determination of seven food additives (sorbic acid, benzoic acid, lactic acid, acetic acid, nitrites, nitrates and phosphates) by ion chromatography coupled with suppressed conductivity detection, was developed, validated and applied for the analysis of different types of cheeses (mozzarella cheese, cheese spread, semi-hard and hard cheese).

The optimisation of sample extraction/purification procedure together with the development of a suitable gradient elution allowed an optimal recovery of each analyte, the removal of the most important interfering compounds, a good resolution of analytes towards endogenous interfering peaks and a good retention times repeatability.

Through a validation procedure, carried out by following an in-house validation model according to European Regulations, it was possible to evaluate the most important validation parameters. These results, expressed in terms of linearity, detection limits (LOD and LOQ), selectivity, accuracy (precision and recovery), ruggedness and measurement uncertainty, demonstrated that the method is well suited to satisfy the demands for accurate confirmation analyses of seven food additives in cheeses, with a simple and fast sample preparation, a rapid analytical response, which is especially valuable in official check analyses and in monitoring schemes.

REFERENCES

1. P. F. Fox, P. L. H. Mc Sweeney and C. M. Lynch, "Significance of Non-Starter Lactic Acid Bacteria in Cheddar Cheese," Australian Journal of Dairy Technology, Vol. 53, 1998, pp. 83-90.

2. Commission Regulation, "Regulation (EC) No. 1129/ 2011 of 11 November 2011 Amending Annex II to Regulation (EC) No 1333/2008 of the European Parliament and of the Council by Establishing a Union List of Food Additives," Official Journal of the European Communities, 2011, pp. L295/1-177.

3. S. Bahruddin, B. Fazlul, I. S. Muhammad, A. Kamarudzaman and M. T. Khairuddin, "Simultaneous Determination of Preservatives (Benzoic Acid, Sorbic Acid, Methylparaben and Propylparaben) in Foodstuffs Using High-Performance Liquid Chromatography," Journal of Chromatography A, Vol. 1073, No. 1-2, 2005, pp. 393- 397.http://dx.doi.org/10.1016/j.chroma.2004.10.105

4. R. F. Witter, E. H. Newcomb and E. Stotz, "The Oxidation of Hexanoic Acid and Derivatives by Liver Tissue in Vitro," The Journal of Biological

Chemistry, Vol. 185, No. 2, 1950, pp. 537-548.

5.	H. J. Deuel, R. Alfin-Slater, C. S. Weil and H. F. Smyth, "Sorbic Acid as a Fungistatic Agent for Foods. I. Harmlessness of Sorbic Acid as a Dietary Component," Food Research, Vol. 19, No. 1, 1954, p. 1.

6.	R. J. Feldmann and H. I. Maibach, "Absorption of Some Organic Compounds through the Skin in Man," Journal of Investigative Dermatology Vol. 54, 1970, pp. 399-404.http://dx.doi.org/10.1111/1523-1747.ep12259184

7.	US FDA (United States Food and Drug Administration), "GRAS (Generally Recognized as Safe) Food Ingredients: Benzoic Acid and Sodium Benzoate," Food and Drug Administration, Washington, DC, 1972.

8.	F. Feillet and J. V. Leonard, "Alternative Pathway Therapy for Urea Cycle Disorders," Journal of Inherited Metabolic Disease, Vol. 21, No. 1, 1998, pp. 101-111.

9.	WHO (World Health Organization), "Concise International Chemical Assessment Document No. 26. Benzoic Acid and Sodium Benzoate," Geneva, Switzerland, 2000.

10.	M. G. El-Ziney, "CG-MS Analysis of Benzoate and Sorbate in Saudi Dairy and Food Products with Estimation of Daily Exposure," Journal of Food Technology, Vol. 7, No. 4, 2009, pp. 127-134.

11.	FAO/WHO (Food and Agriculture Organization of the United Nations/ World Health Organization), "Toxicological Evaluation of Certain Food Additives," 67th Meeting of the 435 Joint FAO/WHO Expert Committee on Food Additives (JECFA), 20-29 June 2006, Rome.

12.	M. Iammarino, A. Di Taranto, M. Muscarella and C. Palermo, "Survey of Benzoic Acid in Cheeses: Contribution to the Estimation of an Admissible Maximum Limit," Food Additives and Contaminants: Part B, Vol. 4, No. 4, 2011, pp. 231-237.http://dx.doi.org/10.1080/19393210.2011.620355

13.	A. M. Fan and V. E. Steinberg, "Health Implications of Nitrate and Nitrite in Drinking Water: An Update on Methemoglobinemia Occurrence and Reproductive and Developmental Toxicity," Regulatory Toxicology and Pharmacology, Vol. 23, No. 1, 1996, pp. 35-43. http://dx.doi.org/10.1006/rtph.1996.0006

14.	G. M. McKnight, C. W. Duncan, C. Leifert and M. H. Golden, "Dietary Nitrate in Man: Friend or Foe?" British Journal of Nutrition, Vol. 81, No. 5, 1999, pp. 349-358.http://dx.doi.org/10.1017/S000711459900063X

15.	O. Matteucci, G. Diletti, V. Principe, E. Di Giannatale, M. M. Marconi and G. Migliorati. "Due casi di Metaemoglobinemia Acuta da Sospetto Avvelenamento da Sodio Nitrito," Veterinaria Italiana, Vol. 44, No. 2, 2008, pp. 439-445.

16. M. C. Archer, "Mechanisms of Action of N-Nitroso Compounds," Cancer Surveys, Vol. 8, 1989, pp. 241-250.

17. Commission Regulation, "Regulation (EC) No. 1169/ 2011 of 25 October 2011 on the Provision of Food Information to Consumers, Amending Regulations (EC) No 1924/2006 and (EC) No 1925/2006 of the European Parliament and of the Council, and Repealing Commission Directive 87/250/EEC, Council Directive 90/496/EEC, Commission Directive 1999/10/EC, Directive 2000/13/ EC of the European Parliament and of the Council, Commission Directives 2002/67/EC and 2008/5/EC and Commission Regulation (EC) No 608/2004. Official Journal of the European Communities, 2011, pp. L304/ 18-63.

18. M. S. Norma, J. Steinhardt Bour, M. S. Barbara, A. Soullier and M. B. Zemel, "Effect of Level and Form of Phosphorus and Level of Calcium Intake on Zinc, Iron and Copper Bioavailability in Man," Nutrition Research, Vol. 4, No. 3, 1984, pp. 371-379.http://dx.doi.org/10.1016/S0271-5317(84)80098-6

19. M. C. Carey, "Pathogenesis of Gallstones," Recenti Progressi in Medicina, Vol. 83, No. 7-8, 1992, pp. 379-391.

20. Commission Regulation, "Regulation (EC) No. 882/2004 of the European Parliament and of the Council of 29 April 2004," Journal of the European Communities, 2004, pp. L165,1-141.

21. Commission Decision, "Decision (EC) No. 657/2002 of 12 August 2002 Implementing Council Directive 96/23 /EC Concerning the Performance of Analytical Methods and the Interpretation of Results," Journal of the European Communities, 2002, L221,8-36.

22. E. J. C. Miller and J. N. Miller, "Statistics for Analytical Chemistry," 3rd Edition, Ellis Horwood PTR Prentice Hall, New York, 1993, p. 115.

23. W. J. Youden and E. H. Steiner, "Statistical Manual of the AOAC— Association of the Official Analytical Chemists," AOAC-I, Washington DC, 1975.

24. E. Hund, D. L. Massart and J. Smeyers-Verbeke, "Operational Definitions of Uncertainty" Trends Analytcal Chemistry, Vol. 20, No. 8, 2001, pp. 394-406.http://dx.doi.org/10.1016/S0165-9936(01)00089-9

25. Dionex Corporation, "Product Manual for IonPac® AS11- HC IonPac® AG11-HC," Document No. 031333-07, Sunnyvale, 2008.

26. F. Toldrà, "Handbook of Meat Processing," John Wiley & Sons, Blackwell Publishing, New York, 2010. http://dx.doi.org/10.1002/9780813820897

27. L. Brüggemann, W. Quapp and R. Wennrich, "Test for Non-Linearity Concerning Linear Calibrated Chemical Measurements," Accreditation and Quality Assurance, Vol. 11, No. 12, 2006, pp. 625-631. http://dx.doi.org/10.1007/s00769-006-0205-x

28. M. Iammarino, A. Di Taranto and M. Cristino, "Endogenous Levels of Nitrites and Nitrates in Wide Consumption Foodstuffs: Results of Five Years of Official Controls and Monitoring," Food Chemistry, Vol. 140, No. 15, 2013, pp. 763-771.http://dx.doi.org/10.1016/j.foodchem.2012.10.094

29. M. Iammarino, A. Di Taranto, M. Muscarella and C. Palermo, "Assessment of Natural Levels of Substances with Preservative Effects in Dairy Products," In: H. Castelli and L. du Vale, Eds., Handbook on Cheese, Production, Chemistry and Sensory Properties, Nova Publisher Inc., New York, 2013, pp. 559-571.

30. L. J. Schuddeboom, "Nitrates and Nitrites in Foodstuffs," Council of Europe, Strasbourg, 1993.

31. A. Topçu, A. A. Topçu, I. Saldamli and M. Yurttagül, "Determination of Nitrate and Nitrite Content of Turkish Cheeses," African Journal of Biotechnology, Vol. 5, No. 15, 2006, pp. 1411-1414.

32. L. Tudor, E. Mitrânescu, L. Tudor and F. Furnaris, "Assessment of Nitrate and Nitrite Content of Romanian Traditional Cheese," Lucrâri Stiintifice Medicinâ Veterinariâ, Timisoara, XL, 2007.

Chapter 4

THE BIOTECHNOLOGICAL DEVELOPMENT OF NEW FOOD PRESERVATIVES

S. Roller

Biotechnology Unit, Leatherhead Food Research Association, Leatherhead, Surrey, UK

INTRODUCTION

The food-poisoning micro-organisms of major concern today are those that contaminate foods and cause illness by infection (Listeria monocytogenes, Yersinia enterocolitica, Vibrio parahaemolyticus, numerous Salmonella species, enteropathogenic strains of Escherichia coli, Campylobacter jejuni and coli and Clostridium perfringens) and by forming toxins (Staphylococcus aureus. Clostridium botulinum, Bacillus cereus, and some strains of Bacillus subtilis and licheiformis). Reported cases of salmonellosis and listeriosis increased steadily during the 1970s and 1980s with an especially sharp rise in some countries during the past 5 years associated with outbreaks in certain foods (Baird-Parker, 1990; Jones. 1990). In addition there is a plethora of spoilage organisms, too numerous to list here that cause wastage and losses of food during processing transport and storage. Thus the use of added preservatives as one of the many options available for extending the shelf-life of foods is likely to remain an important activity in food product formulation. Indeed, while consumer pressure is forcing many food companies to reconsider and reduce the levels of usage of certain food 'additives' the need for effective food preservatives is greater than ever and is fuelling a renewed interest in microbial inhibitors. It is in this search for hitherto unknown, more effective antimicrobials tailor-made to combat specific spoilage or pathogenic organisms, that the role of biotechnological techniques will be most evident in the future.

In the following review the impact of recent advances in traditional biotechnology (fermentation and enzymology) as well as in the newer sciences of molecular biology (genetic manipulation and protein engineering) on the development of new food preservatives will be examined.

Organic Acids

Organic acids such as acetic, lactic, citric, fumaric, malic, tartaric and gluconic acids are used extensively in foods such as soft drinks, alcoholic beverages salad dressings, baked products. Jams. Jellies, ice-cream, processed cheese and confectionery. These edible acidulants are indispensable to the food industry not only for their preserving action but also for the characteristic flavour they give to foods. However, most organic acids are produced by traditional fermentation techniques, some of which have not changed for decades. Although it is conceivable that improvements in strain stability and product yields could be made using recombinant DNA technology, the relatively low cost of food acidulants makes it unlikely that biotechnological techniques will have a major impact on this business sector. Therefore, further discussion of the organic acids as food preservatives will not be undertaken here. The interested reader is referred to reviews by Buchta (1983a, 1983b). Ebner and Follmann (1983), Rohr, Kubicek and Kominek (1983a, 1983b), Ruttloff (1987) and Dziezak (1990).

Microbial Peptides

Small, usually cationic, peptides with pronounced antimicrobial activity are common in nature and many have been described from animal, plant and microbial sources (Sahl, 1985). In spite of their different sources, these peptides display structural similarities including molecular masses of between 3000 and 6000 and isoelectric points of approximately 10. Peptides that are easily degraded in the humall digestive tract hold an important potential safety advantage over chemically synthesized preservatives that may enter the bloodstream. On the other hand, sensitivity to proteolytic enzymes can constitute a disadvantage if the peptide is to be used in foods such as fresh meat. Furthermore, many peptides would not be suitable for use as food preservatives, owing to the pathogenic nature of some of the microbial producers. For example, the best understood group of antimicrobial peptides, the colicins, are produced by E. coli an organism usually associated with poor hygiene in the food industry. Further examples are the peptides epidermin and Pep 5, produced by strains of Staph. epidermidis (Ersfeld-Dressen, Sahl and Brandis, 1984; Horner et al., 1989), which may be useful as antimicrobial agents in topical applications, such as in creams and salves, but are unlikely to receive regulatory approval for use in foods without extensive (and expensive) toxicological evidence that the products are safe to ingest. Nevertheless, several peptides with antimicrobial spectra useful in foods have been identified; some have been commercialized and many more show good potential for future commercialization.

Nisin is produced by Lacto coccus (formerly Streptococcus) lactis and belongs to a class of compounds known as the bacteriocins. These have been defined by Tagg, Dajani and Wannamaker (1976) as proteinaceous compounds with a bactericidal action against a limited range of organisms usually closely related to the producer organism. In addition, bacteriocin production is usually, but not always, governed by plasmid-borne genetic determinants. For recent reviews on the biology chemistry toxicity and food applications of nisin, the reader is referred to papers by Delves-Broughton (1990a, 1990b). Although discovered as long ago as 1928. nisin was not used in foods until 1951 when Hirsch et al. demonstrated the ability of nisin-producing cultures to prevent cheese spoilage by clostridial gas formation. Indeed the ability of nisin to inhibit growth of Gram-positive bacteria such as Staphylococcus and spore outgrowth of several species of Clostridium and Bacillus has been its most useful property. In addition, recent work by Benkerroum and Sandine (1988) suggests that it may be possible to use nisin in the control of the important food pathogen, L. Monocytogenes. Furthermore, the combination of nisin with the chelating agent EDT A andlor the surfactants Triton X-100 and Tween 20 has been shown to be effective against a number of Gramnegative bacteria, including Salmonella typhimurium and E. coli in laboratory media and a model food system based on chicken (Blackburn et al .• 1989). In 1969, nisin received clearance from the Joint FAP/WHO Expert Committee on Food Additives for use in foods and is now positively allowed by regulatory authorities in 47 countries. Nisin has been and is, used as an effective preservative in processed cheese and cheese spreads, dairy desserts canned food and. when refrigeration facilities are inadequate (as in some Middle Eastern countries) in pasteurized milk (Delves-Broughton. 1990a). Attempts have also been made to use nisin as an alternative preservative system to nitrites in cured meats but the levels of addition necessary to inhibit C/ botulinum have been too high to make the process economic (Calderon, Collins-Thompson and Usborne, 1985; Taylor and Somers. 1985; Taylor, Somers and Krueger, 1985; Bell and de Lacy, 1987). Today. nisin remains the only bacitracin produced by lactic acid bacteria that is com metrically available. Nisin has been shown to contain 34 amino acid residues including the unusuallanthionine, and its molecular weight has been determined as 3510 with some evidence of dimer (MW 7000) and tetramer (MW 14000) formation in solution (Gross and Morell, 1971). The antimicrobial activity of nisin is optimal under acidic conditions and is reduced by heat treatment and storage in foods. It has been suggested that the mode of action of nisin is based on the disruption of the cytoplasmic membrane (Morris, Walsh and Hansen. 1984). The possibility that nisin production may be plasmid-mediated was suggested by several investigators in the 1970s to early 19805. For example, Le Blanc. Crow and Lee (1980) reported phenotypic and physical evidence for a 28 MDa nisin plasmid in L. lactis. In 1984. Gasson linked nisin production and resistance to a 30 MDa plasmid and successfully transferred the trait to a non-producing. Plasmid-free strain of L. lactis. Further studies have shown that the final nisin structure was subject to post-translational modification of a precursor polypeptide by enzymic modification of cysteine, threonine and serine residues (Dodd, Horn and Gasson, 1990). The gene sequences of nisin and its structural analogue subtilin (from B. subtilis) have recently been elucidated by scientists at the University of Maryland (Banerjee and Hansen, 1988; Buchman, Banerjee and Hansen. 1988; Hansen. Banerjee and Buchman. 1989) these studies provide the background data essential for the development of a range of useful bacteriocin analogues tailored against the spoilage and pathogenic flora of specific foods. For further details of the genetic work on nisin and other bacteriocins the interested reader is referred to papers by Larry McKay. Todd Klaenhammer and Mike Gasson (for example, Gasson. 1984; Gasson and Anderson, 1985; Klaenhammer. 1988; McKay. 1989).

Bac.'Teriocins Other Than Nisin

Examination of a sufficiently large number of strains (that is. 100 or more) of anyone species of micro-organism is generally rewarded with some evidence of antagonism. A quick glance through the contents pages of any microbiological journal published in the past few years will reveal numerous papers reporting the existence of bacteriocins. Particularly from lactic acid bacteria. However, this group of organisms also produces a host of other metabolites with inhibitory properties and these can be the source of many errors in the screening procedure. For example, the production of hydrogen peroxide by lactic acid bacteria is well documented (Whittenbury, 1964; Anders, Hogg and Jago. 1970; Premi and Bottazzi. 1972; Collins and Aramaki. 1980) as is its inhibitory effect on Staph aureus (Dahiya and Speck. 1968; Attaie el al• 1987), pseudomonads (Price and Lee, 1969) and salmonellae (Mulder, van der Hulst and Bolder, 1987). In addition. many screening assays arc based on active cells, which produce an abundance of lactic and acetic acids; these primary metabolites are the main substances responsible for the antimicrobial properties of lactic acid bacteria. Although many authors have used these primary metabolites as controls individually the effect of a combination of primary metabolites has often been ignored. Correct controls for each individual strain must be constructed following analysis of the fermentation broth for all the primary metabolites. Using a very carefully controlled screening procedure, ten Brink,

Bol and Huis in't Veld (1988) found that only a handful of strains out of WOO screened produced bacteriocins. However, the lessons of the past are slow to be learned and systematic studies of the benefits and pitfalls of different assay methods are only just beginning to appear in the scientific literature (Spelhaug and Harlander, 1989). For a detailed examination of the literature on the biochemistry and genetics of some of the better-characterized bacteriocins reported up to mid-1987, the review by Klaenhammer (1988) is recommended. The bacteriocins reviewed by Klaenhammer included lactocin 27 and helveticin J from strains of Lactobacillus helveticus, lactacin Band F from Lactobacillus acidopltilus, plantaracin A from Lactobacillus plantarum, diplococcin and Las 5 from Streptococcus cremoris, the lactostrepcins from Streptococcus lactis, and pediocin A from Pe(/iococcus pelliosaceus. Other authors have also reported the production of acidophilin from Lactobacillus acidophilus (Shahani. Vakil and Chandan. 1972; Shahani, Vakil and Kilara, 1976. 1977) and bulgarican from Lactobacillus bulgaricus (Reddy and Shahani, 1971; Reddy et al• 1983). However. as acidophilin had not been characterized as proteinaceous it is uncertain whether this compound could be classified as a bacteriocin by Tagg's 1976 definition. Numerous additional bacteriocins have been described since the Klaenhammer review. Including for example, sakacin A and lactocin S from strains of Lactobacillus sake (Schillinger and Lucke, 1989; Mortvedt and Nes, 1990) and leucocin A from a strain of Leuconostoc gelidum isolated from refrigerated meat (Harding and Shaw, 1990; Hastings and Stiles, 1991). Except for pediocin all of the above bacteriocins have been shown to possess a very narrow inhibitory spectrum. The organisms producing these narrow-spectrum inhibitors may find uses in the construction of starter cultures for traditional food fermentations in which they would compete more effectively with the lactic acid bacteria in the natural microflora and therefore produce fermented products with more consistent quality than is currently possible. However, scale-up, isolation and purification of these bacteriocins are unlikely to be undertaken by industry owing to the limited number of foods in which the inhibition of lactic acid bacteria alone would be advantageous. On the other hand the broader-range bacteriocins such as pediocin A from P. pentosaceus with reported activity against the food-borne pathogens Staph aureus. Ct. Botulinum, C/. perfringens. lJ. Cereus and L. monocytogenes and the food-spoilage agents Streptococcus faeca/is and Clostridium sporogelles hold more promise from the industrial point of view (Fleming, Etchells and Costilow. 1975; Daeschel and Klaenhammer, 1985: Daeschel. 19~9). Another broad-spectrum pediocin named pediocin AcH from Pediococcus acidilactici has been shown to be active against food-spoilage bacteria such as Pseudomonas pUlida and Broc!wlhrix rhermosphacla and pathogens such as B. cereus. C/. perfringens. L. monocytogenes and Staph au reus (8hunia. Johnson and Ray.

1988). Furthermore, partially purified pediocin AcH has been shown to be non-immunogenic to eight mice and a rabbit (Bhunia e/ al .. 1990). Mode-of-action studies suggest that pediocin AcH acted on sensitive Gram-positive target cells by binding to specific receptors on the cell envelope. Followed by disruption of the integrity of the cell membrane with consequent cell death (Bhunia et al .. 1991). Although the development of the pediocins as food preservatives is currently hindered by low levels of production. Difficulties in purification and genetic instability (Klaenhammer, 1988; Bamby-Smith et al .. 1989; BarnbySmith and Roller. 1990), some recent publications suggest that it may not be long before these problems are resolved. Thus it has been claimed that a preparation of pediocin PA-1 obtained from P. acidilactici by centrifugation and filtration of the culture, followed by ammonium sulphate precipitation and dialysis, was effective in preventing the spoilage of a salad dressing inoculated with 103 CFU g-I of Lactococcus fermentans and stored at room temperature for 7 days (Gonzales, 1989). The dressings were assessed by smell and taste and were found to be satisfactory. The inoculated control dressing spoiled after 5 days' storage. Using the same producer organism as Gonzales (1989), Nielsen, Dickson and Crouse (1990) tested the efficacy of a crude preparation of pediocin (the microbial culture was centrifuged, neutralized and filtered) on beef surfaces contaminated with L. monocytogenes. The bacteriocin was shown to have a bactericidal mode of action, reducing the number of L. monocytogenes by 0·5-2·5 log cycles, depending on bacteriocin concentration, numbers of listeria present and the time of application of the bacteriocin (before or after contamination). There appeared to be no inactivation of the bacteriocin by the prot eases normally present in fresh meat; however, this may be explained by the low temperature (5°C) of storage. Further work in other food applications is awaited with interest. An alternative approach to extracting and purifying the antagonistic compounds has been to add whole, bacteriocin-producing cultures or centrifuged, 'cell-free' culture preparations to foods. However, this approach has met with limited success (Gibbs, 1987; Pucci et al., 1988; Gombas, 1989; Berry et at., 1990; S. Roller, unpublished). For example. Berry and co-workers (1990) have found only marginal differences in the survival of L. monocytogenes in fermented semi-dry sausage supplemented with 107 CFU g-l of non-producing and bacteriocin-producing pediococci. On the other hand, it has been reported that the addition of several strains of Lactobacillus species at a level of 107 CFU ml[-1] to model food systems inhibited the growth of moulds, psychrotrophic spoilage organisms and pathogens such as Salmonella newport and L. monocytogenes when stored at 7°C (Boudreaux, Matrozza and Leverone, 1989). As in studies with known food preservatives, it would appear that data obtained from inhibitory assays in laboratory media often cannot be extrapolated to food systems. One of the disadvantages of adding whole cultures of lactic acid bacteria to foods for preservation purposes is that this group of organisms produces copious amounts of lactic and acetic acids. Whereas these primary metabolites may be desirable in fermented foods, where they play an important role in flavour formation, in fresh foods (e.g. fresh beefburgers) they have been shown to lead to undesirable organoleptic effects (Gibbs, 1987). Thus, it has been proposed that the antibacterial properties of lactic acid bacteria could be exploited in fresh foods by curing the organisms of their genetic determinants for organic acids. Gonzales, of Microlife Technics, has patented a method for using selected strains of Strep. lactis subsp. diacetylactis that have been cured of the ability to ferment lactose by removal of a 41 MDa plasmid (Gonzales, 1986). The patent claims that the modified organisms added at a level of 10^5-10^6 CFU g^{-1}, were particularly active in cottage cheese against the spoilage organisms Pseudomonas fragi and Pseudomonas fluorescens inocu* lated at a level of 10^3 g. In addition milk inoculated with the cured organisms remained fresh after 12 days' storage at 10°C whereas the control milk spoiled after 10 days (Gonzales, 1986).

The Wisconsin bacon process has been one of the few examples where the addition of a live culture of lactic acid bacteria to a non-fermented food has been a success. However in this process, the preservative action was attributable to the production of acid by the pediococci. Thus, work carried out in the early 1980s at the University of Wisconsin showed that the level of sodium nitrite in bacon could be reduced from the recommended 120 p.p.m. to 80 and 40 p.p.m. without exposing consumers to an increased risk of botulism (Tanaka et al ... 1985a, 1985b). This was accomplished by incorporating P. acidilactici and a fermentable sugar (0.7% sucrose) into the curing mixture. When the bacon was temperature-abused, the lactic acid bacterium utilized the sugar to produce lactic acid, which lowered the pH and thus prevented the growth of CI botulinum (Anon., 1984a, 1984b). By 1986, ABC Research Laboratories were marketing 'Bacon Blend' for reducing nitrites (and thereby nitrosamines) in processed meats (Anon., 1986)

Other Microbial Products

Reuterin, a broad-spectrum antimicrobial substance produced by Lactococcus reuteri has recently been described (Talarico et al., 1988; Talarico and Dobrogosz, 1989). Reuterin has been characterized as a highly soluble, pH-neutral mixture of monomeric and dimeric forms of J3-hydroxypropionaldehyde produced during anaerobic fermentation of the organism on glycerol. The product was reported to be active against several species of Salmonella, Shigella, Clostridium, Staphylococcus, Listeria. Candida and Trypanosoma, and preliminary results suggest promising preservative activity in refrigerated ground beef (Daeschel. 1989).

Engineered Peptides

The immunoproteins of insects are a recently described group of peptides with antimicrobial activity which may have potential for exploitation as food preservatives in the future. In response to bacterial infection, the pupae of the Cecropia moth produce 10-15 immune proteins including lysozyme, the cecropins A. Band D (MW around 4000) and the attacins (MW around 200(0). These compounds have been studied extensively by research groups in Stockholm and Uppsala, Sweden (Hultmark et al .. 1980, 1982, 1983; Boman. 1986). The cecropins have been shown to be active against both Gram-positive and Gram-negative bacteria including pathogens of medical significance such as Pseudomonas aeruginosa. The attacins were active primarily against E. coli and their mode of action was based on the disruption of the outer membrane (Engstrom et al., 1984). The Swedish group concluded that the three immune proteins cecropin. attacin and lysozyme appeared to act in concert, resulting in the effective inactivation of a whole range of structurally different microorganisms.

Clearly only minute quantities of the inhibitory peptides can be isolated from the insect. However, scientists at Ingene Inc. in California have recently reported the successful cloning and expression of a gene encoding for a fusion protein incorporating cecropin A into E. coli (Lai et al., 1989). It has been necessary to construct the gene encoding for a fusion protein rather than a simple cecropin peptide in order to protect the host organism from the product's antimicrobial properties once the gene is expressed. Following a fermentation cycle, the E. coli cells were harvested and homogenized, and the inclusion bodies containing the fusion protein were sedimented. The inclusions were solubilized and the cecropin released by acidification. Further purification by ion-exchange chromatography was required to produce food additive quality cecropin. Limited toxicity tests carried out by the company have shown no adverse effects of cecropin on sheep red blood cells, human B-cells and human fibroblast cell lines at 300 ì g ml^{-1} or on mice following feeding trials of up to 1·5 g kg-I body weight per day in a three-week period. The recombinant cecropin A, however, had a carboxyl group at the C-terminus where the insect-derived cecropin had an amide group; consequently, it was found that the recombinant cecropin also lost its antimicrobial activity against Gram-positive bacteria but not against Gram-negative bacteria. Ingene scientists have suggested chemical methods for attaching amide groups to the recombinant product to produce 'cecromycin' with antibacterial activity against a wide spectrum of Gram-positive and Gram-negative organisms. Thus, it would appear that the application of recombinant DNA technology may, in the future, allow the production of economically feasible quantities of the material.

ENZYMES AS PRESERVATIVES

Lysozyme

Although several forms of lysozyme are known to exist in nature it is the hen albumen lysozyme that has been most studied and that is commercially available (Wilkins and Board, 1989). Lysozyme is a muramidase that cleaves the **β-1,4** -glycosidic bond between N-acetyl-D-glucosamine and N-acetylmuramic acid in cell walls of Gram-positive bacteria. Recently it has been demonstrated that lysozyme was effective against L. monocytogenes in laboratory media (Osa et al., 1990). Used alone, avian lysozyme is inactive against Gram-negative bacteria but milk lysozyme has been reported to lyse E. coli and P. aeruginosa if the outer membrane is first removed by sodium chloride and EDTA, respectively (Vakil et al., 1969). Thus, it is conceivable that the bactericidal spectrum of lysozyme could be extended by design, for example by using it as a preservative in food formulations that already contain agents, such as polyphosphates, known to interfere with the outer membrane of Gram-negative organisms. Similarly, sublethal heat treatments or freeze! Thaw injury may predispose Gram-negative organisms to the bactericidal action of lysozyme (Wilkins and Board, 1989). Recent work on the shelf-lives of mayonnaise, confectionery cream filling and fruit mousse has demon

Strated synergistic effects between lysozyme and calcium sorbate resulting in extension of shelf~lives of up to 50% (Luck et al .• 1988). In addition there is some indication in the literature that lysozyme may attack chitin. An impor Tent constituent of the fungal cell wall (Tokura, 1989). The commercially available lysozyme from hen egg white has a molecular weight of 14400 and contains 129 amino acid residues with four intramolecu· lar disulphide bonds (Scott. Hammer and Szalkucki. 1987). Hen egg white contains approximately 3·5% lysozyme of which about 80% is readily extracted by ion·exchange chromatography. The 'late blowing' of cheese (formation of unsightly holes in hard cheeses during maturation) by Clostridium ryrobutyricum is a recurrent spoilage problem in cheese-producing areas. In France alone the cost of spoilage of Swiss-type cheeses has been estimated at around $US7 million per year (M. Maitenaz, Institut Technique du Gruyere. personal communication). Scott, Hammer and Szalkucki (1987) have estimated that in 1984 100 tonnes of lysozyme were used to prevent the late gas blowing of cheese in Europe. Although nitrate has traditionally been used to prevent late blowing, it is not permitted for use in cheese in the USA, and concern over the possible formation of carcinogenic nitrosamines has encouraged the use of alternative preservatives such as lysozyme. Lysozyme is usually added to milk destined for cheese production at a level of 25 mg I-I to give a final concentration of 300 mg kg-I cheese. The enzyme has been shown to arrest the growth of vegetative cells once germinated from spores (Carminati et al., 1984). Interestingly. Lysozyme has not been reported to interfere with the subsequent growth of lactic starter cultures except on a few occasions. Lysozyme has been shown to be particularly effective in Edam and in the Italian cheeses Provlone, Grana, Emmenthal, Asiago and Montasio (Wasserfall, Voss and Prokopek. 1976; Carini and Lodi. 1982).

There has been some interest recently in the possible application of lysozyme in infant formulae in order to achieve a similar balance of faecal flora to that found in breast-fed babies (Wharton. 1982; Wilkins and Board. 1989). However, results of trials have been inconclusive, probably because hen egg lysozyme, which is less active than milk lysozyme. Was used (Reiter, 1984). It has been suggested that the cost of lysozyme could be reduced if it were produced from a micro-organism by using fermentation. Bacteriophage T4 lysozyme has been a particularly attractive target as it has been shown to be 250 times more active against E. coli than the avian enzyme (Grutter and Matthews. 1982). Although structurally homologous to hen egg lysozyme. T4 lacks the intramolecular disulphide bonds (Hawkes. Grutter and Schellman, 1984). However, scientists at Genentech and Genencor have used protein engineering techniques to introduce a disulphide bond into the T4 lysozyme which increased its heat stability (Perry and Wetzel, 1984). More recently, the genes encoding for hen egg white lysozyme have been successfully transformed into E. coli and Saccharomyces cerevisiae, although yields (1-2 mg l^{-1}) of the recombinant protein have been disappointingly low (Kumagai et aI., 1987; Miki et al., 1987). Higher levels (up to 12 mg l^{-1}) of recombinant lysozyme secreted by Aspergillus niger have been reported by Archer et af. (1990), although this is still too low for a viable industrial process. Nevertheless, these problems do not appear insurmountable and rapid future developments can be anticipated. Several native microbial N-acetylmuramidases similar to lysozyme have been subjected recently to trials in foods. Thus, Hayashi et at. (1989) have shown that 0·11 % of a purified muramidase from Streptomyces rUlgersensis extended the shelf-life of an adzuki bean paste (called 'an' and used for making traditional Japanese sweets) from 48 to 109 h when stored at lO°C. The enzyme was shown to be effective against lactic acid bacteria, which constitute the predominant spoilage flora of an.

Lactoperoxidase

Mammalian secretions such as milk and saliva contain low concentrations of peroxidases, which, coupled with thiocyanate ion and hydrogen peroxide, form hypothiocyanate, an unstable but highly reactive biocide. This rather elegant antimicrobial system has been exploited in a toothpaste that utilizes the endogenous peroxidase and thiocyanate in saliva with exogenous amyloglucosidase (to produce glucose) and glucose oxidase (to produce hydrogen peroxide from glucose and oxygen) (Hoogendoorn, 1974). It has also been recommended that the optimal activity of the endogenous lactoperoxidase in milk can be obtained by the addition of 12 and 8 p.p.m. of thiocyanate and hydrogen peroxide, respectively (Reiter and Harnulv, 1984). Indeed, the lactoperoxidase system has been used successfully in the preservation of raw milk in Sri Lanka, Pakistan, Kenya and Mexico when refrigeration prior to pasteurization was not available.

The lactoperoxidase system was activated by the direct addition of 10 p.p.m. hydrogen peroxide to the raw milk (Banks, Board and Sparks, 1986). Hydrogen peroxide could also be generated in situ by glucose oxidase, although this option tends to be too expensive for application in developing countries. Nevertheless, it has been suggested that a combination of ~-galactosidase and glucose oxidase immobilized on a column could be used as a means of 'cold sterilization' of milk (Bjorck et al., 1975; Banks, Board and Sparks, 1986). However, at levels of peroxide addition used in some countries (300-500 p.p.m.), the lactoperoxidase system is destroyed and the biocidal action stems directly from the peroxide (Bjorck, Claesson and Schulthess, 1979).

Reiter (1984) has shown that the lactoperoxidase system is effective against the Gram-negative E. coli, S. typhimurium and Ps. aeruginosa. Recently, the lactoperoxidase system has also been shown to inhibit growth of L. monocytogenes in skimmed milk subjected to temperature abuse for 6 h at 30°C and 20 h at 20°C (Bibi and Bachmann, 1990). The only commercial sources of lactoperoxidase to date are raw milk and whey (Anon., 1985; Burling, 1989).

The enzyme, however, together with co-produced milk proteins, has already shown some promising results in feeding trials with young calves: lactoperoxidase-fed animals appear to be less prone to diarrhoeal diseases than control animals, with consequent improvements in weight gain (Gudmundsson, 1984 (cited in Scott. Hammer and Szalkucki. 1987». Supplementation of infant milk formulae with commercially available lactoperoxidase has also been shown to delay the onset of growth of S. typhimurium in the reconstituted product for 3 days and E. coli for 1 day at 15°C (Earnshaw et al, 1990); although this represents a useful possible application of lactoperoxidase in countries where infant feed may be prepared with contaminated water. it is unlikely that the supplementation would currently be economic.

Other Enzymes

Lytic enzymes with specific activity against fungi would be particularly attractive for development as food preservatives because many of the antimicrobials in current use are poor inhibitors of this class of organism. Although lysozyme has been shown to have some activity against fungal chitin (Tokura, 1989), its activity has not been sufficiently pronounced to warrant widespread application studies. Chitin a homopolymer of N-acetylglucosamine, is the major component of fungal cell walls. Therefore, it follows that an enzyme specific for fungal chitin may be useful as an inhibitor of common food-borne spoilage and pathogenic moulds. However chitinases have not been tested against foodborne fungi.

Enzymes of the chitinase complex have received substantial attention recently mainly for the degradation of the naturally occurring and abundant polymer chitin, and also because of their involvement in the resistance mechanisms of plants to pathogenic fungi. Thus, a number of chitinaselchitosanase enzymes have been described and, in some cases, isolated and characterized from higher plants (Oishi. Fumiyasu and Masao. 1989) and from Streptococcus olivaceoviridis (Diekman. Tschech and Plattner. 1989). Bacillus circulans (Yabuki. 1989), Mucor rouxi; (Pedraza-Reyes and Lopez-Romero. 1989) and Neurospora crassa (McNab and Glover. 1989), to name but a few. Recently stable chitinase-overproducing mutants of the fungus Aphanocladium album have been isolated following one-step UV mutagenesis (Vasseur et al. 1990).

Two chitinases one from Serratia marcescens and another from Streptococcus griseus are also available commercially. Another enzyme, glucose oxidase which catalyses the oxidation of glucose to gluconate and hydrogen peroxide in the presence of molecular oxygen has frequently been mistaken for a novel antimicrobial compound. Glucose oxidase is ubiquitous in nature and has been found in honey (White, Subers and Schepartz. 1963) and fungi (Kwang-ae Kim. Fravel and Papavizas. 1990). Combinations of glucose oxidase and catalase have also been found effective. with varying degrees of success, in reducing total viable counts and extending shelf-life when sprayed onto the surfaces of hamburger patties, sausages and smoked rainbow trout and packaged under modified atmospheres (Aaltonen. Lehtonen and Karilainen, 1990).

ENZYMICALLY PREPARED PRODUCTS AS PRESERVA-TIVES

Sucrose Esters

The antimicrobial activity of fatty acids and their esterified derivatives has been studied extensively, and useful reviews of the earlier literature can be found in works by Kabara (1979. 1983). Most of these compounds are produced chemically and are used extensively by the food industry as emulsifiers. However. a small selection of these, including sucrose esters, are amenable to enzymic methods of synthesis and are therefore reviewed here. The work of Klibanov at the Massachusetts Institute of Technology has been instrumental in changing our perception of what enzymes can and cannot do. Thus, in 1983, Zaks and Klibanov reported the successful reversal of the degradative action of lipase when used in a hydrocarbon solvent. Synthetic reactions carried out in food solvents (e.g. hexane), low-water systems or supercritical carbon dioxide hold much promise for the future of food ingredient technology. In addition to offering opportunities in the flavourings area (Gatfield, 1988), the reverse action of hydrolytic enzymes could be applied to the synthesis of novel sucrose and carbohydrate esters with antimicrobial as well as emulsifying properties. Although the latter has been a difficult target, recent reports suggest that new breakthroughs are imminent (Poole et al_. 1989; Godtfredsen. 1990; Janssen et al. 1990). Badel et at. (1988) have shown that sucrose esters of mono- and dipalmitate and stearate possess bacteriostatic properties against Staphylococcus spp Strep jaecalis, E. coli and Proteus morganii in laboratory media incubated at $35°C$. The minimum inhibitory concentrations ranged from 0-05 to 1-25 mg ml-I and were lower in the presence of surfactants such as Tween 80 and Triton X-IOO. It is likely that the surfactants played a role not only in increasing the surface area available for the esters to exert their antimicrobial effect but also as potentiators of inhibition by their action on the cell envelope of the Gram-negative bacteria. However control experiments containing the surfactants only were not carried out by the authors. Furthermore, Marshall and Bullerman (1986) have shown that mixtures of sucrose esters substituted with palmitic and stearic acids inhibited the growth of several mould species from Aspergillus, Penicillium, Cladosporium and Alternaria at an addition level of 1 % in laboratory media. Aflatoxin production by Aspergillus parasiticus, however was not inhibited by 0-1 % of sucrose ester. A recently marketed application of sucrose esters has been in the extension of shelf-life of fresh fruit and vegetables (Anon., 1990a). The new product. Semperfresh, has been prepared from a proprietary combination of sucrose esters palm oil and cellulose to form an invisible. tasteless and odourless film_ Semperfresh acts essentially as a barrier to oxygen and water, and somewhat less so to carbon dioxide consequently extending the ripening time of the produce (Anon . 1990b). The shelf-life of pineapples stored at ambient temperatures has been extended from 11 to 28 days, while bananas remained fresh for 4 weeks instead of the more usual 2 weeks following coating of the fruit with Semperfresh. Water loss from refrigerated mange-tout peas has also been inhibited by the coating (Anon. 1990b).

PLANT AND ALGAL PRODUCTS

It has been suggested that plant cell tissue culture could be used to produce valuable chemicals including antimicrobial food ingredients (Evans and Whitaker. 1987). Although plant cell culture cannot at present, compete with bacterial or yeast batch cell culture for cost of production certain unique compounds, synthesized exclusively by plants may become commercially important in the future. Furthermore the potential for public acceptance of food ingredients prepared by plant cell culture is good in the light of an apparent consumer demand for more 'natural' foods or foods to which chemically synthesized preservatives have not been added. Plants produce an impressive array of antimicrobial compounds some of which have been recognized if not necessarily characterized chemically since ancient times. For detailed reviews of the area the reader is referred to Hargreaves et al. (1975). Beuchat and Golden (1989) and Leadbetter (1991). The active ingredients responsible for the antimicrobial activity in herbs and spices have generally been found in the esssential oil fraction and have been identified as mixtures of esters, aldehydes ketones terpenes and phenolic compounds. Much recent interest has centred on the terpene class of compounds-notably eugenol from cloves and thymol from thyme. However although shown to be active in laboratory trials (Jay and Rivers, 1984) the flavours associated with these materials have limited their use in foods. Similarly some plant-derived pigments such as the anthocyanins have been shown to possess antimicrobial activity (Beuchat and Golden. 1989) but are unlikely to find application in foods where high pigmentation is undesirable. Nevertheless it may be possible to enhance the antimicrobial potential of plant-derived flavouring and colouring compounds in foods in which the microbial flora has been subjected to additional stresses such as for example heat. Thus it has been shown that heat-stressed yeasts had increased sensitivity to the essential oils and crude solvent extracts of allspice clove garlic onion oregano thyme and cinnamon (Conner and Beuchat. 1984, 1985).

In response to infection or injury many plants are also known to produce phytoalexins low molecular weight compounds with broad-spectrum antimicrobial activity (Dixon. 1986). Although many phytoalexins have been shown to be toxic to animal cells in tissue culture it is conceivable that once more is known about their structure/activity relationships analogues without the undesirable toxic elements could be prepared (Wilkins and Board. 1990).

The use of plant cell culture for the production of antimicrobial compounds is clearly still in the future. Nevertheless, the first products of plant cell culture, such as shikonin, the purple pigment from Uthospermum erythrorhizon, are already on the market in Japan and more products are likely to follow

Algae

Of the 150000 species of marine algae estimated to exist, approximately 30 000 have been identified, although even the basic taxonomy of some of the commonly harvested species has not been completed (Harvey. 1988). The better-known macroalgae or seaweeds the Gefidium, Gracilaria, Chondrus and Macrocystis are still hand-collected from the world's oceans for their agar, carrageenan and alginate. The microalgae, a large and diverse group of photosynthetic micro-organisms, also comprise several thousand species (Borowitzka, 1988). Of these fewer than 50 species have been studied in detail with respect to their physiology, biochemistry and potential for mass culture and industrial exploitation.

The microalgae macroalgae and cyanobacteria (blue-green algae) produce a host of antibacterial substances, and the reader is referred to a paper by Jones (1986) for a more extended review of the subject. However, the exploitation of many algal antibiotics will be heavily dependent on the ability to mass cultivate these organisms. Between 1974 and 1981, scientists at the Roche Research Institute of Marine Pharmacology (RRIMP) near Sydney, Australia, screened extracts from marine macroalgae for biologically active chemical substances with potential for application in the medical and veterinary fields. Initial screening was carried out on crude extracts obtained from frozen algae using a range of aqueous and non-aqueous solvents. The extracts were tested ill vitro against several human pathogens (three Gram-positive bacteria, four Gram-negative bacteria, two yeasts, two fungi and one protozoan). Those extracts shown to have promising activity were also tested in vivo in laboratory mice. Of the 159 species of marine plants screened, 118 (74%) showed in vitro activity against one or more of the microbes used in the screen. However, of the crude extracts subsequently tested in vivo, only nine (5%) showed activity (Reichelt and Borowitzka, 1984)

In the Reichelt and Borowitzka study (1984) several interesting compounds active in vitro were isolated from the Rhodophyta (red algae) and found to be generally halogenated. However a number of these were toxic to mice.

By far the highest proportion of inhibitory activity was found in the genus Cystophora of the Phaeophyta class (brown algae) when tested against Gram-positive bacteria. Much of this activity was attributed to phenolic compounds that commonly occur in brown algae. Some of the compounds with in vivo activity were identified e.g. an alkylated resorcinol from Cystophora torulosa a phloroglucinol from Cystophora scalars and a ä — tocotrienol from Cystophora expansa. Phenolics are generally considered to be toxic compounds; however the presence of an alkyl group, such as in the algal phenolics, is known to render them less toxic. In addition, the Australian group observed that many of the algal antimicrobials identified in their study were rapidly eliminated in the mouse by either excretion or metabolism, a property that is undesirable in pharmaceuticals but may be highly beneficial in food applications. Studies on the in vivo activities of these compounds were not completed, owing to the closure of the Institute by Roche (Reichelt and Borowitzka. 1984).

Mabrouk el al. (1985) have shown that five marine macroalgae of the Phaeophyta class (Sargassum despiense. Turbinaria decurrense. Dilophus ligulatus, Cystoseira myrica and Padina pavonia) inhibited mycelial growth and aflatoxin production by a toxigenic strain of Aspergillus flavus. The algae were prepared by washing fresh biomass with tap-water drying at 40°C until constant weight was reached and finely grinding the product. Concentrations of algal powder ranging from 10 to 200 g 1-1 of culture medium were tested in shake-flask culture. The volatile constituents of the Phaeophyta have been shown to contain fatty acids from C1 to CIO including acetic and propionic acids: lower terpenes, such as linalool and geraniol; and phenolic compounds such as p-cresol. The preservative activity of all of these compounds is well known. Other compounds that may also contribute to the preservative potential of algal extracts include fatty acids with longer chain lengths (CIO-C20), sterols and triterpenes.

It has been claimed that both cell extracts and extracts of the spent growth media of the microalgae Chlorella vulgaris and Chlamydomonas pyrenoidosa have antimicrobial activity against Gram-positive and Gram-negative bacteria. Extracts from Dunaliella tertiolecta, Rivularia firma and Gomphosphaeria aponina have been reported to produce a range of pharmacological activities (Pabon de Majid and Martin, 1983; de Pauw and de la Noue, 1986; Borowitzka. 1988; Regan, 1988). However the cyanobacterium Lyngbya majuscula is known to produce a toxin that is the causative agent of 'swimmer's itch' (Cohen. 1986). Clearly as with all novel food preservatives, potential toxicological effects should not be ignored. The major obstacle to the exploitation of algal products is the difficulty in adapting these marine plants and micro-organisms to mass cultivation methods. Microalgae are much more amenable than macroalgae to strictly control mass culture conditions and can be manipulated to produce the desired product in abundance. Thus, a potentially interesting biological activity will only be useful if access to future supplies of the same organism is assured. Several biotechnology companies in California, Australia. Israel and the UK are known to be working in this area.

CONSTRAINTS

A number of factors can be expected to impede the rapid development of biotechnological production methods in the food industry in the near future. First, most foods are of a very complex nature and the interrelationships between the molecular components in food are still poorly understood. Consequently, the relationship between the functional properties of a food ingredient tested in isolation and the performance of that ingredient in the finished food product is often difficult to establish. Thus, an antimicrobial that works well ill vitro will not necessarily perform well in a food and vice versa.

Secondly, there is often a weak relationship between the market value of a food ingredient and its technical quality. Food is an emotional issue, which affects everyone personally. The properties of appearance, taste, freshness, texture, odour, 'naturalness', convenience, etc. are all judged subjectively by the consumer. Consumer acceptance of biotechnoiogically produced food preservatives will be critical in the success of these new products. In this respect, the association in the public mind of genetic engineering with biotechnology has already had adverse effects on the penetration of biotechnology in the food industry. In the future, it will be important to maintain an open flow of information between the food industry, health educators, consumer organizations and the media in order to avoid a reaction of the sort recently experienced with the introduction of food irradiation into the UK

Thirdly, the regulatory clearances necessary for the manufacture of any new additive frequently involve the investment of substantial resources (Korwek, 1987; McNamara, 1989). The requirements for toxicological testing of new food ingredients are now so stringent that some food manufacturers are avoiding any new developments. Furthermore, many countries now require that a case of 'need' for any new additive be established. Nevertheless, precedents are being set with several biotechnologically derived products entering the market in 1989 and 1990: FDA approval for use in selected foods of gellan gum from Pseudomonas elodea; the mammalian chymosins from recombinant Kluyveromyces lactis and E. coli; and a genetically engineered bakers' yeast. All of these events will undoubtedly help to make the biotechnological production of food ingredients more acceptable to the public in the future.

ACKNOWLEDGEMENTS

The author wishes to thank Dr Paul Gibbs of the Applied Microbiology Section and Professor lain Dea of the Technology Group for encouragement, advice and a critical review of the manuscript

REFERENCES

1.   AALTONEN, P.K., LEHTONEN, P.O. AND KARILAINEN, U. (1990). GhlCose oxidase food treatment and storage method. International Patent Application (peT), WP 90/04336.

2.   ANDERS, R.F., HOGG, D.M. AND JAGO, G.R. (1970). Formation of hydrogen peroxide by Group N streptococci and its effect on their growth and metabolism. Applied Microbiology 19(4),608-612.

3.   ANONYMOUS (1984a). Bacon nitrite can be greatly reduced with no botulism risk, USDA says. Food Chemical News 22, 33-34.

4.   ANONYMOUS (1984b). USDA proposal to permit 213 reduction in bacon nitrite levels. Food Chemical News 24, 50-51

5.   ANONYMOUS (1985). Purified milk proteins preserve foods. Food Processing July, 8.

6.   ANONYMOUS (1986). Biotechnology creates natural preservatives. Food Engineering August, 95.

7.   ANONYMOUS (1990a). Extended shelf life at no extra cost'? The Grocer July, 41-43.

8.   ANONYMOUS (1990b). Semperfresh taking root on a worldwide basis. Fresh Produce Journal 20 July, 16

9.   ARCHER, D.B., JEENES, D.J., McKENZIE, D.A., BRIGHTWELL, G., LAMBERT, N., LOWE. G., RADFORD, S.E. AND DOBSON. M. (1990). Hen egg white lysozyme expressed in. and secreted from. Aspergillus niger is correctly processed and folded. Bio/Technology 8, 74[-745.

10. ATTAIE. R.. WHALEN, P.J.. SHAHAN', K.M. AND AMER. M.A (1987). Inhibition of growth of Staphylococcus aureus during production of acidophilus yogurt. Journal of Food Protection 50(3). 224-228.

11. BADEL. .. DESCOTES, G . MENTECH, J. ANDTHIRIET. B. (1988). Nouveauxproduits derives du saccharose. European Patent Application 0 349 431 Al.

12. BAIRD·PARKER. A.e. (1990). Foodborne salmonellosis. Lancet 336. 1231-1235.

13. BANERJEE, S. AND HANSEN. J.N. (1988). Structure and expression of a gene encoding the precursor of subtilin. a small protein antibiotic. Journal of Biological Chemistry 263. 9508-9514.

14. BANKS, J.G .• BOARD. R.G. AND SPARKS. N.H.e. (1986). Natural antimicrobial systems and their potential in food preservation of the future. Biotechnology and Applied Biochemistry 8, 103--147.

15. BARNBy·SMITH. F.M. AND ROLLER. S.D. (1990). Production of antimicrobial compounds by lactic acid bacteria. Proceedings of the 5th European Congress on

16. Biotechnology, volume 1 (e. Christiansen, L. Munck and J. Villadsen, Eds). pp. 302-305. Munksgaard, Denmark.

17. BARNBy·SMIHI, F.M .. ROLLER, S.D., WOODS, L.F.J .• BARKER. M .. NIGHTINGALE. M. AND GIBBS. P. A. (1989). Production of antimicrobial compounds by lactic acid bacteria. Leatherhead Food Research Association Report No. 662.

18. BELL. RG. AND DE LACY, K.M. (1987). The efficacy of nisin, sorbic acid and monolaurin as preservatives in pasteurized cured meat products. Food Microbiology 4, 2n-283.

19. BENKERROUM. N. AND SANDINE, W.E. (1988). Inhibitory action of nisin against Listeria monocytogenes. Journal of Dairy Science 71. 3237-3245.

20. BERRY. E.D .• LIEWEN, M.B .• MANDIGO, R.W. AND HUTKINS. R.W. (1990). Inhibition of Listeria monocytogenes by bacteriocin-producing Pediococcus during the manufacture of fermented semidry sausage. Journal of Food Protection 53(3}. 194-197.

21. BEUCHAT. LR. AND GOLDEN. D.A (1989). Antimicrobials occurring naturally in foods. Food Technology 43(1). 134-150.

22. BHUNJA, AK., JOHNSON. M.e. AND RAY, B. (1988). Purification. Characterization and antimicrobial spectrum of a bacteriocin produced by Pediococcus acidilactici. Journal of Applied Bacteriology 65,261-268.

23. BHUNIA. AK .. JOHNSON. M.e.. RAY. B. AND BELDEN. E.L (1990). Antigenic property of pediocin AcH produced by Pediococcus acidilactici H. Journal of Applied Bacteriology 69. 211-215.

24. BHUNIA, A.K .. JOHNSON. M.e., RAY. B. AND KALCHAYANAND. N. (1991). Mode of action of pediocin AcH from Pediococcus acidilactici H. on sensitive bacterial strains. Journal of Applied Bacteriology 70, 25-33.

25. BIBI. W. AND BACHMANN. M.R. (1990). Antibacterial effect of the

lactoperoxidase thiocyanate hydrogen peroxide system on the growth of Listeria spp. in skim milk. Milchlllissenschaft 45(1). 26-28.

26. BJORCK, L.. CLAESSON. O. AND SCHULTHESS. W. (1979). The lactoperoxidase thiocyanate hydrogen peroxide system as temporary preservative for raw milk in developing countries. Milchlllissenschaft 4,726-729.

27. BJORCK, L., ROSEN. C.-G., MARSHALL, V. AND REITER. B. (1975). Antibacterial activity of the lactoperoxidase system in milk against pseudomonads and other gram-negative bacteria. Applied Microbiology 30(2), 199-204.

28. BLACKBURN. P., POLAK, J., GUSIK. S.-A. AND RUBINO, S.D. (1989). Nisin compositions for use as enhanced, broad range bactericides. International Patent Application (PCT), WO 89/12399.

29. BOMAN, H.G. (1986). Structure and function of attacins and cecropins. Two classes of antibacterial proteins from insects. In Natural Antimicrobial Systems, Part J, Antimicrobial Systems in Plants and Animals, FEMS Symposium No. 35, pp.116-130. Bath University Press.

30. BOROWITZKA, M. (1988). Vitamins and fine chemicals from microalgae. In Microalgal Biotechnology (M.A. Borowitzka and L.J. Borowitzka. Eds). Cambridge University Press.

31. BOUDREAUX. D.P .• MATROZZA. M.A. AND LEVERONE. M.F. (1989). Method for inhibiting food-borne human pathogens and preventing microbial spoilage in refrigerated foods using Lactobacillus. US Patent 4.874.704.

32. BUCHMAN, G.W., BANERJEE, S. AND HANSEN. J.N. (1988). Structure. expression and evolution of a gene encoding the precursor of nisin. a small protein antibiotic. Journal of Biological Chemistry 263, 16260-16266.

33. BUCHTA. K. (1983a). Lactic acid. In Biotechnology. A Comprehensive Treatise in 8 Volumes, volume 3 (H. HeJlweg, Ed.), pp.409-417. Verlag Chemic, Weinheim.

34. BUCHTA. K. (1983b). Organic acids of minor importance. In Biotechnology. A Comprehensive Treatise in 8 Volumes. volume 3 (H. Hellweg. Ed.). pp.467-478. Verlag Chemie, Weinheim.

35. BURLING. H. (1989). Process for extracting pure fractions of lactoperoxidase and lactoferrin from milk serum. International Patent Application (per) WO 89/04608.

36. CALDERON, c., COLLINS.THOMPSON, D.L. AND USBORNE. W.R.

(1985). Shelf life studies of vacuum packaged bacon treated with nisin. Journal of Food Protection 48(4).330-333.

37. CARINI, S. AND LaDI. R. (1982). Inhibition of germination of clostridial spores by lysozyme. Industria del Latle 18, 35-48.

38. CARMINATI, D., MUCCHETTI. G .• NEVIANI, E. AND EMALDI. G.c. (1984). La germinazione di Clostridium tyrobutyricum in presenza di lisozima. Lalle 9, 897-904.

39. COHEN, Z. (1986). Products from microalgae. In CRC Handbook of Microalgal Culture (A. Richmond, Ed.). CRC Press. Boca Raton. Florida.

40. COLLINS, E.B. AND ARAMAKI. K. (1980). Production of hydrogen peroxide by Lactobacillus acidophilus. Journal of Dairy Science 63. 353-357.

41. CONNER. D.E. AND BEUCHAT, L.R. (1984). Sensitivity of heat-stressed yeasts to essential oils of plants. Applied and Environmental Microbiology 47.229-233.

42. CONNER, D .E. AND BEUCHAT, L. R. (1985). Recovery of heat-stressed yeasts in media containing plant oleoresins. Journal of Applied Bacteriology 59, 49-55.

43. DAESCHEL, M.A. (1989). Antimicrobial substances from lactic acid bacteria for use as food preservatives. Food Technology Jan., 164-167.

44. DAESCHEL, M.A. AND KLAENHAMMER. T.R. (1985). Association of a 13·6· megadalton plasmid in Pediococcus pentosaceus with bacteriocin activity. Applied and Environmental Microbiology SO, 1538-1541.

45. DAHIYA, R.S. AND SPECK, M.L. (1968). Hydrogen peroxide formation by lactobacilli and its effect on Staphylococcus aureus. Journal of Dairy Science 51, 1568-1572.

46. DELVES-BROUGHTON, J. (l990a). Nisin and its uses as a food preservative. Food Technology 44(11), 100-117.

47. DELVES-BROUGHTON, J. (J990b). Nisin and its application as a food preservative. Journal of the Society of Dairy Technology 43(3). 73-76.

48. DE PAUW, N. AND DE LA NOUE, J. (1986). Production and utilisation of microalgae: The potential of microalgal biotechnology. [n Proceedings of the International Symposium on Food and Biotechnology. Quebec. Universite Laval, Canada.

49. DIEKMANN, H .• TSCHECH, A. AND PLATTNER, H. (1989). Purification of Streptomyces olivaceoviridis chitinase by fast protein liquid chromatography. In Chitin and Chitosan (G. Skjak-Braek. T. Anthonsen and P. Sandford, Eds), pp. 207-214. Elsevier Applied Science,

London.

50.   DIXON, R.A. (1986). The phytoalexin response: elicitation. signalling and control of host gene expression. Biological Reviews 61,239-291.

51.   DODD, H.M., HORN, N. AND GASSON. M.J. (1990). Analysis of the genetic determi nant for production of the peptide antibiotic nisin. Journal of General Microbiology 136, 555-566.

52.   DZIEZAK. J.D. (1990). Acidulants: Ingredients that do more than meet the acid test. Food Technology 44(1). 76-83.

53.   EARNSHAW. R.G .. BANKS, J.G.. FRANCOTIE. C. AND DEFRISE. D. (1990). Inhibition of Salmonella typhimurium and Escherichia coli in an infant milk formula by an activated lactoperoxidase system. Journal of Food Protection 53(2). 170-172.

54.   EBNER. H. AND FOl.LMANN. H. (1983). Acetic acid. In Biotechnology. A Comprehensive Treatise in 8 Volumes. volume 3 (H. Hellweg. Ed.). pp.387-407. Verlag Chemie, Weinheim.

55.   ENGSTROM, P., CARLSSON. A.. ENGSTROM. A .. TAO, Z.-J. AND BENNICH. H. (1984). The antibacterial effect of attacins from the silk moth Hyalophora cecropia is directed against the outer membrane of Escherichia coli. EMBO Journal 3, 3347-3351.

56.   ERSFELD·DRESSEN. H .• SABL. H.-G. AND BRANDIS. H. (1984). Plasmid involvement in production of and immunity to the staphylococcin-like peptide Pep 5. Journal of Microbiology]30.3029-3035.

57.   EVANS, D.A. AND WHITAKER, R.J. (1987). Technology for the development of new breeding lines and plant varieties for the food industry. In Food Biotechnology (D. Knorr. Ed.). pp.323-345. Marcel Dekker. New York.

58.   FLEMING, H.P .. ETCHELLS. J.L. AND COSTILOW. R.N. (1975). Microbial inhibition by an isolate of Pediococcus from cucumber brines. Applied Microbiology 30, 1040--1042.

59.   GASSON. M.J. (1984). Transfer of sucrose-fermenting ability, nisin resistance and nisin production into Streptococcus lactis. FEMS Microbiology Letters 21. 7-10.

60.   GASSON. M.J. AND ANDERSON. P.H. (1985). High copy number plasmid vectors for use in lactic streptococci. FEMS Microbiology Letters 30, 193-197.

61.   GATFIELD. I.L. (1988). Production of flavor and aroma compounds by biotechnology. Food Technology October, 110-122.

62.   GIBBS. P.A. (1987). Novel uses for lactic acid fermentation in food

preservation. JOllrnal of Applied Bacteriology. Symposium Supplement. 515-58S.

63. GODTFREDSEN. S.E. (1990). Application of Iipases for synthesis of new chemicals. In Opportunities in Biotransformatiorzs (L.G. Copping. R.E. Martin. I.A. Pickett. C. Bucke and A. W. Bunch. Eds). Elsevier Applied Science, London.

64. GOMBAS. D.E. (1989). Biological competition as a preserving mechanism. Journal of Food Safety 10. 107-117.

65. GNZALES, C. F. (1986). Preservation of foods with non-lactose fermenting Streptococcus lactis subsp. diacetylactis. US Patent 4,599,313.

66. GONZALES, C.F. (1989). Method for inhibiting bacterial spoilage and resulting compositions. US Patent 4,883.673.

67. GROSS, E. AND MORELL, J.L. (1971). The structure of nisin. Journal of the American Chemical Society 93, 4634-4637.

68. GRUTTER, M.G. AND MATmEws. B. W. (1982). Amino acid substitutions far from the active site of bacteriphage T4 lysozyme reduce catalytic activity and suggest that the C-terminal lobe of the enzyme participates in substrate binding. Journal of Moleclliar Biology 154.525-535.

69. GUDMUNDSSON. B. (1984). Lactoperoxidase System; A Defense System of Nature with Importance for Allimal Health. EWOS. Sodertalja, Sweden.

70. HANSEN, J.N. BANERJEE. S. AND BUCHMAN, G.W. (1989). Potential of small ribosomally synthesised bacteriocins in design of new food preservatives. Journal of Food Safety 10, 119-130.

71. HARDING. C.D. AND SHAW. B.G. (1990). Antimicrobial activity of Leuconostoc gelidum against closely related species and Listeria monocytogenes. Journal of Applied Bacteriology 69,648-654.

72. HARGREAVES. L.L., JARVIS, B. RAWLINSON, A.P. AND WOOD, J.M. (1975). The antimicrobial effects of spices, herbs and extracts from food plants. Leatherhead Food Research Association Scientific and Technical Surveys No. 88.

73. HARVEY. W. (1988). Cracking open marine algae's biological treasure chest. Biotechnology 6. 488-492.

74. HASTINGS,J.W. ANOSTILES, M.E. (1991). Antibiosis of Leuconostocgelidum isolated from meat. Journal of Applied Bacteriology 70. 127-134.

75. HAWKES, R., GRUTTER, M.G. AND SCJiELLMAN, J. (1984).

Thermodynamic stability and point mutations of bacteriophage T4 lysozyme. Journal of Molecular Biology 175, 195-212.

76. HAYASHI. K., KASUMI. T• KUBO, N., HARAGUCHI. K. AND TSUMURA. N. (1989). Effects of N-acetylmuramidase from Streptomyces rurgersensis H-46 as a food preservative. Agricultural and Biological Chemistry 53(12). 3173-3177.

77. HIRSCH. A. GRINSTED. E., CHAPMAN. H.R. AND MATTICK. A. (1951). A note on the inhibition of an anaerobic spore former in Swiss-type cheese by a nisin-producing Streptococcus. Journal of Dairy Research 18(2).205-207.

78. HOOGENDOORN. H. (1974). The effect of lactoperoxidase thiocyanate hydrogen peroxide on the metabolism of cariogenic microorganisms in vitro and in the oral cavity. In Industrial Enzymology (T. Godfrey and 1. Reichelt. Eds), pp.433-443. Nature Press, New York.

79. HORNER, T., ZAHNER, H KELLNER. R. AND JUNG. G. (1989). Fermentation and isolation of epidermin. a lanthionine containing polypeptide antibiotic from Staphylococcus aureus. Applied Microbiology and Biotechnology 30. 219-225.

80. HULTMARK. D STEINER. H., RASMUSON. T. AND BOMAN. H.G. (1980). Insect immunity. Purification and properties of three inducible bactericidal proteins from hemolymph of immunized pupae of Hyalophora cecropia. European Journal of Biochemistry 106,7-16.

81. HULTMARK. D., ENGSTROM, A., BENNICH. H• KAPUR, R. AND BOMAN. H.O. (1982). Insect immunity. Isolation and structure of cecropin D and four minor antibacterial components from Cecropia pupae. European Journal of Biochemistry 127,207-217.

82. HULTMARK, D., ENGSTROM, A., ANDERSSON. K• STEINER, H .• BENNIC~I. H. AND BOMAN. H.G. (1983). Insect immunity. Attacins. a family of antibacterial proteins from Hyalophora cecropia. EMBO Journal 2, 571-576.

83. JANSSEN, A.E.M., KLABBERS, c., FRANSSEN. M.C.R. AND VAN'T RIET, K. (1990). Enzymatic synthesis of carbohydrate esters in 2-pyrrolidone, Proceedings of the 5th European Congress on Biotechnology, volume 1 (C. Christiansen. L. Munck and J. Villadsen, Eds), pp.441-444. Munksgaard. Denmark.

84. JAY, 1.M. AND RIVERS, G.M. (1984). Antimicrobial activity of some food flavoring compounds. Journal of Food Safety 6, 129-139.

85. JONES, A.K. (1986). Eukaryotic algae - Antimicrobial systems. In Natural Antimicrobial Systems, Part 1, Antimicrobial Systems in Plants

and Animals. FEMS Symposium No. 35, pp. 1 16-130. Bath University Press.

86. JONES, D. (1990). Foodbome listeriosis. Lancet 336, 1171-1174.

87. KABARA, I.J. (Ed.) (1979). The Pharmacological Effects of Lipids. American Oil Chemists Society. Monograph No.5, Urbana, Champaign, II., USA. .

88. KABARA, 1.J. (1983). Medium-chain fatty acids and esters. In Antimicrobials in Food (A. Branen and M. Davidson, Eds). Pp.109-140. Marcel Dekker, New York.

89. KATAYAMA, T. (1967). In Physiology and Biochemistry of Algae (R.A. Lewin. Ed.). pp.467-473. Academic Press. New York.

90. KLAENHAMMER. T.R. (1988). Bacteriocins of lactic acid bacteria. Biochimie 70. 337-349.

91. KORWEK. E.L. (1987). Regulatory aspects of the use of modern biotechnological methods in the food industry. In Food Biotechnology (D. Knorr, Ed.), pp.559-578. Marcel Dekker, New York.

92. KUMAGAI,I., KOJIMA, S . TAMAKI. E. AND MIURA, K. (1987). Conversion ofTrp 62 of hen egg white lysozyme to Tyr by site-directed mutagenesis. Journal of Biochemistry 102.733-740.

93. KWANG-AE KIM. K . FRAVEL. D.R. AND PAPAVIZAS. G.e. (1990). Glucose oxidase as the antifungal principle of tala ron from Taloromyces flavus. Canadian Journal of Microbiology 36.760 ... 764.

94. LAI. J.S LEE. J.-H. LEI. S.-P. LIN. Y.-L WEICKMANN. J.L. AND BLAIR. L.e. ((989). Production of food additives and food processing enzymes by recombinant DNA technology. In Biotechnology and Food Quality (5.-0. Kung. D.O. Bills and R. Quatrono. Eds). Pp.337-354. Butterworths. Boston.

95. LEADBETIER. S. (1991). Natural antimicrobial agents - A literature survey. Leatherhead Food Research Association Food Focus No. 13 May.

96. LE BLANC, D.J .. CROW. V.L. AND LEE. L.N. (1980). Plasmid mediated carbohydrate catabolic enzymes among strains of Streptococcus JaClis. In Plasmids and TransPOSOllS: Environmental Effects and Maintenance Mechanisms (c. Stuttard and K. Rozee. Eds). Pp.31-41. Academic Press. New York.

97. LOCK. E.. KLUG. C LOTZ. A. AND WOfiNER. G. (1988). Zubereitung zur Verkmgerullg der Haltbarkeit VOll Lebensmiueln. Arzneimitteln und Kosmetischen Produkten. International Patent Application (PCT). WO 88/05992.

98. MABROUK. S.S. El.-SHAYEB. N.M.A. EL-REFAIR. A.H. SALLAM. L.A.R. AND HAMDY. A.A. (1985). Inhibitory activities of some marine algae on aflatoxin accumulation. Applied Microbiology and Biotechnology 22. 152-155.

99. MCKA Y. L.L. (1989). Genetic engineering of lactic starter cultures. In Biotechnology and Food Quality (S.-D. Kung. D.O. Bills and R. Quatrano. Eds). pp.317-335. Butterworths. Boston.

100. McNAB. R. AND GLOVER. A. (1989). The cytosolic chitinase of Neurospora crassa. In Chitin and Chitostlll (G. Skjak-Braek. T. Anthonsen and P. Sandford. Eds). 255-265. Elsevier Applied Science. London.

101. MCNAMARA. S.H. (1989). FDA regulation of food substances produced by new techniques of biotechnology. In Biotechnology Challenges in the Flavor and Food Industry (R.C. Lindsay and BJ. Willis. Eds). Pp.65-89. Elsevier Applied Science. London.

102. MARSHALL. D.L. AND BULLERMAN. L.B. (1986). Antimicrobial activity of sucrose fatty acid ester emulsifiers. Journal of Food Science 51(2).468-470.

103. MIKI. T YASlJKOCIII. . NAGATANI. H. FURUNO. M., ORITA. T YAMADA, H .lMOTO. T. AND HORlunll. T. (1987). Construction of a plasmid vector for the regulatable high level expression of eukaryotic genes in Escherichia coli: an application to overproduction of chicken lysozyme. Protein Engineering I. 327-332.

104. MORRIS, S.L. WALSH. R.C. AND HANSEN, J.N. (1984). Identification andcharacteTisation of some bacterial membrane sulphydryl groups which are targets of bacteriostatic and antibiotic action. Journal of Biological Chemistry 259, 13590-13598.

105. MORTVEDT. C.I. AND NEs. I.F. (1990). Plasmid-associated bactreriocin production by a Lactobacillus sake strain. Journal of General Microbiology 136. 1601-1607.

106. MULDER. R.W.A.W... VAN DER HULST, M C. AND BOLDER, N.M. (1987). Research Note: Salmonella decontamination of broiler carcasses with lactic acid. L-cysteine and hydrogen peroxide. Poultry Science 66, 1555-1557.

107. NIELSEN. J.W DICKSON. J.S. AND CROUSE, J.D. (1990). Use of a bacteriocin produced by Pediococcus addilactld to inhibit Listeria monocYlogenes associated with fresh meat. Applied and Environmelllal Microbiology 56(7), 2142-2145.

108. OISHI. K. FUMIYA$U. I. AND MASAO, N. (1989). Chitinolytic and lysozymic activities in plants. In Chitin and Chitosan (G. Skyjak-Braek,

T. Anthonsen and P. Sandford, Eds). Pp.185-195. Elsevier applied Science. London.

109. OSA. J.M._ BENEZET. A., BOTAS, M.• OLMO, N. AND PEREZ. FLOREZ, F. (1990). Accion de la Iisozima commercial (c1orhidrato), frente a distinas especies bacterianas. Alimentaria 210,49-59.

110. PABON DE MAJID. L. AND MARTIN. D.F. (1983). Induction of sessile-stage formation of the red tide organism Ptychodiscus brevis by materials elaborated by Gomphosphaeria aponina. Microbios Letters 22,59-65.

111. PEDRAZA-REYES, M. AND LOPEZ-ROMERO, E. (1989). Purification and some properties of two forms of chitinase from mycelial cells of Mucor rouxii Journal of General Microbiology 135, 211-218.

112. PERRY, L.J. AND WETZEL, R. (1984). Disulfide bond engineered into T4 lysozyme: stabilisation of the protein toward thermal inactivation. Science 226,555-557.

113. POOLE, S., WYETH, L.J., JAMES, M.J., PATEL, P.D., HART, R.J. AND CLARK, S.A. (1989). Chemical and enzymic modification of biopolymers. Leatherhead Food Research Association Research Report No. 645.

114. PREMI, L. AND BOITAZZI, V. (1972). Hydrogen peroxide formation and hydrogen peroxide splitting activity in lactic acid bacteria. Milchwissenschaft 27(12), 762-765.

115. PRICE, R.J. AND LEE, J.S. (1969). Inhibition of Pseudomonas species by hydrogen peroxide producing lactobacilli. Journal of Milk and Fermentation Technology 33, 13-18.

116. PUCCI, M.J., VEDAMUTHU, E.R., KUNKA, B.S. AND VANDENBERGH, P.A. (1988). Inhibition of Listeria monocylOgenes by using bacteriocin PA-l produced by Pediococcus acidilactici PAC 1.0. Applied and Environmental Microbiology 54(10), 2349-2353.

117. REDDY, G.V. AND SHAHANI, K.M. (1971). Isolation of an antibiotic from Lactobacillus bulgaricus. Journal of Dairy Science 54,748-752.

118. REDDY, G.V., SHAHANI, K.M., FRIEND, B.A. AND CHANDAN, R.C. (1983). Natural antibiotic activity of Lactobacillus acidophulus and bulgaricus III: production and partial purification of bulgarican from Lactobacillus bulgaricus. Cultured and Dairy Products jounal18, 15-17.

119. REGAN, D.L. (1988). Other micro-algae. In Microalgal Biotechnology (M.A. Borowitzka and L.J. Borowitzka, Eds), pp. 52-78. Cambridge University Press.

120. REICHELT, 1.L. AND BOROWITZKA, M.A. (1984). Antimicrobial

activity from marine algae: Results of a large-scale screening programme. Hydrobiologia 116/117, 158-168.

121. REITER, B. (1984). The biological significance and exploitation of some of the immune systems in milk: a review. Microbiologie Aliments Nutrition 2, 1-20.

122. REITER, B. AND HARNULV, G. (1984). Lactoperoxidase antibacterial system: Natural occurrence, biological functions and practical applications. }ournal of Food

123. Protection 47(9), 724-732.

124. ROHR, M., KUBICEK, C.P. AND KOMINEK, J. (1983a). Citric acid. In Biotechnology. A Comprehensive Treatise in 8 Volumes, volume 3 (H. Hellweg, Ed.), pp.419-454. Verlag Chemie, Weinheim.

125. ROHR, M., KUBICEK, C.P. AND KOMINEK, J. (l983b). Gluconic acid. In Biotechnology. A Comprehensive Treatise in 8 Volumes, volume 3 (H. Hellweg. Ed.), pp.455-465. Verlag Chemie, Weinheim.

126. RUITLOFF, H. (1987). Impact of biotechnology on food and nutrition. In Food Biotechnology (D. Knorr, Ed.), pp.37-95. Marcel Dekker, New York.

127. SAHL, H.-G. (1985). Bactericidal cationic peptides involved in bacterial antagonism and host defence. Microbiological Sciences 2(7),212-217.

128. SCHILLINGER, U. AND LUCKE, F.-K. (1989). Antibacterial activity of Lactobacillus sake isolated from meat. Applied and Environmental Microbiology 55, 1901-1906.

129. ScOIT, D., HAMMER, F.E. AND SZALKUCKI, T.J. (1987). Bioconversions: Enzyme technology. In Food Biotechnology (D. Knorr, Ed.), pp.413-440. Marcel Dekker, New York.

130. SHAHANI, K.M., VAKIL, J.R. AND CHANDAN, R.C. (1972). Antibiotic acidophilin and process of preparing the same. US Patent No. 3,689,640.

131. SHAHANI, K.M., VAKIL, J.R. AND KILARA, A. (1976). Natural antibiotic activity of Lactobacillus acidophilus and bulgaricus. I. Cultural conditions required for the production of antibiosis. Cultured and Dairy Products Jouma/ll(4). 1417-1422.

132. SHAHANI. K.M .• VAKIL. J.R. AND KILARA. A. (19n). Natural antibiotic activity of Lactobacillus acidophilus and bulgaricus. II. Isolation of acidophilin form L. acidophilus. Cultured Dairy Products Journalll, 8-11.

133. SPELliAUG. S.R. AND HARLANDER. S.K. (1989). Inhibition of

foodborne bacterial pathogens by bacteriocins from Lactococcus lactis and Pediococcus pentosaceus. Journal of Food Protection 52(12).856-862.

134. TAGG, J.R., DAJANI, A.S. AND WANNAMAKER. L.W. (1976). Bacteriocins of gram-positive bacteria. Bacteriological Reviews 40. 722-756.

135. TALARICO. T.L. AND DOBROGOSZ. W.J. (1989). Chemical characterization of an antimicrobial substance produced by Lactobacillus reuteri. Antimicrobial Agents al!d Chemotherapy 33. 67~79.

136. TALARICO, T.L... CASAS, LA., CHUNG, T.C. AND DOBROGOSZ, W.J. (1988). Production and isolation of reuterin, a growth inhibitor produced by Lactobacillus reuteri. Antimicrobial Agents and Chemotherapy 32, 1854-1858.

137. TANAKA, N., GORDON. N., LINDSAY, R.e., MESKE, L., DOYLE, M.P. AND TRAIS· MAN. E. (1985a). Sensory characteristics of reduced nitrite bacon manufactured by the Wisconsin Process. Journal of Food Protection 48(8). 687-692.

138. TANAKA, N., MESKE, L., DOYLE, M.P., TRAISMAN, E., THAYER. D.W. AND JOHNSTON. R.W. (1985b). Plant trials of bacon made with lactic acid bacteria sucrose and lowered sodium nitrite. Journal of Food Protection 48(8). 679-686.

139. TAYLOR. S. AND SOMERS, E. (1985). Evaluation of the antibotulinal effectiveness of nisin in bacon. Journal of Food Protection 48. 949-952.

140. TAYLOR. S., SOMERS. E. AND KRUEGER. L. (1985). Antibotulinal effectiveness of nisin-nitrite combinations in culture medium and chicken frankfurter emulsions. Journal of Food Protection 48. 234-239.

141. TEN BRINK. B., BOL. J. AND HUIS IN'T VELD, J.H.J. (1988). Antimicrobial compounds produced by lactic acid bacteria: The importance of a good screening procedure. Paper presented at the SAB Colloquium on Lactic Acid Bacteria in Food Preservation, Guildford, 18-21 July. Journal of Applied Bacteriology 65, 261-268

142. TOKURA, S. (1989). Structure and chemical modification of chitin and chitosan. In Chitil! and Chitosan (G. Skjak-Braek, T. Anthonsen and P. Sandford. Eds). Pp.45-50. Elsevier Applied Science, London.

143. VAKIL. J.R., CHANDAN, R.C., PARRY. R.M. AND SHAHANI, K.M. (1969). Susceptibility of several microorganisms to milk Iysozymes. Journal of Dairy Science 52. 1192-1195.

144. VASSEUR. V. ARIGONI, F., ANDERSEN, H. DEFAGO. G.

BOMPEIX. G. AND SENG. J.-M. (1990). Isolation and characterization of Aphanocladium album chitinaseoverproducing mutants. Journal of General Microbiology 136, 2561-2567.

145. WASSERFALL. F., VOSS. E. AND PROKOPEK, D. (1976). Experiments on cheese ripening: the use of lysozyme instead of nitrite to inhibit late blowing of cheese. Kielische Milchwirlschaft Forschungsbereite 28.3-16.

146. WHARTON, B.A. (Ed.) (1982). Food for the suckling. Acta Paediatrica Scandinavica, Suppl. 299.

147. WHITE. J.W. JR. SUBERS, M.H. AND SCHEPARTZ, A.1. (1963). the identification of inhibine, the antibacterial factor in honey as hydrogen peroxide and its origin in a honey glucose-oxidase system. Biochimica et Biophysica Acta 73, 57-70.

148. WHITTENBURY. R. (1964). Hydrogen peroxide formation and catalase activity in the lactic acid bacteria. Journal of General Microbiology 35, 13-26.

149. WILKINS. K.M. AND BOARD. R.G. (1989). Natural antimicrobial systems. In Mechanisms of Action of Food Preservation Procedures (G.W. GOUld, Ed.). Pp.285-362. Elsevier Applied Science, London.

150. WORLD, HEALTH, ORGANISATION, (1969). Specification for the identity and purity of food additives and their toxicological evaluation: Some antibiotics. 12th Report of the Joint F APIWHO Expert Committee on Food Additives. WHO Technical Report Series No. 430. World Health Organisation. Geneva, Switzerland.

151. YABUKJ, M. (1989). Characterisation of chitosanase produced by Bacillus circulans. In Chitin and Chitosan (G. Skjak-Braek, T. Anthonsen and P. Sandford. Eds), pp.197-206. Elsevier Applied Science.

152. ZAKS, A. AND KLINABOV, A.M. (1983). Enzymic catalysis in organic media at 100°c. Science 244, 1249-1251.

Chapter 5

OXYGEN SCAVENGERS: AN APPROACH ON FOOD PRESERVATION

Renato Souza Cruz[1], Geany Peruch Camilloto[2] and
Ana Clarissa dos Santos Pires[2]

[1]Technology Department, State University of de Feira de Santana, Feira de Santana,
BA,, Brazil
[2]Food Tecnhology Department, Federal University of Viçosa, Viçosa, MG,, Brazil

INTRODUCTION

Many foods are very sensitive for oxygen, which is responsible for the deterioration of many products either directly or indirectly. In fact, in many cases food deterioration is caused by oxidation reactions or by the presence of spoilage aerobic microorganisms. Therefore, in order to preserve these products, oxygen is often excluded.

Oxygen (O_2) presence in food packages is mainly due to failures in the packaging process, such as mixture of gases containing oxygen residues, or inefficient vacuum. Vacuum packaging has been widely used to eliminate oxygen in the package prior to sealing. However, the oxygen that permeates from the outside environment into the package through the packaging material cannot be removed by this method (Byun et al., 2011).

Modified atmosphere packaging (MAP) is often used as an alternative to reduce the O_2 inside food packaging. However, for many foods, the levels of residual oxygen that can be achieved by regular (MAP) technologies are too high for maintaining the desired quality and for achieving the sought shelf-life (Damaj et al., 2009). The use of oxygen scavenging packaging materials means that oxygen dissolved in the food, or present initially in the headspace, can potentially be reduced to levels much lower than those achievable by modified atmosphere packaging (Zerdin et al., 2003). In this context, research and developments in the food packaging area have been conducted, aiming to eliminate residual O_2. One of the most attractive subjects is the active packaging concept. Active packaging includes oxygen and ethylene scavengers, carbon dioxide scavengers and emitters, humidity controllers, flavor emitters or absorbers and films incorporated with antimicrobial and antioxidant agents (Santiago-Silva et al., 2009).

The most used active packaging technologies for food are those developed to scavenge oxygen and were first commercialized in the late 1970s by Japan's Mitsubishi Gas Chemical Company (Ageless®). In the case of gas scavengers, reactive compounds are either contained in individual sachets or stickers associated to the packaging material or directly incorporated into the packaging material (Charles et al., 2006).

The first patent of an absorber was given in 1938 in Finland. This patent was developed to remove the residual oxygen in headspace of metallic packaging. The method of introduction of hydrogen gas in the packaging to react with oxygen in palladium presence was commercialized in 1960s however this method has never been popularized and well accepted because the hydrogen was unstable during manipulation and storage and, furthermore, it is expensive and unwholesome (Abe and Kondoh, 1989). Recently, more than 400 patents were recorded, mainly in EUA, Japan and Europe, due the great interest by absorbers use (Cruz et al., 2005).

Oxygen scavengers are becoming increasingly attractive to food manufacturers and retailers and the growth outlook for the global market is bullish. Pira International Ltd estimated the global oxygen scavenger market to be 12 billion units in Japan, 500 million in the USA and 300 million in Western Europe in 2001. This market was forecast to grow to 14.4 billion in Japan, 4.5 billion in the USA and 5.7 billion in Western Europe in 2007 (Anon., 2004). In addition, Pira International Ltd. estimated the global value of this market in 2005 to be worth $588 million and has forecast this market to be worth $924 million in 2010. The increasing popularity of oxygen scavenging polyethylene terephthalate (PET) bottles, bottle caps and crowns for beers and other beverages has greatly contributed to this impressive growth (Anon., 2005).

Overall, oxygen absorbing technology is based on oxidation or combination of one of the following components: iron powder, ascorbic acid, photosensitive polymers, enzymes, etc. These compounds are able to reduce the levels of oxygen to below 0.01%, which is lower than the levels typically found (0.3-3%) in the conventional systems of modified atmosphere, vacuum or substitution of internal atmosphere for inert gas (Cruz et al., 2007). A summary of the most important trademarks of oxygen scavenger systems and their manufacturers is shown in Table 1.

An appropriate oxygen scavenger is chosen depending on the O_2-level in the headspace, how much oxygen is trapped in the food initially and the amount of oxygen that will be transported from the surrounding air into the package during storage. The nature of the food (e.g. size, shape, weight), water activity and desired shelf-life are also important factors influencing the choice of oxygen absorbents (Vermeiren et al., 2003).

Oxygen scavengers must satisfy several requirements such as to be harmless to the human body, to absorb oxygen at an appropriate rate, to not produce toxic substances or unfavorable gas or odor, to be compact in size and are expected to show a constant quality and performance, to absorb a large amount of oxygen and to be economically priced (Nakamura and Hoshino, 1983; Abe, 1994; Rooney, 1995).

The most well known oxygen scavengers take the form of small sachets containing various iron based powders containing an assortment of catalysts. However, non-metallic oxygen scavengers have also been developed to alleviate the potential for metallic taints being imparted to food products and the detection of metal by in-line detectors. Non-metallic scavengers include those that use organic reducing agents such as ascorbic acid, ascorbate salts or catechol. They also include enzymatic oxygen scavenger systems using either glucose oxidase or ethanol oxidase (Day, 2003).

Table 1: Some manufacturers and trade names of oxygen scavengers (Ahvenainen and Hurme, 1997; Day, 1998; Vermeiren et al., 1999)

Company	Trade Name	Type	Principle/Active substances
Mitsubishi Gas Chemical Co., Ltd. (Japan)	Ageless	Sachets and Labels	Iron based
Toppan Printing Co., Ltd. (Japan)	Fresilizer	Sachets	Iron based
Toagosei Chem. Ind. Co. (Japan)	Vitalon	Sachets	Iron based
Nippon Soda Co., Ltd. (Japan)	Seaqul	Sachets	Iron based
Finetec Co., Ltd. (Japan)	Sanso-cut	Sachets	Iron based
Toyo Pulp Co. (Japan)	Tomatsu	Sachets	Catechol
Toyo Seikan Kaisha Ltd. (Japan)	Oxyguard	Plastic Trays	Iron based
Dessicare Ltd. (US)	O-Buster	Sachets	Iron based

Multisorb technologies Inc. (US)	FreshMax	Labels	Iron based
	FreshPax	Sachets	Iron based
Amoco Chemicals (US)	Amosorb	Plastic film	Unknown
Ciba Specialty chemicals (Switzerland)	Shelfplus O2	Plastic film	Iron based
W.R. Grace and Co. (US)	PureSeal	Bottle crowns	Ascorbate/metallic salts
	Darex	Bottle crowns, bottle	Ascorbate/sulphite
CSIRO/Southcorp Packaging (Australia)	Zero2	Plastic film	Photosensitive dye/organic compound
Cryovac Sealed Air Co. (US)	OS1000	Plastic film	Light activated scavenger
CMB Technologies (UK)	Oxbar	Plastic bottle	Cobalt catalyst/nylon polymer
Standa Industrie (France)	ATCO	Sachets	Iron based
	Oxycap	Bottle crowns	Iron based
	ATCO	Lables	Iron based
Bioka Ltd. (Finland)	Bioka	Sachets	Enzyme based

Structurally, the oxygen scavenging component of a package can take the form of a sachet, label, film (incorporation of scavenging agent into packaging film) (Figure 1), card, closure liner or concentrate (Suppakul et al., 2003).

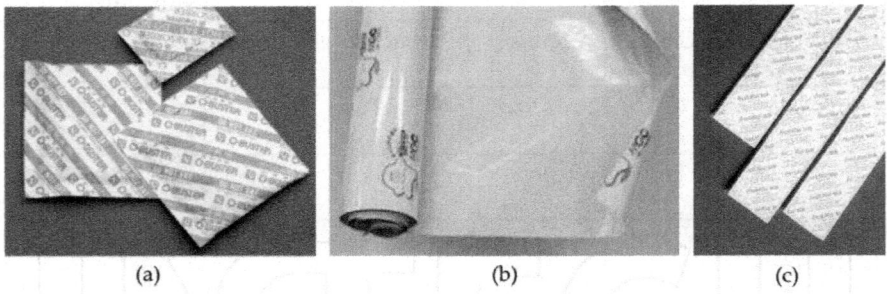

(a)   (b)   (c)

Figure 1: Oxygen scavengers: (A) O-Buster® sachet, (B) OMAC® film and (C) Fresh-Max™ SLD label.

Although the performance of oxygen-absorbing sachets was quite satisfactory for a wide range of food storage conditions, a number of limitations to their use in practice were recognized. The esthetics of inserts, coupled with a concern about possible ingestion or rupture, as well as their unsuitability for use with liquid foods, drove researchers to seek package-based solutions (Rooney, 2005). The incorporation of scavengers in packaging films is a better way of resolving sachet-related problems. Scavengers may either be imbedded into a solid, dispersed in the plastic, or introduced into various layers of the package, including adhesive, lacquer, or enamel layers (Ozdemir and Floros, 2004). In general, the speed and capacity of oxygen-scavenging systems incorporated in the packaging materials are considerably lower than those of (iron-based) oxygen scavenger sachets and labels (Kruijf et al., 2002).

For an oxygen scavenger sachet to be effective, some conditions have to be fulfilled (Nakamura and Hoshino, 1983; Abe, 1994; Smith, 1996). First of all, packaging containers or films with a high oxygen barrier must be used, otherwise the scavenger will rapidly become saturated and lose its ability to trap O_2. Films with an oxygen permeability not exceeding 20 ml/m^2.d.atm are recommended for packages in which an oxygen scavenger will be used. Secondly, for flexible packaging heat sealing should be complete so that no air invades the package through the sealed part. Finally, an oxygen scavenger of the appropriate type and size must be selected. The appropriate size of the scavenger can be calculated using the following formulae (Roussel, 1999; ATCO® technical information, 2002). The volume of oxygen present at the time of packaging (A) can be calculated using the formula:

$$A = \frac{(V - P) \times [O_2]}{100}$$

where V is the volume of the finished pack determined by submission in water and expressed in ml, P is the weight of the finished pack in g and $[O_2]$ is the initial O_2 concentration in package (= 21% if air).

In addition, it is necessary to calculate the volume of oxygen likely to permeate through the packaging during the shelf-life of the product (B). This quantity in ml may be calculated as follows:

$$B = S \times P \times D$$

where S is the surface area of the pack in m^2, P is the permeability of the packaging in ml/m^2/24h/atm and D is the shelf-life of the product in days.

The volume of oxygen to be absorbed is obtained by adding A and B. Based on these calculations, the size of the scavenger and the number of sachets can be determined.

According Cruz et al. (2005), the scavengers may be used alone or combined with modified atmosphere. This association requires the equipments to apply the modified atmosphere and decreases the filling velocity. However, this technique is generally used in the market to reduce the oxygen to desirable levels.

Oxygen scavengers have attracted interest of food researchers, and then in this chapter we will discuss the principles involved in scavenge of O_2, as well the main applications and researches in this field of active food packaging.

OXYGEN SCAVENGERS SYSTEMS

Nowadays, there are many systems of oxygen scavengers based on metallic and non-metallic coumpounds. The mechanism of each system is described below.

Iron Powder Oxidation

The commercially oxygen scavengers available are in form of small sachets containing metallic reducing agents, such as powder iron oxide, ferrous carbonate and metallic platinum. The majority of these scavengers are based on the principle of iron oxidation in water presence. A self-reacting type contains moisture in the sachet and as soon as the sachet is exposed to air, the reaction starts. In moisture-dependent types, oxygen scavenging takes place only after moisture has been taken up from the food. These sachets are stable in open air before use because they do not react immediately upon exposure to air therefore they are easy to handle if kept dry (Vermeiren et al., 1999; Cruz et al., 2005). The action mechanism of oxygen scavenger based on iron oxidation is very complicated and is described by the following reactions.

$$Fe \rightarrow Fe^{2+} + 2\,e^-$$

$$\frac{1}{2}\,O_2 + H_2O + 2\,e^- \rightarrow 2\,OH^-$$

$$Fe^{2+} + 2\,OH^- \rightarrow Fe(OH)_2$$

$$Fe(OH)_2 + \frac{1}{4}\,O_2 + \frac{1}{2}\,H_2O \rightarrow Fe(OH)_3$$

According Shorter (1982), if the oxidation rate of the food product and the oxygen permeability of the packaging were known, it is possible to calculate the required iron amount to maintain the desirable oxygen level during the storage time. A rule of thumb is that 1 g of iron will react with 300 ml of O_2(Labuza, 1987; Nielsen, 1997; Vermeiren *et al.*, 1999). The LD_{50} (lethal dose that kills 50% of the population) for iron is 16 g/kg body weight. The largest commercially available sachet contains 7 grams of iron so this would amount to only 0.1 g/kg for a person of 70 kg, or 160 times less than the lethal dose (Labuza and Breene, 1989).

Cruz et al. (2007) evaluate the efficiency of O-Buster® oxygen - absorbing sachets at relative humidity of 75%, 80% and 85% and different temperatures, 10 ± 2 °C and 25 ± 2 °C. They observed that oxygen absorption by the sachet increased as the relative humidity increased for both temperature. Therefore the oxygen - absorbing sachets were most active under 25 ± 2 °C and 85 % relative humidity. At ambient condition (25 ± 2 °C/75 % RH) the rate of oxygen absorbed was 50 ml/day and 18.5 ml/day for 10 ± 2°C.

Some important iron-based O_2 absorbent sachets are Ageless® (Mitsubishi Gas Chemical Co., Japan), ATCO® O2 scavenger (Standa Industrie, France), Freshilizer® Series (Toppan Printing Co., Japan), Vitalon (Toagosei Chem. Industry Co., Japan), Sanso-cut (Finetec Co., Japan), Seaqul (Nippon Soda Co., Japan), FreshPax® (Multisorb technologies Inc., USA) and O-Buster® (Dessicare Ltd., USA).

Ascorbic Acid Oxidation

The ascorbic acid is another oxygen scavenger component which action based on ascorbate oxidation to dehydroascorbic acid. Most of these reactions is slow and can be accelerated by light or a transition metal which will work as catalyst, e.g., the copper (Cruz et al, 2005).

The ascorbic acid reduce the Cu^{2+} to Cu^+ to form the dehydroascorbic acid (Equation I). The cuprous ions (Cu^+) form a complex with the O_2 originating the cupric ion (Cu^{2+}) and the superoxide anionic radical (Equation II). In copper presence, the radical leads to formation of O_2 and H_2O_2 (Equation III). The copper-ascorbate complex quickly reduces the H_2O_2 to H_2O (Equation IV) without the OH⁻formation, a highly reactive oxidant. The following reactions show the process of oxygen absorber by ascorbic acid.

$$AA + 2\,Cu^{2+} \rightarrow DHAA + 2\,Cu^+ + 2\,H^+ \tag{1}$$

$$2\,Cu^+ + 2\,O_2 \rightarrow 2\,Cu^{2+} + 2\,O_2^- \tag{2}$$

$$2\,O_2^- + 2\,H^+ + Cu^{2+} \rightarrow O_2 + H_2O_2 + Cu^{2+} \tag{3}$$

$$H_2O_2 + Cu^{2+} + AA \rightarrow Cu^{2+} + DHAA + 2\,H_2O \tag{4}$$

These equations can be summarized as described below:

$$AA + \tfrac{1}{2}O_2 \rightarrow DHAA + H_2O,$$

where AA is the ascorbic acid and DHAA is the dehydroascorbic acid.

The total capacity of the O_2 absorption is determined by the amount of ascorbic acid. The complete reducing of 1 mol of O_2 requires 2 moles of ascorbic acid (Cruz et al., 2005).

Ascorbic acid and ascorbate salts are being used in the design of scavengers in both sachet and film technologies. A patent from Pillsbury describes the oxygen-reducing properties of these substances. The active film may contain a catalyst, commonly a transition metal (Cu, Co), and it is activated by water; therefore, this technology is specially indicated for aqueous food products, or when the packaged product is sterilized because the water vapor inside the autoclave is capable of triggering the scavenging process (Brody et al., 2001a).

Enzymatic oxidation (e.g., glucose oxidase and alcohol oxidase)

Some O_2-scavengers use a combination of two enzymes, glucose oxidase and catalase, that would react with some substrate to scavenge incoming O_2. The glucose oxidase transfers two hydrogens from the -CHOH group of glucose, that can be originally present or added to the product, to O_2 with the formation of glucono-delta-lactone and H_2O_2. The lactone then spontaneously reacts with water to form gluconic acid (Labuza and Breene, 1989; Nielsen, 1997). A negative factor of this process is the catalase presence, a natural contaminant found in the glucose oxidase preparation, since the catalase reacts with the H_2O_2 forming H_2O and O_2 and, therefore, decreasing the system efficiency. However, the glucose oxidase production without catalase is so expensive. The reactions can be expressed as follows:

$2glucose + 2O_2 + 2H_2O \rightarrow 2gluconic\ acid + 2H_2O_2$

Where glucose is the substrate.

Since H_2O_2 is an objectionable end product, catalase is introduced to break down the peroxide (Brody and Budny, 1995):

$2H_2O_2 + catalase \rightarrow 2H_2O + O_2$

According the reaction, 1 mol of glucose oxidade reacts with 1 mol of O_2. So, in an impermeable packaging with 500 ml of headspace only 0.0043 mol of glucose (0.78 g) is necessary to obtain 0 % of O_2. The enzymatic efficiency depends on the enzymatic reaction velocity, the substrate amount and the oxygen permeability of the packaging.

Coupled enzyme systems are very sensitive to changes in pH, a_w, salt content, temperature and various other factors. Additionally, they require the addition of water and, therefore, cannot be effectively used for low-water foodstuffs (Graff, 1994). One application for glucose oxidase is the elimination of O_2 from bottled beer or wine. The enzymes can either be part of the packaging structure or put in an independent sachet. The immobilization occurs by different process, such as, adsorption and encapsulation. Both polypropylene (PP) and polyethylene (PE) are good substrates for immobilizing enzymes (Labuza and Breene, 1989). A commercially available O_2-removing sachet based on reactions catalyzed by food-grade enzymes is the Bioka O_2-absorber (Bioka, Finland). It is claimed that all components of the reactive powder and the generated reaction products are food-grade substances safe for both the user and the environment (Bioka technical information, 1999). The oxygen scavenger eliminates the oxygen in the headspace of a package and in the actual product in 12–48 hours at 20 °C and in 24–96 hours at 2–6 °C. With certain restrictions, the scavenger can also be used in various frozen products. When introducing the sachet into a package, temperature may not exceed 60°C because of the heat sensitivity of the enzymes (Bioka technical information, 1999). An advantage is that it contains no iron powder, so it presents no problems for microwave applications and for metal detectors in the production line.

Besides glucose oxidase, other enzymes have potential for O_2-scavenging, including ethanol oxidase which oxidises ethanol to acetaldehyde. It could be used for food products in a wide aw range since it does not require water to operate. If a lot of oxygen has to be absorbed from the package, a great amount of ethanol would be required, which could cause an off-odour in the package. In addition, considerable aldehyde would be produced which could give the food a yoghurt-like odour (Labuza and Breene, 1989).

Unsaturated Hydrocarbon Oxidation

The oxidation of polyunsaturated fatty acids (PUFAs) is another technique to scavenge oxygen. It is an excellent oxygen scavenger for dry foods. Most known oxygen scavengers have a serious disadvantage: when water is absent, their oxygen scavenging reaction does not progress. In the presence of an oxygen scavenging system, the quality of the dry food products may decline rapidly because of the migration of water from the oxygen scavenger into the food. Mitsubishi Gas Chemical Co. holds a patent that uses PUFAs as a reactive agent. The PUFAs, preferably oleic, linoleic or linolenic, are contained in carrier oil such as soybean, sesame or cottonseed oil. The oil and/or PUFA are compounded with a transition metal catalyst and a carrier substance (for example calcium carbonate) to solidify the oxygen scavenger composition. In this way the scavenger can be made into a granule or powder and can be packaged in sachets (Floros et al., 1997).

In many patent applications (Ackerley et al., 1998; Akkapeddi and Tsai, 2002; Barski et al., 2002;Cahill and Chen, 2000; Goodrich et al., 2003; Kulzick et al., 2000; Mize et al., 1996; Morgan et al., 1992; Roberts et al., 1996; Speer and Roberts, 1994; Speer et al., 2002), it was disclosed that ethylenic-unsaturated hydrocarbons, such as squalene, fatty acids, or polybutadiene, had sufficient commercial oxygen scavenging capacity to extend the shelf-life of oxygen-sensitive products. These unsaturated hydrocarbons, after being functionally terminated with a chemical group to make them compatible with the packaging materials, can be added during conventional mixing processes to thermoplastics such as polyesters, polyethylene, polypropylene, or polystyrene, and the films can be obtained using most conventional techniques for the plastic processing such as coinjection or coextrusion. 1,2-Polybutadiene is specially preferred because it exhibits transparency, mechanical properties, and processing characteristics similar to those of polyethylene. In addition, this polymer is found to retain its transparency and mechanical integrity, and exhibits a high oxygen-scavenging capacity (Roberts et al., 1996). Transition metal catalysts, such as cobalt II neodecanoate or octoate (Barski et al., 2002; Mize et al., 1996; Speer et al., 2002), are also included in the oxygen scavenger layer in order to accelerate the scavenging rate. Photoinitiators can also be added to further facilitate and control the initiation of the scavenging process. Adding a photoinitiator or a blend of photoinitiators to the oxygen-scavenging composition is a common practice, especially where antioxidants were added to prevent premature oxidation of the composition during processing and storage.

The main problem of this technology is that during the reaction between these polyunsaturated molecules and oxygen, by-products such as organic acids, aldehydes, or ketones can be generated that affect the sensory quality of the food or raise food regulatory issues (Brody et al., 2001a). Indeed, some of these compounds are used to determine the quality and shelf-life of fatty foodstuffs because they are intrinsically related to rancidity (Jo et al., 2002; Van Ruth et al., 2001). This problem can be minimized by the use of functional barriers that impede migration of undesirable oxidation products. This functional layer must provide a high barrier to organic compounds, but allow oxygen to migrate, and it has to be inserted between the food product and the scavenger layer. Another solution comes from the use of adsorber materials. Some polymers present inherent organic compound-scavenging properties. Others incorporate adsorbers within the polymer structure (i.e., silica gel, zeolites, etc). It has also been found that when the ethylenic unsaturation is contained within a cyclic group, substantially fewer by-products are produced upon oxidation as compared with analogous noncyclic materials. The Oxygen Scavenging Polymer developed by Chevron Chemical is an example of this kind of technology. This system is reported to scavenge oxygen without degrading into smaller, undesirable compounds. Ten percent of the polymer is a concentrate that contains a photoinitiator plus a transition metal catalyst that maintains the polymer in a nonscavenging state until triggered by ultraviolet (UV) radiation (Rooney, 1995).

Oxbar™ is a system developed by Carnaud-Metal Box (now Crown Cork and Seal) that involves cobalt-catalyzed oxidation of a MXD6 nylon that is blended into another polymer. This system is used especially in the manufacturing of rigid PET bottles for packaging of wine, beer, flavored alcoholic beverages, and malt-based drinks (Brody et al., 2001b).

Another O_2 scavenging technology involves using directly the closure lining. Darex® Container Products (now a unit of Grace Performance Chemicals) has announced an ethylene vinyl alcohol with a proprietary oxygen scavenger developed in conjunction with Kararay Co. Ltd. In dry forms, pellets containing unsaturated hydrocarbon polymers with a cobalt catalyst are used as oxygen scavengers in mechanical closures, plastic and metal caps, and steel crowns (both PVC and non-PVC lined). They reportedly can prolong the shelf life of beer by 25% (Brody et al., 2001b).

Immobilization of Microorganisms in Solid Holders

At least two patents from the 1980s and 1990s describe the use of yeast to remove oxygen from the headspace of hermetically sealed packages. One patent, from enzyme manufacturer Gist Brocades, focused on the incorporation of immobilized yeast into the liner of a bottle closure (Edens et al., 1992). The other patent used the yeast in a pouch within the package (Nezat, 1985). The concept of the patents was that, when moistened, the yeast is activated and respires, consuming oxygen and producing carbon dioxide plus alcohol. In the bottleclosure application, any carbon dioxide and alcohol produced would enter the contents, in this case beer, without causing measurable changes in the product.

Other researchers proposed an alternative approach: the use of entrapped aerobic microorganisms, capable of consuming oxygen (Tramper et al., 1983; Doran and Bailey, 1986; Gosmann and Rehem, 1986 and Gosmann & Rehem 1988). Natural and biological oxygen scavengers, based on the use of microorganisms entrapped in a polymeric matrix, effective in preserving foods, safe to use, agreeable to consumer, inexpensive, environment friendly, could be a very interesting concept to modern food technology. In fact, the possibility to create a new package, having many desirable characteristics, is very promising, also taking into account the new consumers' demand for mildly preserved convenience foods, having fresh-like qualities and being environmental friendly. In the field of biotechnology, immobilization of whole cells is gaining increasing importance (Gosmann and Rehem, 1988). Alginate, agar, and gelatin (Tramper et al., 1983; Doran and Bailey, 1986; Gosmann and Rehem, 1986 andGosmann and Rehem, 1988) have been successfully used. Unfortunately, the above study cannot be used for the development of a biological oxygenscavenger. In fact, the cycle life of a biological oxygen-scavenger film includes the entrapment of the microorganisms in an appropriate polymeric matrix (film manufacturing), the maintenance of the desiccated film till its use (film storage and distribution), and the re-hydration (film usage, obtained by putting the film in contact with the food).

Altiere et al. (2004) develop an environmental friendly oxygen-scavenger film using microorganisms as the active component. In particular, hydroxyethyl cellulose (HEC) and polyvinyl alcohol (PVOH) were used to entrap two different kinds of microorganisms: *Kocuria varians* and *Pichia subpelliculosa*. In this work a new method is proposed to produce oxygen-scavenger films using aerobic microorganisms as the "active compound". The manufacturing cycle of the investigated oxygen-scavenger film was optimized both to prolong the microorganism's viability during storage and to improve the efficiency of the film to remove oxygen from the package headspace. It was found that it is possible to store the desiccated film over a period of 20 days without monitoring any appreciable decrease of microorganism viability. It was also pointed out that the highest respiratory efficiency of the proposed active film is obtained by entrapping the microorganisms into polyvinyl alcohol, and by using the active film as a coating for a high humidity food.

Photosensitive Dye Oxidation

Another technique of oxygen absorption is a photosensitive dye impregnated onto a polymeric film. When the film is irradiated by ultraviolet (UV) light, the dye activates the O_2 to its singlet state, making the oxygen-removing reaction much faster (Ohlsson and Bengtsson, 2002).

Australian researchers have reported that reaction of iron with ground state O_2 is too slow for shelf-life extension. The singletexcited state of oxygen, which is obtained by dye sensitisation of ground state oxygen using near infra-red, visible or ultraviolet radiation, is highly reactive and so its chemical reaction with scavengers is rapid. The technique involves sealing of a small coil of ethyl cellulose film, containing a dissolved photosensitising dye and a singlet oxygen acceptor, in the headspace of a transparent package. When the film is illuminated with light of the appropriate wavelength, excited dye molecules sensitise oxygen molecules, which have diffused into the polymer, to the singlet state. These singlet oxygen molecules react with acceptor molecules and are thereby consumed. The photochemical reaction can be presented as follows (Vermeiren et al., 2003, Cruz et al., 2005).

photon + dye $\rightarrow$ dye*

dye* + O_2 $\rightarrow$ dye + O_2^*

O_2^* + acceptor $\rightarrow$ acceptor oxide

This scavenging technique does not require water as an activator, so it is effective for wet and dry products. Its scavenging action is initiated on the processor's packaging line by an illumination-triggering process (Vermeiren et al., 2003).

Cryovac® 0S2000™ polymer based oxygen scavenging film has been developed by Cryovac Div., Sealed Air Corporation, USA. This UV light-activated oxygen scavenging film (Figure 2), composed of an oxygen scavenger layer extruded into a multilayer film, can reduce headspace oxygen levels from 1% to ppm levels in 4–10 days and is comparable in effectiveness with oxygen scavenging sachets. The OS2000™ scavenging films have applications in a variety of food products including dried or smoked meat products and processed meats (Butler, 2002).

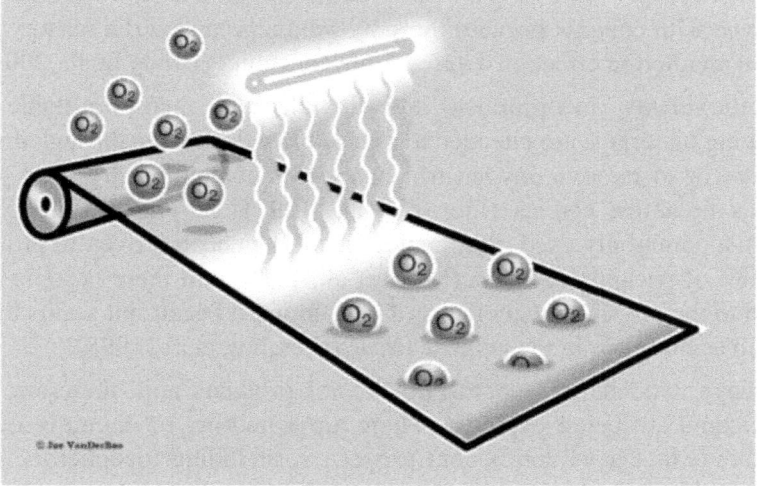

Figure 2: Light-activated oxygen scavenging films Cryovac® OS Films (Cryovac Food Packaging, Sealed Air Corporation, USA).

A similar UV light-activated oxygen scavenging polymer ZERO$_2$®, developed by CSIRO, Division of Food Science Australia in collaboration with Visy Pak Food Packaging, Visy Industries, Australia, forms a layer in a multilayer package structure and can be used to reduce discoloration of sliced meats. The active ingredient of the ZERO$_2$® is integrated into the polymer backbones of such common packaging materials as PET, polyethylene, polypropylene and EVA. The active ingredient is nonmetallic and is activated by UV light once it is incorporated into packaging material (Graff, 1998).

Another successful commercial example for use with meat is the OSP™ system (Chevron Philips Chemical Company, USA). The active substances of OSP™ systems are ethylene methacrylate and cyclohexene methacrylate, which need to be blended with a catalyst or photoinitiator in order to activate the oxygen scavenging mechanism.

Others

Sulphites have also been proposed as active substances for use, not only in sachets, but also in plastic gasket liners of bottle closures, as liquid trapped between sheets of flexible packaging material, or directly incorporated into plastic film structures to pack products such as wine or ketchup. For example, potassium sulfite is cited as an O_2 scavenger that can be readily triggered by the moist high temperature of the retorting process, and it also has enough thermal stability to pass unchanged through thermoplastic processes. However, any oxygen scavenger producing an end-product compound such as sulfur dioxide is viewed with concern because these by-products can exert a sensory change, or even an allergic effect on a susceptible consumer (Brody et al., 2001a).

Antioxidants, incorporated into flexible and thermoformable plastic packaging materials, are intended to reduce oxygen passage through the plastic structure or to remove oxygen from packages containing dry food products such as breakfast cereals (Floros et al., 1997). Butylated hydroxytoluene (BHT), a commonly used plastic antioxidant, has been proven to prolong the shelf-life of packed oat flakes (Miltz et al., 1989), but there is some concern related to the physiological effects of consuming it because it seems that BHT tends to accumulate in the adipose tissue (Wessling et al., 1998).

Nowadays, tendencies lead toward natural products and, therefore, natural antioxidants are being explored. There are a number of naturally occurring compounds that have antioxidant properties, including tocopherols, lecithin, organic acids and rosemary extracts. Among them, there is a growing interest in the use of vitamin E (also known as a-tocopherol) and vitamin C to be in-corporated into polymers. Vitamin E has been marketed as a food-grade odor remover in packaging materials. For example, Laermer et al. (1996) showed that addition of vitamin E to high density polyethylene (HDPE)-ethylene/vinyl acetate (EVA)-HDPE flexible packaging system could reduce the "plastic" taste and preserve the fresh taste of breakfast cereals. Ho et al. (1994) showed that vitamin E was effective in reducing off-flavor compounds released from HDPE bottles. Vitamin E has somehow superior antioxidant behavior than

BHT related to the off-flavor generation, stability and solubility, in polyolefins. The incorporation of vitamins E or C into the plastic material presents another advantage when compared with the addition of synthetic antioxidants because the possible migration of these compounds into the food not only does not produce adverse effects, but also improves the nutritional characteristics of the food product. However, being a bigger molecule than BHT, it is less mobile (Wessling et al., 1998). The amount of antioxidant added to the polymer must be controlled, as high levels of antioxidant incorporated into films can alter the polymer properties. Oxygen permeability of the film would increase and somemechanical properties of the film would change (Wessling, 2000).

Many patents have been issued for UV light activated oxygen scavengers, however, these UV activation steps reduces packaging line speeds and resulting in reduced profitability. In addition, there is a significant cost increases for oxygen scavenging films due to the high cost of photoinitiators and the operation and maintenance costs of the UV machine. Therefore the development of new oxygen scavenging systems that don't require a UV activation step should be valuable to the food packaging industry.

Oxygen scavenging systems that utilise natural compounds as the basis for the oxygen scavenger may provide added benefit. One such potential compound is α-tocopherol which is a natural free radical scavenger with a positive consumer perception (Hamilton et al., 1997). It has been incorporated into the polymer materials as a stabilizer (Al-Malaika et al., 1999) and as an antioxidant in controlled release packaging to reduce the oxidation in food products (Byun et al., 2010, Lacoste et al., 2005, Siro et al., 2006 and Wessling et al., 2000).

The oxygen scavenging principle for the use of α-tocopherol was that oxygen free radicals can be produced by a transition metal. Oxygen free radicals are derived from the non-enzymatic reactions of oxygen along with transition metals (Bagchi and Puri, 1998). The transition metal activates oxygen to the singlet electron state oxygen. Then, this activated oxygen undergoes subsequent reduction to reactive oxygen species (ROS), which is an oxygen free radical. α-Tocopherol is a strong free radical scavenger which can also react irreversibly with singlet oxygen and produce tocopherol hydroperoxydienone, tocopherylquinone, and quinine epoxide (Choe and Min, 2006). α-Tocopherol can donate its electrons to scavenge the oxygen free radical. When the free radical gains the electron from α-tocopherol, it returns to its ground state and the free radical is eliminated.

There are two chemical reaction steps in this oxygen scavenging reaction as follow. In first step, oxygen free radicals are produced in the presence of a transition metal. In the second step, the oxygen free radicals are eliminated by receiving electrons from α-tocopherol (Smirnoff, 2005). Therefore, the presence of both the transition metal and α-tocopherol are essential conditions for the oxygen scavenging system. Furthermore, thermal processing can accelerate oxygen scavenging reaction.

$$\text{Initiation step: Oxygen + transition metal} \xrightarrow{\Delta} \text{oxygen free radical}$$

$$\text{Scavenging step: } \alpha - \text{tocopherol + oxygen free radical} \rightarrow \text{dimer or tocopherylquinone}$$

PRACTICAL APPLICATION AND RESEARCHES

Oxygen scavengers have been studied for many researchers. There are many different types of oxygen scavengers that have been successfully applied to reducing food spoilage. In this section, we will discuss about the main and recent studies involving this technology.

Acid ascorbic is degraded to dehidroascorbic acid in the presence of oxygen, and the rate at which dehydroascorbic acid is formed is approximately first order with respect to the concentrations of ascorbic acid, oxygen, and metal catalysts. To evaluate the ascorbic acid loss in orange juice due to oxygen presence, the product was packed in oxygen scavenging film and oxygen barrier film. The initial concentration of ascorbic acid in the orange juice was 374 mg/l and this decreased by 74 and 104 mg/l after 3 days of storage at 25 °C in the O_2 scavenger film and O_2 barrier film, respectively. The rapid loss in ascorbic acid was related to the high oxygen content initially present in the headspace and that dissolved in the juice. This content of oxygen could not be eliminated by O_2 barrier film. The authors concluded that the rapid removal of oxygen is an important factor to maintain the ascorbic acid content in orange juice over long storage times (Zerdin et al., 2003).

Altieri et al. (2004) purposed a new method to produce oxygen-scavenger film based on aerobic microorganisms (*Kocuria varians* and *Pichia subpelliculosa*). These microorganisms were entrapped into hydroxyethyl cellulose and polyvinyl alcohol and maintained their viability over 20 days. Both films were able to reduce oxygen content present into vials, however the highest respiratory efficient was obtained by entrapping the microorganism into polyvinyl alcohol.

Mohan et al. (2009) studied the effect of commercial oxygen scavenger in reducing the formation of biogenic amines during chilled storage of fish. It was observed that the O_2 scavenger was able to reducing the oxygen content of the pack up to 99.95% within 24 h and it extended the fish shelf-life up to 20 days compared to only 12 days for air packs. The biogenic amine content was significantly higher in air packs compared to the O_2 scavenger packs. Inhibition of enzymatic activity of food or bacterial decarboxylase activity and prevention of bacterial growth are essential to control the production of biogenic amine. The authors verified that the use of oxygen scavengers associated to chilled storage temperature helps in reducing the formation of biogenic amines in fish. In conclusion, the authors believe that by using O_2 scavengers, use of vacuum packing machine can be avoided.

The health benefits of the Mediterranean diet are often related to the consumption of olive oil. The container material has been related to influence the oi quality and sensorial characteristics. Glass is the most used material, however the use of polyethylene terephthalate (PET) bottle have increased, since it is transparent, recyclable, unbreakable, inexpensive and it has demonstrated the ability to preserve the characteristics of olive oil during its shelf-life. In the other hand, the permeability of the PET bottle to gases and vapour, such as oxygen limits the use of these containers to olive oil, since rancidity is the main cause of oil spoilage. In this context, Cecchi et al. (2010) evaluated the quality of extra-virgin olive oil packed into PET bottles containing or not commercial oxygen scavenger. Results of the 13-months experimental study indicate that the presence of the O_2 scavenger in the plastic matrix was able to better maintain the quality and authenticity attributes of the oil. A reduced flux of oxygen through the PET bottle keeps the level of primary and secondary oxidation products lower than that obtained in simple PET bottles stored under the same conditions. The active barrier reduces the olive oil antioxidant activity decline during storage. The chlorophylls content decay can only be prevented via the storage of the sample in the dark, while the active barrier is able to diminish the carotenes loss at the end of the shelf-life. On the whole, the performance of the tested innovative packageing proved to better preserve the extra virgin characteristics of the oil during its shelf-life.

A variety of oxygen scavengers have been commercialised for use in the food packaging industry. These oxygen scavenging system are used in various forms such as; sachets, plastic films, labels, plastic trays, and bottle crowns. The most used O_2 scavengers are based on the principle of iron oxidation.

Cruz et al. (2006) evaluated an O_2 absorbent system on the inhibition of microorganisms growth in fresh lasagna pasta during storage at $10 \pm 2°C$. Fresh lasagna pasta was produced with and without potassium sorbate and acondicionated in high O_2 barrier bags containing an O_2-absorber sachet in the headspace. Three treatments were obtained: pasta with potassium sorbate, pasta without potassium sorbate packed with sachet and pasta without potassium sorbate packed without sachet (Figure 3). Oxygen absorbers were efficient in controlling the growth of filamentous fungi and yeasts,*Staphylococus* spp, total coliforms and *E. coli* in lasagna type fresh pasta without the addition of potassium sorbate, vacuum-packed in O_2-absorbent sachets, stored at 10 ± 2 °C. Therefore, the O_2-absorber sachet can be used as a hurdle technology, associated with vacuum packaging and applying the good manufacturing practices, to preserve lasagna pasta without additives.

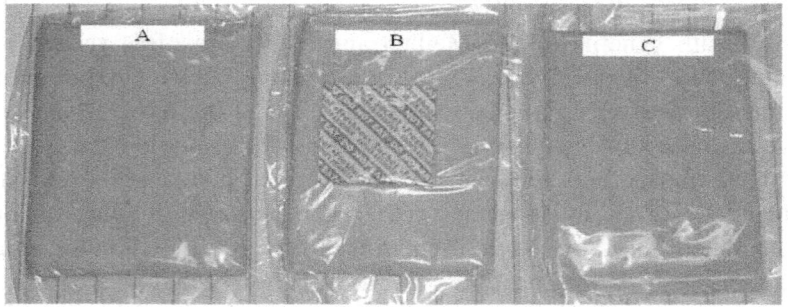

Figure 3: Fresh lasagna pasta vacuum-packed with potassium sorbate (A), without potassium sorbate and with oxygen scavenger (B) and without potassium sorbate.

However, nowadays, many consumers have a negative view of the term "iron-based." Therefore, Byun et al. (2011) studied the development of na oxygen scavenger using a natural compound: α-tocopherol. A natural free radical scavenger, α-tocopherol, and a transition metal in an oxygen scavenging system were evaluated as a possible oxygen scavenger. An initial, cup headspace oxygen content (%) of 20.9% was decreased to 18.0% after thermal processing and 60 days of storage at room temperature when the oxygen scavenging system containing α-tocopherol (500 mg) and transition metal (100 mg) was utilised. The oxygen content (%) decreased further to 17.1% when the amount of transition metal increased from 100 to 150 mg. The authors concluded that α-tocopherol (500 mg) and transition metal (150 mg) had an oxygen scavenging capacity of 6.72 ml O_2/g and an oxygen scavenging rate of 0.11 ml O2/g•day.

Others authors also research alternative systems able to scavenging oxygem. Anthierens et al. (2011) developed an O_2 scavenger using an endospore-forming bacteria genus *Bacillus amyloliquefaciens* as the "active ingredient". Spores were incorporated in poly(ethylene terephthalate, 1,4-cyclohexane dimethanol) (PETG), na amorphous PET copolymer having a considerable lower processing temperature and higher moisture absorption compared to PET (Figure 4). The work showed that endospores were able to survive incorporation in PETG at 210 °C, and the spores could consume oxygen for minimum 15 days, after na activation period of 1-2 days at 30 °C under high moisture conditions. According to the authors, the usse of viable spores as oxygen scavengers could have advantages towards consumer perception, recyclability, safety, material compatibility and production costs compared to currently available chemical oxygen scavengers.

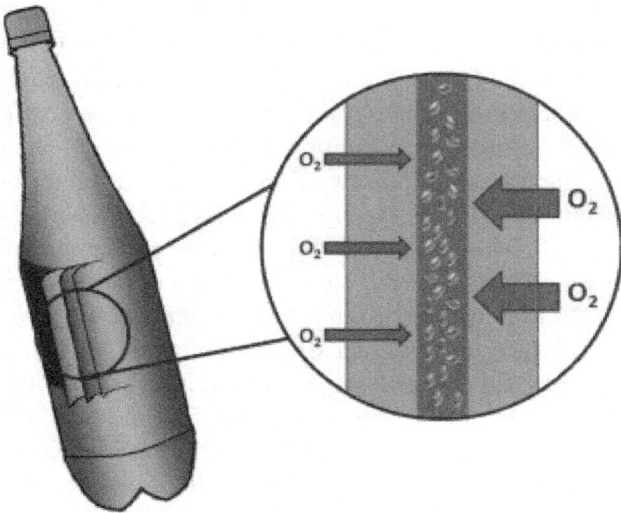

Figure 4: Schematic representation of a multilayer PET bottle consisting of a PETG middle layer containing bacterial spores surrounded by two outer PET layers. The inside of the bottle is in contact with the product, allowing moisture uptake of the bottle needed for spore germination. The system allows scavenging of residual oxygen from the in-bottle environment and scavenging from atmospheric oxygen permeating through the bottle wall (Anthierens et al., 2011)

An ascorbyl palmitate-β-cyclodextrin inclusion complex was produced and used as oxygen scavenger by Byun and Whiteside (2012). Cyclodextrin inclusion complex is one microencapsulation technique that has a significant potential for oxygen scavenging technology. Cyclodextrins (CDs) are cyclic oligosaccharides with a hydrophilic exterior and a hydrophobic central cavity. Its molecular dimensions allow total or partial inclusion of guest compounds. Among conventional microencapsulation methods, β–cyclodextrin inclusion is the most effective for protecting flavors. Production of off-flavors is a common problem of conventional oxygen scavenging sachets and films. Therefore, eliminating or reducing these potential off-flavors is a major concern for developing new oxygen scavenger. Cyclodextrin has other advantages, such as its thermal and chemical stability. The new O_2 scavenger based on ascorbyl palmitate-β-cyclodextrin inclusion complex was able to reduce oxygen content under 4 and 23 °C more than iron powder based sachet. In addition, the effect of thermal processing on oxygen scavenging capability was also investigated, and the O_2 scavenger developed maintained good oxygen scavenging capability after thermal processing. The results indicated that ascorbyl palmitate-β-cyclodextrin inclusion complex is an effective O_2 scavenger.

Gibis and Rieblinger (2011) incorporate the oxygen scavenger into the packaging material aiming to achieve better quality preservation and longer shelf-life of the chilled food. First investigations concentrated on defining the influence of temperature to the oxygen consumption of an oxygen scavenger film. Reducing the temperature from 23 °C to 5 °C caused a decrease (factor 3.0) in the oxygen consumption rate of the oxygen scavenger multilayer film PE /AL(SP2400; PE) within the first four days (RH 100 %; 0.5 % initial headspace-oxygen). Moreover the influence of using a polymer with a higher oxygen permeation rate than PP (commonly used) to the oxygen consumption of the scavenger film was investigated. Thus the masterbatch SP2500 was mixed with EVA that shows higher oxygen permeability than PP (by factor 2.3). Consequently the oxygen scavenger multilayer film PE/AL(SP2500; EVA) showed a faster oxygen consumption than the film PE/AL(SP2500; PP) (by factor 2.3). Finally the oxygen concentration in measuring cells with scavenger film PE/AL(SP2500; EVA) and with sausage were compared at 5 °C (initial oxygen concentration in headspace: 0.5 %). The combination (calculated) with oxygen scavenger film showed a faster oxygen decrease in the headspace of the measuring cell than the sausage alone. This leads to the assumption of a certain protection of the sausage against oxygen deterioration. Better protection of the sausages might be achieved by storing the food sample in combination with the scavenger film in darkness for the first few days. This would allow the scavenger to absorb the oxygen much faster than the sausage because the fast photo-oxidation processes in the food do not appear without light-exposure.

Absorption kinetics of two commercial O_2 and CO_2 scavengers (ATCO® LH-100 and ATCO® CO-450, respectively) commonly used in active modified atmosphere packaging (MAP), were studied. Individual scavenger sachets were placed in polyvinylidene chloride pouches filled with air or modified atmosphere at 0% or 100% relative humidity and at 5, 20 and 35 °C. The headspace gas composition was measured as a function of time. Absorption kinetics were described by a first-order reaction with an Arrhenius type behaviour. The absorption capacity, absorption rate constant, energy of activation, Arrhenius constant and variation of all these parameters were evaluated This study illustrated the importance to take into account the temperature effect and the variation of the scavenger absorption kinetics to understand gas kinetics inside pouches, as well as to predict the product quality in modified atmosphere packaging (Charles et al., 2006).

Rodrigues et al. (2012) evaluated the antioxidant capacities of gum arabic and maltodextrin microcapsules containing antioxidant molecules (trolox, α-tocopherol, β-carotene, apo-8'-carotenal and apo-12'-carotenal) against reactive oxygen and nitrogen species. The scavenging capacities were influenced by the wall material, the reactive species, namely ROO•,H2O2,HO•, HOCl and ONOO⁻, and the antioxidant molecule. In general, a more pronounced enhancement of the antioxidant capacity due to incorporation of antioxidant molecules was observed in gum arabic microcapsules. The empty microcapsules showed capacity to scavenge reactive oxygen species (ROS) and reactive nitrogen species (RNS), being gum arabic a more potent antioxidant than maltodextrin. Apo-8'-carotenal incorporation promoted the highest increase in the scavenging capacities among the evaluated antioxidants, varying from 50% to 132% and from 39% to 85% for gum arabic and maltodextrin microcapsules, respectively, suggesting that this carotenoid presented the best balance between the molecule localization inside the microcapsules and the reactivity against the specific reactive species. These results contribute to the development of multi-functional microcapsules that are able to scavenge a broad range of reactive species of biological relevance, serving as a dietary supplement or as antioxidants for food products, and can also be used as colourants in hydrophilic matrices, such as foods and drugs, without raising the fat content.

Zeolites (mostly faujasites) with adsorbed terpenes ((R)-(+)-limonene or D-pinene) or phenol derivatives (thymol, resorcin, pyrocatechol) have been applied as effective oxygen scavengers of oxygen in packing bags. Their efficiency depends on type of zeolite and on cation modification. Na- and Cu-forms of zeolites X and Y accelerate the oxidation of terpenes greatly, whereas the H-forms retard the reaction with oxygen. The reactivity of phenol derivatives with oxygen is also affected by the zeolite support markedly. Although the reactivity of phenols does not increase after adsorption on zeolites, the oxidation products remain adsorbed and do not affect the packing system (Frydrych et al., 2007).

An oxygen scavenging system (OSS), composed of oxygen scavenging nanoparticles α-tocopherol and iron chloride (II), was incorporated into warm-water fish gelatin film and their oxygenvscavenging capability was investigated. The initial oxygen content (%) in the cup headspace, 20.90%, was decreased to 4.56% after 50 days of storage. The oxygen scavenging fish gelatin (OSFG) film had good oxygen scavenging capacity,1969.08 cc $O_2/m^2/$

mil, and moisture was used as the activator to trigger the oxygen scavenging reaction (Byun et al., 2012).

The researches briefly presented above show that there is an increasing interest in the oxygen scavengers field, and that the role of packaging in food preservation is more active, contributing for extending food shelf-life

REFERENCES

1.   Y. Abe, Kondoh, 1989, Kondoh Y (1989) CA/MA vacuum packaging of foods. In: Oxygen absorbers. Trumbull, Westport, CT, 149158.

2.   Abe 1994 (1994) Active packaging with oxygen absorbers. In Ahvenainen R, Nattila-Sandholm T, Ohlsson T. Minimal processing of foods. VTT symposium 142, Espoo, 209233.

3.   Ackerley DF, Barski RFJr, Cashhill PJ, Chen SY, Johnson DC, Nyderek WM, Rotter GE 1998 Zero Oxygen Permeation Plastic Bottle for Beer and Other Applications. WO Patent 9,812,127.

4.   R. Ahvenainen, Hurme, 1997 (1997) Active and smart packaging for meeting consumer demands for quality and safety. Food addit contam. 14 753763.

5.   Akkapeddi 2002K, Tsai M (2002) Ultra High Oxygen Barrier Films and Articles Made Therefrom. US Patent 6,479,160.

6.   S. Al-Malaika, C. Goodwin, S. Issenhuth, Burdick, 1999 (1999) The antioxidant role of α-tocopherol in polymers II. Melt stabilizing effect in polypropylene. Polym Degrad Stabil. 64 145156 .

7.   C. Altieri, M. Sinigaglia, M. R. Corbo, G. G. Buonocore, P. Falcone, Del Nobile, 2004 Use of entrapped microorganisms as biological oxygen scavengers in food packaging applications. LWT 37 915 .

8.   Anon. 2004 US invasion- it dominates the US, but Multisorb is pushing its oxygen scavengers in Europe. Active & intelligent pack news. 2: 5.

9.   Anon. 2005 Up and active- Pira's latest market report plots a healthy future for active packaging. Active & intelligent pack news. 3: 5.

10.   T. Anthierens, P. Ragaert, S. Verbrugghe, A. Ouchchen, B. G. De Geest, B. Noseda, J. Mertens, L. Beladjal, D. De Cuyper, W. Dierickx, Prez. F. Du, Devlieghere, 2011, Devlieghere F (2011) Use of endospore-forming bacteria as an active oxygen scavenger in plastic packaging materials. Innov. Food Sci. Emerg. Tech. 12 594599 .

11.   ATCO® Technical Information 2002 ATCO® oxygen absorbers, Standa Industrie, France.

12.   K. Bagchi, Puri, 1998 (1998) Free radicals and antioxidants in health

and disease. East Mediterr Health J. 4 350360 .

13.   Barski RFJr, Richardson JA, Cahill PJ, Rotter GE, Wass RV, Nyderek WM, Smyser GL 2002 Active Oxygen Scavenger Compositions and Their Use in Packaging Articles. US Patent 6,346,308.

14.   Bioka Technical Information 1999 Bioka oxygen absorber, Bioka Oy, Finland.

15.   Brody AL, Budny JA 1995 Enzymes as active packaging agents. In: Rooney ML (Ed) Active Food Packaging. London, Blackie Academic Professional. 174192 .

16.   Brody AL, Strupinsky ER, Kline LR 2001a Oxygen scavenger systems. In: Active Packaging for Food Applications. Pennsylvania, USA, Technomic Publishing Company.

17.   Brody AL, Strupinsky ER, Kline, LR 2001b Active packaging for food applications. Lancaster: Technomic Publishing Co., Inc. 218 p.

18.   B. L. Butler, 2002 Cryovac® OS2000TM Polymeric oxygen scavenging systems. Presented at Worldpak 2002.http://www.sealedair.com/library/articles/article-os2000.html

19.   Y. Byun, H. J. Bae, Whiteside, 2012 (2012) Active warm-water fish gelatin film containing oxygen scavenging system. Food Hydrocol 27 250255 .

20.   Y. Byun, D. Darby, K. Cooksey, P. Dawson, Whiteside, 2011 (2011) Development of oxygen scavenging system containing a natural free radical scavenger and a transition metal. Food Chem. 124 615619 .

21.   Y. Byun, Y. Kim, Whiteside, 2010 (2010) Characterization of an antioxidant polylactic acid (PLA) film prepared with α-tocopherol, BHT and polyethylene glycol using film cast extruder. J Food Eng. 100 239244 .

22.   Y. Byun, Whiteside, 2012 (2012) Ascorbyl palmitate-β-cyclodextrin inclusion complex as an oxygen scavenging microparticle. Carbohydrate Pol. 87 21142119 .

23.   Cahill PJ, Chen SY 2000 Oxygen Scavenging Condensation Copolymers for Bottles and Packaging Articles. US Patent 6,083,585.

24.   T. Cecchi, P. Passamonti, P. Cecchi, 2010 Study of the quality of extra virgin olive oil stored in PET bottles with or without an oxygen scavenger. Food Chem. 120 730735 .

25.   F. Charles, J. Sanchez, Gontard, 2006 (2006) Absorption kinetics of oxygen and carbon dioxide scavengers as part of active modified atmosphere packaging. J. Food Eng. 72 17 .

26. E. Choe, D. B. Min, 2006 Chemistry and reactions of reactive oxygen species in foods. Crit Rev Food Sci Nutr. 46 122 .

27. Cruz RS, Soares NFF, Andrade NJ 2005 Absorvedores de oxigênio na conservação de alimentos: Uma revisão. Rev Ceres. 52 191206 .

28. Cruz RS, Soares NFF, Andrade NJ 2006 Evaluation of oxygen absorber on antimicrobial preservation of lasagna-type fresh pasta under vacuum packed. Cienc. Agrotec. 30 11351138 .

29. Cruz RS, Soares NFF, Andrade NJ 2007 Efficiency of oxygen- absorbing sachets in different relative humidities and temperatures. Cienc. Agrotec. 31 18001804 .

30. Z. Damaj, A. Naveau, L. Dupont, E. Hénon, G. Rogez, Guillon, 2009, Rogez G, Guillon E (2009) Co(II)(L-proline)2(H2O)2 solid complex: Characterization, magnetic properties, and DFT computations. Preliminary studies of its use as oxygen scavenger in packaging films. Inorg. Chem. Commun. 12 1720 .

31. Day BPF 1998 Active packaging of foods. CCFRA New technologies bulletin. 17 123 .

32. Day BPF 2003 Active packaging. In: Coles R, McDowell D, Kirwan M. (Eds) Food Packaging Technologies. Boca Raton, FL, USA: CRC Press, 282302 .

33. Doran PM, Bailey JE 1986 Effects of immobilization on growth, fermentation properties, and macromolecular composition of Saccharomyces cerevisiae attached to gelatin. Biot Bioeng. 28 7387 .

34. L. Edens, F. Farin, A. F. Ligtvoet, J. B. Van Der Platt, 1992 Dry yeast immobilized in wax or paraffin for scavenging oxygen. US Patent 5106633.

35. Floros JD, Dock LL, Han JH 1997 Active packaging technologies and applications. Food Cosmet Drug Packag. 20 1017 .

36. E. Frydrycha, Z. Foltynowicz, S. Kowalakb, Janiszewskab, 2007, Foltynowicz Z, Kowalakb S, Janiszewskab E (2007) Oxygen scavengers for packing system based on zeolite adsorbed organic compounds. In: Xu R, Gao Z, Chen J, Yan W (Eds) Zeolites to Porous MOF Materials-the 40th Anniversary of International Zeolite Conference. Elsevier B.V., 15971604 .

37. D. Gibis, Rieblinger, 2011 (2011) Oxygen scavenging films for food application. Proc Food Sci 1 229234 .

38. Goodrich JL, Schmidt RP, Jerdee GD, Ching TY, Leonard JP, Rodgers BD 2003 Oxygen Scavenging Packaging. US Patent 6569506.

39.   B. Gosmann, H. J. Rehem, 1986 Oxygen uptake of microrganisms entrapped in Ca-alginate. Appl Microbiol Biot. 23 163167 .

40.   B. Gosmann, H. J. Rehem, 1988 Influence of growth behaviour and physiology of alginate-entrapped microorganisms on the oxygen consumption. Appl Microbiol Biot. 29 554559 .

41.   E. Graff, 1994 Oxygen removal. US patent 5284871

42.   1998raff G (1998) O2 scavengers give "smart" packaging a new lease on shelflife. Mod Plast. 75 6972 .

43.   Hamilton RJ, Kalu C, Prisk E, Padley FB, Pierce H (1997) Chemistry of freeradicals in lipids. Food Chem. 60: 193-199.

44.   Ho YC, Yam KL, Young SS, Zambetti PF 1994 Comparison of vitamin E, IRGANOX 1010 and BHT as antioxidants on release of off-flavors from HDPE bottles. J. Plast Film Sheet. 10 194212 .

45.   C. Jo, D. U. Ahn, M. W. Byun, 2002 Irradiation-induced oxidative changes and production of volatile compounds in sausages prepared with vitamin E-enriched commercial soybean oil. Food Chem. 76 299305 .

46.   Kulzick MA, Cahill PJ, Barnes TJ, Bauer CW, Johnson DC, Lynch T-Y, Rotter GE, Nyderek WM 2000 Improved Active Oxygen Scavenger Packaging. WO Patent 0037321.

47.   N. Kruijf, M. van Beest, R. Rijk, T. Sipiläinen-Malm, Losada. P. Paseiro, De Meulenaer, 2002eest M, Rijk R, Sipiläinen-Malm T, Paseiro Losada P, De Meulenaer B (2002) Active and intelligent packaging: applications and regulatory aspects. Food Addit Contam. 19 144162 .

48.   Labuza TP 1897 Oxygen absorber sachets. Food Res. 32 276277 .

49.   Labuza TP, Breene WM 1989 Applications of active packaging for improvement of shelf-life and nutritional quality of fresh and extended shelf-life foods. J Food Process Pres. 13 169 .

50.   A. Lacoste, K. M. Schaich, D. Zumbrunnen, K. L. Yam, 2005 Advancing controlled release packaging through smart blending. Packag Tech Sci. 18 7787 .

51.   Laermer SF, Young SS, Zambetti PF 1996 Could your packaging use a dose of vitamin E? Converting Magazine. 8082 .

52.   J. Miltz, P. Hoojjat, J. Han, J. R. Giacin, B. R. Harte, I. J. Gray, 1989 Loss of antioxidants from high density polyethylene-its effect on oatmeal cereal oxidation. In: Food and Packaging Interactions. Washington D.C. ACS Symposium 365.

53.   Mize JAJr, Stockley HWIII, Logan RH, Miranda NR 1996 Peelable

Package with Oxygen Scavenging Layer. EP Patent 0698563.

54. C. O. Mohan, C. N. Ravishankar, Gopal. T. K. Srinivasa, Kumar. K. Ashok, K. V. Lalitha, 2009 Biogenic amines formation in seer fish (Scomberomorus commerson) steaks packed with O2 scavenger during chilled storage. Food Res. Int. 42 411416 .

55. Morgan CR, Speer DV, Roberts WP 1992 Methods and Compositions for Oxygen Scavenging. EP Patent 0520257 A3 B1.

56. H. Nakamura, Hoshino, 1983 (1983) Techniques for the preservation of food by employment of an oxygen absorber. Mitsubishi Gas Chemical Co., Tokyo, Ageless® Division, 145 .

57. Nezat JW 1985 Composition for absorbing oxygen and carrier thereof. US Patent 4510162.

58. Nielsen 1997 (1997) Active packaging-a literature review, The Swedish Institute for Food and Biotechnology, 631, Sweden.

59. T. Ohlsson, Bengtsson, 2002 (2002) Active and intelligent packaging: oxygen absorbing packaging materials. In: Minimal Processing Technologies in the Industry. Cambridge, Woodhead Publishing Limited and CRC Press LLC.

60. M. Ozdemir, Floros, 2004 Active Food Packaging Technologies. Crit Rev Food Sci Nut. 44 185193 .

61. Roberts WP, Vanputte AW, Speer DV, Morgan CR 1996 Multilayer Structure for A Package for Scavenging Oxygen. US Patent 5350622.

62. E. Rodrigues, L. B. R. Mariutti, A. F. Faria, A. Z. Mercadante, 2012 Microcapsules containing antioxidant molecules as scavengers of reactive oxygen and nitrogen species. Food Chem (article in press).

63. Rooney ML 1995 Active packaging in polymer films. In: Active Food Packaging. London, Blackie Academic Professional.

64. Rooney ML 2005 Introduction to active food packaging technologies. In: Han JH (Ed) Innovations in Food Packaging. London, UK: Elsevier Ltd., 6369 .

65. Roussel 1999 Les emballages absorbeurs d'oxygène. In: Gontard N (Ed) Les emballages actifs. Paris, Tec and Doc, 3137 .

66. P. Santiago-Silva, N. F. F. Soares, J. E. Nóbrega, M. A. W. Júnior, K. B. F. Barbosa, A. C. P. Volp, E. R. M. A. Zerdas, N. J. Würlitzer, 2009 Antimicrobial efficiency of film incorporated with pediocin (ALTA® 2351) on preservation of sliced ham. Food Control 20 8589 .

67. Shorter AJ 1982 Evaluation of rapid methods for scavenging oxygen in flexible pouches. LWT. 5 380381 .

68.   I. Siro, E. Fenyvesi, L. Szente, Devlieghere. F. Meulenaer, J. Orgovanyi, J. Sényi, Barta, 2006, Sényi J, Barta, J (2006) Release of alpha-tocopherol from antioxidative low density polyethylene film into fatty food simulant: influence of complexation in beta-cyclodextrin. Food Addit Contam 23 845853 .

69.   Smith JP 1996 Improving shelf life of packaged baked goods by oxygen absorbents. AIB research department technical bulletin. 18 27 .

70.   Smirnoff 2005 (2005) Ascorbate, tocopherol and carotenoids: metabolism, pathway engineering and functions. In: Antioxidants and reactive oxygen species in plants. New York: Blackwell Publishing, 5386 .

71.   Speer DV, Cotterman RL, Kennedy TD 2002 Process for Pasteurizing an Oxygen Sensitive Product and Triggering an Oxygen Scavenger and the Resulting Package. US Patent 2002142168.

72.   Speer DV, Roberts WP 1994 Oxygen Scavenging Compositions for Low Temperature Use. US Patent 5310497.

73.   P. Suppakul, J. Miltz, K. Sonneveld, S. W. Bigger, 2003 Active packaging technologies with an emphasis on antimicrobial packaging and its applications. J Food Sci. 68 408420 .

74.   J. Tramper, K. C. A. M. Luyben, W. J. J. van der Tweel, 1983 Kinetic aspects of glucose oxydation by Gluconobacter oxydans cells immobilized in Ca-alginate. Eur J Appl Microbiol. 17 1318

75.   van Ruth SM, Shaker ES, Morrissey PA 2001 Influence of methanolic extracts of soybean seeds and soybean oil on lipid oxidation in linseed oil. Food Chem. 75 177184 .

76.   L. Vermeiren, F. Devlieghere, M. van Beest, N. de Kruijf, Debevere, 1999 (1999) Developments in the active packaging of foods. Trends Food Sci. Tech. 10 7786 .

77.   L. Vermeiren, L. Heirlings, F. Devlieghere, Debevere, 2003 (2003) Oxygen, ethylene and other scavengers. In: Ahvenainen R (Ed.) Novel food packaging techniques. Boca Raton, USA, CRC Press.

78.   Wessling 2000 (2000) Antioxidant-Impregnated Polymers Antioxidant Ability and Interactions with Food. Ph.d Thesis Cornell University Ithaca NY USA.

79.   C. Wessling, T. Nielsen, Leufven, 2000 (2000) The influence of α-tocopherol concentration on the stability of linoleic acid and the properties of low density polyethylene. Packag Tech Sci. 13 1928 .

80.   C. Wessling, T. Nielsen, A. Leufven, Jagerstad, 1998 (1998) Mobility of

tocopherol and BHT in LDPE in contact with fatty food simulants. Food Addit Contam. 15 709715 .

81.   K. Zerdin, M. L. Rooney, Vermuë, 2003 (2003) The vitamin C content of orange juice packed in an oxygen scavenger material. Food Chem. 82 387395 .

Chapter 6

DEVELOPMENT AND VALIDATION OF HPLC METHOD FOR THE SIMULTANEOUS DETERMINATION OF FIVE FOOD ADDITIVES AND CAFFEINE IN SOFT DRINKS

Bürge Aşçı, Şule Dinç Zor, and Özlem Aksu Dönmez

Department of Chemistry, Faculty of Science and Arts, Yildiz Technical University, Davutpasa, 34220 Istanbul, Turkey

ABSTRACT

Box-Behnken design was applied to optimize high performance liquid chromatography (HPLC) conditions for the simultaneous determination of potassium sorbate, sodium benzoate, carmoisine, allura red, ponceau 4R, and caffeine in commercial soft drinks. The experimental variables chosen were pH (6.0–7.0), flow rate (1.0–1.4 mL/min), and mobile phase ratio (85–95% acetate buffer). Resolution values of all peak pairs were used as a response. Stationary phase was Inertsil OctaDecylSilane- (ODS-) 3V reverse phase column (250 × 4.6 mm, 5 µm) dimensions. The detection was performed at 230 nm. Optimal values were found 6.0 pH, 1.0 mL/min flow rate, and 95% mobile phase ratio for the method which was validated by calculating the linearity (r^2> 0.9962), accuracy (recoveries ≥ 95.75%), precision (intraday variation ≤ 1.923%, interday variation ≤ 1.950%), limits of detection (LODs), and limits of quantification (LOQs) parameters. LODs and LOQs for analytes were in the range of 0.10–0.19 µg/mL and 0.33–0.63 µg/mL, respectively. The proposed method was applied successfully for the simultaneous determination of the mixtures of five food additives and caffeine in soft drinks.

INTRODUCTION

Food additives are widely used in foodstuffs to prevent from spoilage and improve color, flavor, and texture of foods. However, these additives in foods may affect individuals who are sensitive with some type of allergy, asthma, and hay fever. Consequently, authorities have set threshold values for acceptable daily intake, varying from country to country. For instance, the list of authorised food additives and maximum permitted levels in European Union are laid down in the annexes of council directive [1, 2].

To ensure food safety from farm to fork, it is also essential to develop effective and reliable analytical methods for the monitoring of the additive levels in food [3]. Therefore, various analytical methods have been reported for the simultaneous determination synthetic food additives, such as thin layer chromatography [4], UV-visible spectrophotometry [5, 6], voltammetry [7, 8], differential pulse polarography [9], capillary electrophoresis [10], HPLC-DAD [11–14], HPLC-MS [15], and HPLC-MS-MS [16, 17]. Until now, although many analytical techniques have been developed for the determination of various food additives in foods, there is no report about simultaneous determination of this combination in food samples. Among these analytical methods, HPLC coupled with UV/Vis or diode array detectors (DADs) are the most commonly used methods due to their sensitivity, selectivity, and high resolution. So, development of effective chromatographic separation method involves judicious selection of experimental conditions that is suitable for the separation of interested components at an adequate resolution with reasonable run time. In this regard, experimental design is a useful tool to simplify the laborious work [18]. It not only is a timesaving method but also it has an ability to reveal possible interactions between variables [19, 20]. Hence, experimental designs have been increasingly used to determine the optimum conditions of chromatographic separation of some analytes in food, drug, and biological fluid samples with a minimum number of experiments for over the past decade [21–28].

In this paper, a new RP-HPLC method was developed, using experimental design, for simultaneous determination of five synthetic food additives in soft drinks, including three synthetic colorants (carmoisine, allura red, and ponceau 4R), two preservatives (potassium sorbate and sodium benzoate), and caffeine. For the optimization procedure, Box-Behnken design (BBD) was used to construct mathematical models that predict how changes input or controlled by variables (pH, flow rate, and mobile phase ratio) affected the resolution in defined experimental region. Further, the method validation has been carried out according to the International Conference on Harmonization guidelines. The optimized and validated method was successfully applied to some commercial soft drinks containing potassium sorbate, sodium benzoate, carmoisine, allura red, ponceau 4R, and caffeine.

EXPERIMENTAL

Apparatus

Chromatographic analyses were performed using a Shimadzu HPLC system (Kyoto, Japan) consisting of a model LC20 AT pump unit, SPD-20A UV-Vis detector, 7725 20 μL sample injection, a computer, and an Inertsil OctaDecylSilane- (ODS-) 3V column (5 μm, 250 mm × 4.6 mm; GL Sciences, Tokyo, Japan). The statistical analysis for the analytical responses and validation data was evaluated with Microsoft Excel 2000 software. The statistical software Statgraphics Centurion XV (StatPoint Inc., VA, USA) was used for the graph plotting and for estimating the responses of experimental variables.

Chemicals and Reagents

All chemicals and solvents were of analytical reagent grade and used without further purification. Milli-Q water was used to prepare the solutions and mobile phases (Millipore, Milford, MA, USA). Sodium acetate trihydrate, glacial acetic acid, and HPLC-grade acetonitrile were acquired from Merck (Darmstadt, Germany). Potassium sorbate (≥99.0% purity), sodium benzoate (≥99.0%, purity), carmoisine (≥98.0% purity), allura red (≥98.0% purity), ponceau 4R (≥99.0% purity), and caffeine (100.0% purity) were purchased from Sigma-Aldrich (St. Louis, Missouri, USA).

Preparation of Standard Solutions

Standard stock solutions of potassium sorbate, sodium benzoate, and caffeine were prepared at a concentration of 250 μg/mL. Standard stock solutions of carmoisine, allura red, and ponceau 4R were prepared at a concentration of 100 μg/mL. Fresh working solutions in the concentration range of 2–10 μg/mL for carmoisine, allura red, and ponceau 4R and 5–25 μg/mL for caffeine, potassium sorbate, and sodium benzoate were prepared by the dilution of the standard stock solutions in Milli-Q water.

Sample Preparation

Soft drink samples were purchased from local supermarkets in Istanbul, Turkey, and were degassed in an ultrasonic bath for 5 min. Then, 1 mL of the sample was transferred to a 10 mL volumetric flask and diluted to the volume with Milli-Q water. Prior to the analysis, both soft drink samples and standard solutions were filtered through 0.45 μm Millipore filters and then injected into HPLC system.

Chromatographic Procedure

The optimum separation of all analytes was achieved with 0.025 M sodium acetate/acetic acid buffer, pH 6.0, acetonitrile gradient that follows 0–5 min, 95 : 75 (v/v); 5–10 min, 70 : 30 (v/v). The mobile phase flow rate was 1.0 mL/min and the injection volume was 20 μL in all the chromatographic runs. The detection was made with a variable ultraviolet-visible detector fixed at 230 nm.

Optimization Procedure

A Box-Behnken design (BBD) using three variables at three levels (coded levels: −1, 0, and +1) was used for the optimization of simultaneous determination of potassium sorbate, sodium benzoate, carmoisine, allura red, ponceau 4R, and caffeine by HPLC. This design was selected due to the small number of experiments required. The variables and levels selected for optimization procedure were pH (A; 6.0, 6.5, and 7.0), flow rate (B; 1.0, 1.2, and 1.4), and mobile phase ratio (in terms of acetate buffer) (C; 85, 90, and 95) (Table 1). The proposed HPLC method analyzed the compounds in two steps as mentioned above. While the first step has an effect on the chromatographic separation, the second step has an effect on the run time of the method. Therefore, experimental variables of the first step of HPLC method were taken into account. 15 experimental runs were performed at random and overall resolution (R) was chosen as the response for the separation of the compounds [19]. Experimental design matrix used and the results obtained by BBD were listed in Table 2.

Table 1: The experimental variables and levels of BBD

Run	A	B	C	R
1	6.0	1.0	90	1.041
2	7.0	1.0	90	0.622
3	6.0	1.4	90	0.553
4	7.0	1.4	90	0.128
5	6.0	1.2	85	0.000
6	7.0	1.2	85	0.027
7	6.0	1.2	95	3175.373
8	7.0	1.2	95	1792.207
9	6.5	1.0	85	0.046
10	6.5	1.4	85	0.000
11	6.5	1.0	95	2487.823
12	6.5	1.4	95	1257.455
13	6.5	1.2	90	0.759
14	6.5	1.2	90	0.327
15	6.5	1.2	90	0.918

Table 2: Experimental design matrix and the responses for BBD.

Variable	Level		
	−1	0	+1
pH (A)	6.0	6.5	7.0
Flow rate (B) (mL min^{-1})	1.0	1.2	1.4
Mobile phase ratio (C) (%)	85	90	95

Validation Procedure

In-house validation of the method was performed according to International Conference on Harmonization guidelines (ICH Q2R1) [29]. Evaluated parameters are linearity of calibration curve, limit of detection (LOD), limit of quantification (LOQ), and precision, accuracy, and stability. The linearity of the HPLC method for the determination of five food additives and caffeine was evaluated in a concentration range of 2–10 µg/mL for carmoisine, allura red, and ponceau 4R and 5–25 µg/mL for potassium sorbate, sodium benzoate, and caffeine covering the normal range of concentrations obtained when analyzing soft drinks. Calibration equations were calculated by the least squares treatment of the peak area of the food additives and caffeine. The limit of detection (LOD) and limit of quantitation (LOQ) were calculated as LOD $^{3x\sigma/S}$ and LOQ $10x\sigma/S$, , where σ is the standard deviation of intercept and S is the slope. In order to test the prediction performance of the proposed methods, intraday (three times in a day operation under the same conditions) and interday (four different days) studies were performed at three different concentrations (Level 1: 10 µg/mL; Level 2: 15 µg/mL; Level 3: 20 µg/mL for potassium sorbate, sodium benzoate, and caffeine; Level 1: 4 µg/mL; Level 2: 6 µg/mL; Level 3: 8 µg/mL for carmoisine, allura red, and ponceau 4R). Accuracy of the method was ascertained by a recovery study by adding a known amount of reference standards to the soft drink samples. Firstly, 0.5 mL of the soft drink sample was transferred to a 10 mL volumetric flask and the reference standards were added on it at three different concentration levels. Then, added samples were diluted to the volume with Milli-Q water, filtered, and analyzed.

RESULTS AND DISCUSSION

Optimization of the HPLC Method

Chromatographic optimization requires selecting suitable criteria for the evaluation of the resultant chromatograms in order to choose the optimum conditions. BBD is an independent, rotatable, or nearly rotatable second-order design based on three-level incomplete factorial designs. It is more efficient compared to other response surface designs, such as central composite designs. It can also provide sufficient information to test the lack of fit, and therefore it is one of the best quadratic models for response surface method and has been widely used in analytical fields. Because of the nonlinearity of the model, a polynomial function to contain second-order model is postulated to describe the evolution phenomenon:

$$y_i = b_0 + \sum_{i=1}^{n} b_i x_i + \sum_{i=1}^{n} b_{ii} x_i^2 + \sum_{1 \le i \le j} b_{ij} x_i x_j + \varepsilon_i,$$

(1)

where n is the number of variables b_0, is the constant term bi, b_{ii}, and b_i represent the coefficient of the first-order terms, quadratic terms, and interaction terms, respectively, and ε_i is a term that represents other sources of variability not accounted for the estimation, such as background noise [30].

The experimental results are shown in Table 2. The regression model for the response was tested through analysis of variance (ANOVA). From the results of ANOVA (Table 3), it can be deduced that linear contribution of mobile phase ratio (C) and quadratic contribution of mobile phase ratio (CC) influence the resolution significantly. Interactions of the individual variables in this study are not significant to resolution in the selected range. Fitted quadratic model equation is also presented in (2). Figure 1 shows the analysis of individual variables of experimental design. From Figure 1, it can be seen how the value of the resolution may increase if we take higher mobile phase ratio (C). Also, we can infer that although pH (A) and flow rate (B) do not greatly influence the resolution better resolutions are obtained for low values of pH and flow rate

$$\hat{R} = 0.67 - 172.99A - 153.92B + 1089.10C + 152.74A^2 - 345.80AC$$

$$- 152.83B^2 - 307.58BC + 1088.49C^2.$$

(2)

Table 3: ANOVA results for optimization by BBD.

Effect	SS	D.f.	MS	F-ratio	P value
A	239422	1	239422	1.95	0.2216
B	189543	1	189543	1.54	0.2694
C	9489050	1	9489050	77.20	0.0003*
AA	86144	1	86144	0.70	0.4407
AB	0	1	0	0.00	1.0000
AC	478304	1	478304	3.89	0.1056
BB	86239	1	86239	0.70	0.4404
BC	378422	1	378422	3.08	0.1397
CC	4374670	1	4374670	35.59	0.0019*
Total error	614539	5	122908		
Total (corr.)	15999393	14			

SS: sum of squares; MS: mean squares; F-ratio: MS/MS_{error}; P value: probability level; D.f.: degree of freedom. $R^2 = 0.961$, R^2 (adjusted for D.f.) = 0.892.
*Significant factor at $\alpha = 0.05$.

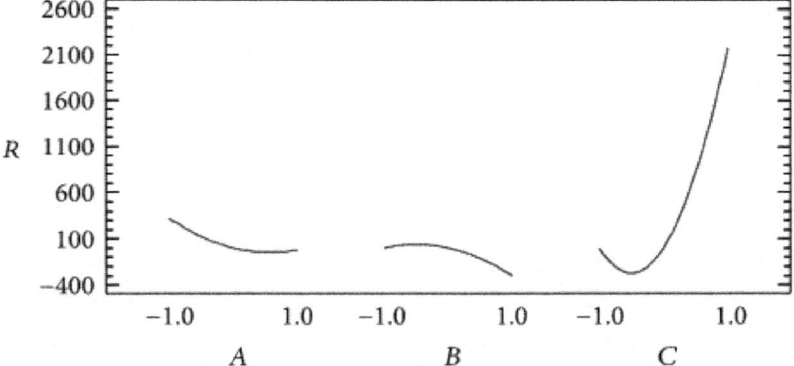

Figure 1: Analysis of the main variables in BBD.

The regression models obtained were used to calculate the response surface for each variable separately. Figure2 illustrates the response surface plots for the resolutions. In particular, the effect of pH (A) and mobile phase ratio (C) on resolution is shown in Figure 2(b). This plot shows that the highest resolution is obtained at greater values of the mobile phase ratio. The relation between the effects of the other variables on the resolution is also plotted in Figures 2(a)–2(c).

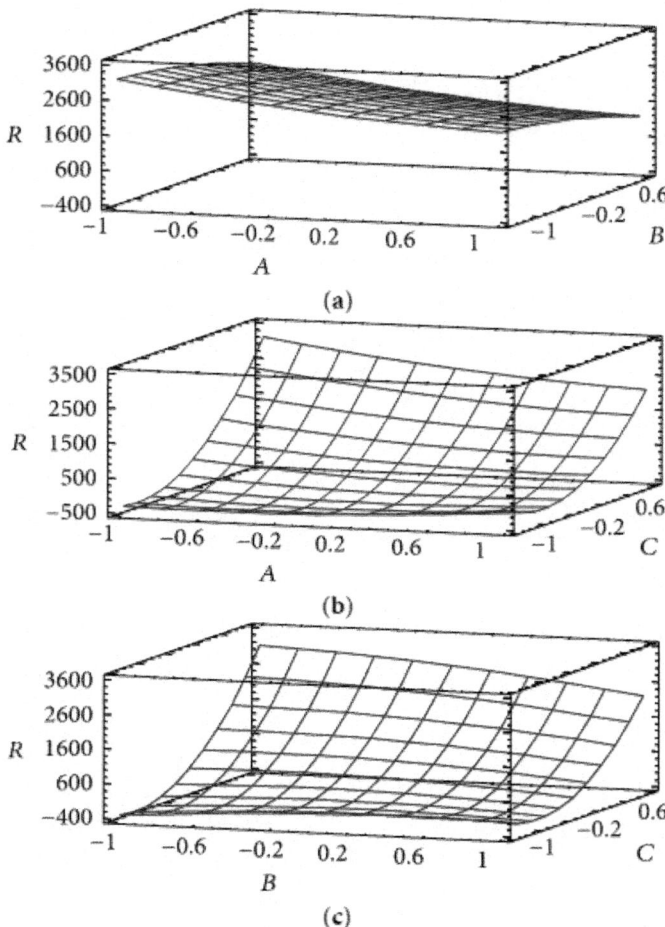

Figure 2: Response surface plots for BBD: (a) pH (A) versus flow rate (B) (mobile phase ratio: 95%); (b) pH (A) versus mobile phase ratio (B) (flow rate: 1.0 mL/min); (c) flow rate (C) versus mobile phase ratio (B) (pH: 6.0).

According to the results of the optimization procedure, the optimum variables corresponded to pH, 6.0; flow rate, 1.0 mL/min; mobile phase ratio, 95%. A typical chromatogram obtained under optimum conditions is shown in Figure 3.

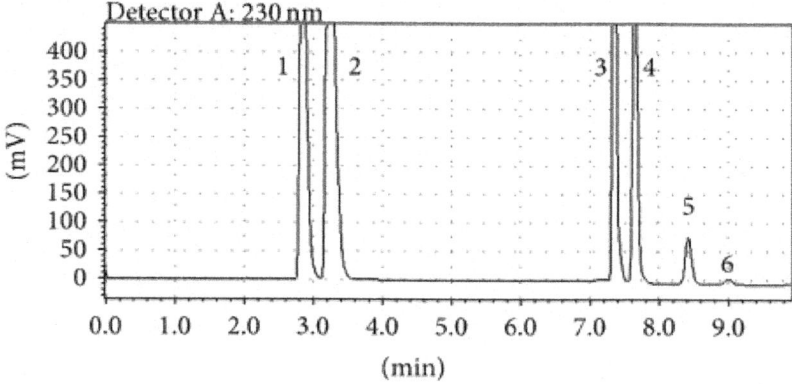

Figure 3: Chromatogram of synthetic standard mixture containing five food additives and caffeine recorded under optimized analysis conditions (1: sodium benzoate (15 μg/mL), 2: potassium sorbate (15 μg/mL), 3: caffeine (15 μg/mL), 4: ponceau 4R (6 μg/mL), 5: allura red (6 μg/mL), and 6: carmoisine (6 μg/mL)).

Validation of the HPLC Method

The results of the linearity, LODs, and LOQs are summarized in Table 4. A good linear relationship is displayed between the corresponding peak areas and the concentrations of the compounds based on the correlation coefficients (r^2 > 0.9962). The LODs of the six compounds were in the range of 0.10–0.19 μg/mL, and the LOQs of the six compounds were in the range of 0.33–0.63 μg/mL. So, these values demonstrated that the proposed analytical method was sufficiently sensitive. A summary of intraday (the RSD of the recoveries of the nine samples) and interday precision (the RSD of the recoveries of the twelve samples) are listed in Table 5. The RSD values ranged from 0.310% to 1.950% for the HPLC method. These results show that the proposed method is precise for the simultaneous determination of these compounds. The recoveries of the six compounds to determine the accuracy of the method are summarized in Table 6. The proposed method resulted in satisfactory recoveries for all additives and caffeine, ranging from 97.67% to 105.56%. The recoveries demonstrated that the matrixes have negligible effect on the quantification of these compounds and the method is accurate within the desired range. Under refrigerated and room temperature conditions, all food additives and caffeine in mobile phase and water were stable for at least 1 month.

Table 4: The important parameters of the calibration equations for the proposed HPLC method for simultaneous determination of potassium sorbate (SOR), sodium benzoate (BEN), carmoisine (CAR), allura red (ALU), ponceau 4R (PON), and caffeine (CAF).

Compounds	Calibration range (µg/mL)	Regression equation ($Y = aX + b$)	S_a	S_b	r^2	LOD (µg/mL)	LOQ (µg/mL)
SOR	5–25	$Y = 7.8097 \times 10^4 X + 4.6484 \times 10^4$	1.391×10^3	2.3076×10^4	0.9990	0.12	0.40
BEN	5–25	$Y = 3.3313 \times 10^4 X + 1.2421 \times 10^4$	7.52×10^2	1.2469×10^4	0.9985	0.10	0.33
CAR	2–10	$Y = 4.3383 \times 10^4 X + 1.8834 \times 10^4$	6.86×10^2	4.555×10^3	0.9992	0.16	0.53
ALU	2–10	$Y = 5.5741 \times 10^4 X - 7.427 \times 10^3$	9.22×10^2	1.5292×10^4	0.9992	0.11	0.35
PON	2–10	$Y = 6.043 \times 10^3 X + 5.0389 \times 10^4$	2.150×10^3	1.4264×10^4	0.9962	0.17	0.56
CAF	5–25	$Y = 2.4885 \times 10^4 X + 4.6613 \times 10^4$	4.81×10^2	3.1891×10^4	0.9989	0.19	0.63

Y: peak area; X: concentration (µg/mL); S_a: standard deviation of the slope; S_b: standard deviation of the intercept.

Table 5: Validation of the simultaneous determination of food additives and caffeine by developed method potassium sorbate (SOR), sodium benzoate (BEN), carmoisine (CAR), allura red (ALU), ponceau 4R (PON), and caffeine (CAF).

Levels	SOR (µg/mL)			BEN (µg/mL)			CAR (µg/mL)			ALU (µg/mL)			PON (µg/mL)			CAF (µg/mL)		
	10.0	15.0	20.0	10.0	15.0	20.0	4.0	6.0	8.0	4.0	6.0	8.0	4.0	6.0	8.0	10.0	15.0	20.0
Intra-assay	9.97	15.22	19.76	9.70	15.37	20.18	4.05	6.17	7.92	3.90	6.02	7.86	3.79	5.72	7.75	9.96	15.15	20.28
	9.80	15.30	20.01	10.02	15.09	20.21	4.15	6.33	8.01	3.83	5.97	7.82	3.89	5.92	7.91	9.99	15.50	20.02
	9.91	15.31	19.86	10.03	14.79	20.38	4.18	6.20	8.11	3.87	6.06	7.70	3.92	5.87	8.04	10.03	15.16	20.20
Mean	9.89	15.28	19.88	9.92	15.08	20.26	4.13	6.23	8.01	3.87	6.02	7.79	3.87	5.84	7.91	9.99	15.27	20.16
RSD[a]%	0.869	0.321	0.634	1.895	1.923	0.533	1.646	1.364	1.186	0.904	0.747	1.060	1.757	1.781	1.643	0.350	0.310	0.689
Recovery%	98.90	101.87	99.40	99.20	100.53	101.30	103.25	103.83	100.13	96.75	100.33	97.83	96.75	97.33	98.88	99.90	101.80	100.80
Interassay	9.92	15.09	19.66	9.84	14.76	20.71	4.14	6.01	7.81	3.91	6.01	7.96	4.10	6.01	8.20	10.35	15.10	20.09
	9.84	15.23	19.59	9.85	14.95	20.74	4.18	6.06	8.14	3.85	5.84	7.90	4.08	6.18	8.26	10.01	15.32	20.43
	9.77	15.21	19.69	9.81	14.79	20.25	4.05	6.19	7.96	3.81	5.86	8.04	3.83	6.20	8.24	10.17	14.88	20.07
	9.68	15.40	19.89	10.02	14.82	20.11	4.19	6.22	7.98	3.76	6.08	7.82	4.01	5.98	7.99	10.03	15.16	20.20
Mean	9.80	15.23	19.71	9.88	14.83	20.45	4.14	6.12	7.97	3.83	5.95	7.93	4.03	6.09	8.14	10.14	15.11	20.20
RSD%	1.041	0.840	0.624	0.961	0.566	1.565	1.546	1.650	1.694	1.644	1.950	1.173	1.919	1.855	1.518	1.548	1.204	0.817
Recovery%	98.00	101.53	98.55	98.80	98.87	102.25	103.50	102.00	99.62	95.75	99.17	99.13	100.75	101.50	102.12	101.40	100.73	101.00

[a]Relative standard deviation, RSD (%), standard deviation/mean × 100.

Table 6: Results of accuracy studies (mean value ± standard deviation, n=5).

Sample	Food additive	Sample concentration (µg/mL)	Added (µg/mL)	Found (µg/mL)	Recovery (%)
Energy drink	Potassium sorbate	10.68	3	13.70 ± 0.048	100.67
			6	16.90 ± 0.057	103.67
			9	20.00 ± 0.065	103.56
	Caffeine	7.11	3	10.18 ± 0.045	102.33
			6	13.25 ± 0.054	102.33
			9	16.61 ± 0.042	105.56
	Allura red	3.26	1	4.31 ± 0.038	105.00
			2	5.34 ± 0.037	104.00
			3	6.39 ± 0.044	104.33
Pomegranate juice	Potassium sorbate	12.71	3	15.64 ± 0.059	97.67
			6	18.61 ± 0.103	98.33
			9	21.74 ± 0.166	100.34
	Sodium benzoate	7.24	3	10.34 ± 0.040	103.60
			6	13.30 ± 0.054	101.20
			9	16.45 ± 0.063	102.50
	Carmoisine	1.21	1	2.25 ± 0.012	104.00
			2	3.30 ± 0.017	104.50
			3	4.34 ± 0.022	104.40
	Ponceau 4R	4.08	1	5.11 ± 0.036	102.66
			2	6.16 ± 0.039	104.16
			3	7.15 ± 0.045	102.22

These results show that the proposed method is precise, accurate, and sensitive for the simultaneous determination of the six compounds and can be used for routine analysis of potassium sorbate, sodium benzoate, carmoisine, allura red, ponceau 4R, and caffeine in soft drinks.

Application of the Method

The proposed HPLC method was applied to the simultaneous determination of potassium sorbate, sodium benzoate, carmoisine, allura red, ponceau 4R, and caffeine in different soft drinks. Five replicates determination was made and the results are summarized in Table 7. The concentration of food additives in soft drinks ranged from 24.26 ± 0.47 $\mu g/mL$ to 254.13 ± 1.24 $\mu g/Ml$. The amounts of food additives and caffeine in all soft drink samples were below the limit value defined in the legislation on the food additives [1, 2].

Table 7: Analysis of soft drinks (mean value ± standard deviation, n=5).

Food additive	Energy drink ($\mu g/mL$)	Pomegranate juice ($\mu g/mL$)	Mandarin juice ($\mu g/mL$)
Potassium sorbate	213.62 ± 0.34	254.13 ± 1.24	246.23 ± 1.76
Sodium benzoate	—	144.71 ± 2.38	148.67 ± 1.99
Caffeine	142.20 ± 1.17	—	—
Allura red	65.28 ± 0.59	—	—
Carmoisine	—	24.26 ± 0.47	—
Ponceau 4R	—	81.58 ± 1.51	—

CONCLUSION

An efficient, accurate, and reliable method for the simultaneous determination of five food additives and caffeine in soft drinks was developed using HPLC. Box-Behnken design was applied to the optimization of the chromatographic separation conditions and this design reduced to the number of experiments required. It can be concluded that a slight change in mobile phase ratio has a direct effect on the resolution. All the validation parameters were within the acceptance range. High percentage recovery data also shows that the proposed method is free from the interference. Consequently, this study will provide a sensitive and rapid method for the detection of potassium sorbate, sodium benzoate, carmoisine, allura red, ponceau 4R, and caffeine in soft drinks.

CONFLICT OF INTERESTS

The authors declare that there is no conflict of interests regarding the publication of this paper.

ACKNOWLEDGMENT

This work was financially supported by Yildiz Technical University Research Foundation (Project no. 2010-01-01-GEP04).

REFERENCES

1. Regulation (EC) no 1333/2008 of the European Parliament and of the Council on Food Additives, 2008.

2. European Commission, Commission Directive 2008/84/EC, Laying down Specific Purity Criteria on Food Additives Other Than Colours and Sweeteners, 2008.

3. A. K. Malik, C. Blasco, and Y. Pico, "Liquid chromatography-mass spectrometry in food safety," Journal of Chromatography A, vol. 1217, no. 25, pp. 4018–4040, 2010.

4. I. Baranowska, M. Zydroń, and K. Szczepanik, "TLC in the analysis of food additives," Journal of Planar Chromatography—Modern TLC, vol. 17, no. 1, pp. 54–57, 2004.

5. E. Dinç, E. Baydan, M. Kanbur, and F. Onur, "Spectrophotometric multicomponent determination of sunset yellow, tartrazine and allura red in soft drink powder by double divisor-ratio spectra derivative, inverse least-squares and principal component regression methods," Talanta, vol. 58, no. 3, pp. 579–594, 2002.

6. S. Altinöz and S. Toptan, "Simultaneous determination of Indigotin and Ponceau-4R in food samples by using Vierordt's method, ratio spectra first order derivative and derivative UV spectrophotometry,"Journal of Food Composition and Analysis, vol. 16, no. 4, pp. 517–530, 2003.

7. A. H. Alghamdi, "Determination of allura red in some food samples by adsorptive stripping voltammetry," Journal of AOAC International, vol. 88, no. 5, pp. 1387–1393, 2005.

8. X. Yang, H. Qin, M. Gao, and H. Zhang, "Simultaneous detection of Ponceat 4R and tartrazine in food using adsorptive stripping voltammetry on an acetylene black nanoparticle-modified electrode," Journal of the Science of Food and Agriculture, vol. 91, no. 15, pp. 2821–2825, 2011.

9. S. Chanlon, L. Joly-Pottuz, M. Chatelut, O. Vittori, and J. L. Cretier, "Determination of carmoisine, allura red and ponceau 4R in sweets and soft drinks by differential pulse polarography," Journal of Food Composition and Analysis, vol. 18, no. 6, pp. 503–515, 2005.

10. M. A. Prado, L. F. V. Boas, M. R. Bronze, and H. T. Godoy, "Validation of methodology for simultaneous determination of synthetic dyes in alcoholic beverages by capillary electrophoresis," Journal of Chromatography A, vol. 1136, no. 2, pp. 231–236, 2006.

11. N. Dossi, R. Toniolo, S. Susmel, A. Pizzariello, and G. Bontempelli, "Simultaneous RP-LC determination of additives in soft drinks,"

Chromatographia, vol. 63, no. 11-12, pp. 557–562, 2006.

12. N. O. Can, G. Arlı, and Y. Lafçı, "A novel RP-HPLC method for simultaneous determination of potassium sorbate and sodium benzoate in soft drinks using C18-bonded monolithic silica column,"Journal of Separation Science, vol. 34, no. 16-17, pp. 2214–2222, 2011.

13. Q. H. Yan, L. Yang, H. R. Zhang, and L. Y. A. Niu, "A sensitive and validated method for determination of four additives in ham sausage by HPLC-DAD method," Journal of Liquid Chromatography & Related Technologies, vol. 35, no. 2, pp. 268–279, 2012.

14. K. Ma, Y. N. Yang, X. X. Jiang, M. Zhao, and Y. Q. Cai, "Simultaneous determination of 20 food additives by high performance liquid chromatography with photo-diode array detector," Chinese Chemical Letters, vol. 23, no. 4, pp. 492–495, 2012.

15. M. Ma, X. Luo, B. Chen, S. Su, and S. Yao, "Simultaneous determination of water-soluble and fat-soluble synthetic colorants in foodstuff by high-performance liquid chromatography-diode array detection-electrospray mass spectrometry," Journal of Chromatography A, vol. 1103, no. 1, pp. 170–176, 2006.

16. H. Gao, M. Yang, M. Wang, Y. Zhao, Y. Cao, and X. Chu, "Determination of 30 synthetic food additives in soft drinks by HPLC/electrospray ionization-tandem mass spectrometry," Journal of AOAC International, vol. 96, no. 1, pp. 110–115, 2013.

17. F. Feng, Y. Zhao, W. Yong, L. Sun, G. Jiang, and X. Chu, "Highly sensitive and accurate screening of 40 dyes in soft drinks by liquid chromatography–electrospray tandem mass spectrometry," Journal of Chromatography B, vol. 879, no. 20, pp. 1813–1818, 2011

18. R. Noguerol-Cal, J. M. López-Vilariño, M. V. González-Rodríguez, and L. F. Barral-Losada, "Development of an ultraperformance liquid chromatography method for improved determination of additives in polymeric materials," Journal of Separation Science, vol. 30, no. 15, pp. 2452–2459, 2007.

19. R. Gheshlaghi, J. M. Scharer, M. Moo-Young, and P. L. Douglas, "Application of statistical design for the optimization of amino acid separation by reverse-phase HPLC," Analytical Biochemistry, vol. 383, no. 1, pp. 93–102, 2008.

20. D. C. Montgomery, Design and Analysis of Experiments, John Wiley & Sons, New York, NY, USA, 2004.

21. S. L. C. Ferreira, R. E. Bruns, E. G. P. da Silva et al., "Statistical designs and response surface techniques for the optimization of chromatographic

systems," Journal of Chromatography A, vol. 1158, no. 1-2, pp. 2–14, 2007.

22. D. B. Hibbert, "Experimental design in chromatography: a tutorial review," Journal of Chromatography B: Analytical Technologies in the Biomedical and Life Sciences, vol. 910, no. 1, pp. 2–13, 2012.

23. F. van de Velde, M. E. Pirovani, M. S. Cámara, D. R. Güemes, and C. M. D. H. Bernardi, "Optimization and validation of a UV–HPLC method for vitamin C determination in strawberries (Fragaria ananassaDuch.), using experimental designs," Food Analytical Methods, vol. 5, no. 5, pp. 1097–1104, 2012.

24. N. García-Villar, J. Saurina, and S. Hernández-Cassou, "High-performance liquid chromatographic determination of biogenic amines in wines with an experimental design optimization procedure," Analytica Chimica Acta, vol. 575, no. 1, pp. 97–105, 2006.

25. T. Sivakumar, R. Manavalan, C. Muralidharan, and K. Valliappan, "Multi-criteria decision making approach and experimental design as chemometric tools to optimize HPLC separation of domperidone and pantoprazole," Journal of Pharmaceutical and Biomedical Analysis, vol. 43, no. 5, pp. 1842–1848, 2007.

26. M. Medenica, B. Jancic, D. Ivanovic, and A. Malenovic, "Experimental design in reversed-phase high-performance liquid chromatographic analysis of imatinib mesylate and its impurity," Journal of Chromatography A, vol. 1031, no. 1-2, pp. 243–248, 2004.

27. G. Iriarte, N. Ferreirós, I. Ibarrondo et al., "Optimization via experimental design of an SPE-HPLC-UV-fluorescence method for the determination of valsartan and its metabolite in human plasma samples," Journal of Separation Science, vol. 29, no. 15, pp. 2265–2283, 2006.

28. E. Nemutlu, S. Kır, D. Katlan, and M. S. Beksaç, "Simultaneous multiresponse optimization of an HPLC method to separate seven cephalosporins in plasma and amniotic fluid: application to validation and quantification of cefepime, cefixime and cefoperazone," Talanta, vol. 80, no. 1, pp. 117–126, 2009.

29. International Conference on Harmonization (ICH) of Technical Requirements for Registration of Pharmaceuticals for Human Use, Topic Q2 (R1): Validation of Analytical Procedures: Text and Methodology, International Conference on Harmonization (ICH), Geneva, Switzerland, 2005.

30. J.-Z. Song, C.-F. Qiao, S.-L. Li, Y. Zhou, M.-T. Hsieh, and H.-X. Xu,

"Rapid optimization of dual-mode gradient high performance liquid chromatographic separation of Radix et Rhizoma Salviae Miltiorrhizae by response surface methodology," Journal of Chromatography A, vol. 1216, no. 42, pp. 7007–7012, 2009.

Chapter 7

PHOSPHATE ADDITIVES IN FOOD-A HEALTH RISK

Eberhard Ritz[1], Kai Hahn[2], Markus Ketteler[3] Martin K. Kuhlmann[4], Johannes Mann[5]

[1]Nierenzentrum Heidelberg

[2]Nephrologische Gemeinschaftspraxis/Dialyse, Dortmund

[3] Nephrologische Klinik, Klinikum Coburg GmbH, Coburg

[4]Vivantes Klinikum im Friedrichshain, Berlin

[5]Klinik für Nieren-, Hochdruck- und Rheumakrankheiten, Städtisches Klinikum München Schwabing

ABSTRACT

Background

Hyperphosphatemia has been identified in the past decade as a strong predictor of mortality in advanced chronic kidney disease (CKD). For example, a study of patients in stage CKD 5 (with an annual mortality of about 20%) revealed that 12% of all deaths in this group were attributable to an elevated serum phosphate concentration. Recently, a high-normal serum phosphate concentration has also been found to be an independent predictor of cardiovascular events and mortality in the general population. Therefore, phosphate additives in food are a matter of concern, and their potential impact on health may well have been underappreciated.

Methods

We reviewed pertinent literature retrieved by a selective search of the PubMed and EU databases (www.zusatzstoffe-online.de, www.codexalimentarius.de), with the search terms "phosphate additives" and "hyperphosphatemia."

Results

There is no need to lower the content of natural phosphate, i.e. organic esters, in food, because this type of phosphate is incompletely absorbed; restricting its intake might even lead to protein malnutrition. On the other hand, inorganic phosphate in food additives is effectively absorbed and can measurably elevate the serum phosphate concentration in patients with advanced CKD. Foods with added phosphate tend to be eaten by persons at the lower end of the socioeconomic scale, who consume more processed and "fast" food. The main pathophysiological effect of phosphate is vascular damage, e.g. endothelial dysfunction and vascular calcification. Aside from the quality of phosphate in the diet (which also requires attention), the quantity of phosphate consumed by patients with advanced renal failure should not exceed 1000 mg per day, according to the guidelines.

Conclusion

Prospective controlled trials are currently unavailable. In view of the high prevalence of CKD and the potential harm caused by phosphate additives to food, the public should be informed that added phosphate is damaging to health. Furthermore, calls for labeling the content of added phosphate in food are appropriate.

The dietary intake of phosphate and the serum phosphate concentration are important matters not just for persons with renal disease, but for the general public as well. It has recently been determined that phosphate additives in food may harm the health of persons with normal renal function (1, e1). This judgment has been made on the basis of large-scale epidemiological studies and is supported by the latest findings of basic research.

It was first recognized in patients with renal disease that a high serum phosphate concentration is a major risk factor for elevated cardiovascular and overall mortality (2,3). Block et al. studied a cohort of 40 538 hemodialysis patients and determined, after multivariate adjustment, that 12% of the 10 015 deaths occurring over the period of observation were associated with hyperphosphatemia (2). Dietary phosphate restriction has been a standard recommendation for decades for patients with chronic renal failure (4). This approach is also supported by the findings of a prospective five-year observational study of patients receiving chronic hemodialysis, in which a low dietary intake of phosphate was found to confer a significant survival advantage. Patients whose dietary phosphate intake was above the 99th percentile died at a rate 2.37 times higher than those whose dietary phosphate intake was below the first percentile (5). In a controlled interventional trial, persons who were informed about food additives containing phosphates and their presence in different types of food went on to avoid food with phosphate additives and to have lower serum phosphate concentrations.

More recent studies have shown that the association between high phosphate concentrations and higher mortality is not restricted to persons with renal disease; it can also be observed in persons with cardiovascular disease and even in the general population. High-normal serum phosphate concentrations are associated with coronary calcification in young, healthy men (6) and were found to be a predictor of cardiovascular events in the Framingham study (7). Elevated mortality in association with high-normal serum phosphate concentrations was seen mainly among persons with cardiovascular disease who had normal renal function (8) *(Figure 1)*. In the Framingham study, 375 of the 4127 subjects died within 60 months; the adjusted mortality risk was 22% for each 1 mg/dL elevation of the serum phosphate concentration.

Percentiles of dietary phosphate intake and overall mortality in 224 hemodialysis patients observed for up to 5 years; the hazard ratio (after full adjustment) in the highest percentile, compared to the lowest percentile, was 2.37

(From: Noori N, et al.: Association of dietary phosphorus intake and phosphorus to protein ratio with mortality in hemodialysis patients. CJASN 2010; 5: 683–92 [5]. Reproduced with the kind permission of the American Society of Nephrology)

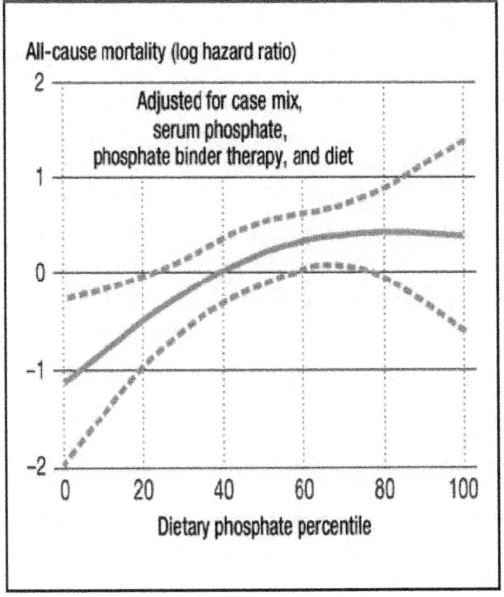

Figure 1: Percentiles of dietary phosphate intake and overall mortality in 224 hemodialysis patients observed for up to 5 years; the hazard ratio (after full adjustment) in the highest percentile, compared to the lowest percentile, was 2.37

Thus, the problem of high serum phosphate concentrations impairing survival is not merely a matter for nephrologists, though admittedly these observational studies do not yet firmly establish a direct cause-and-effect relationship. For the present article, the authors reviewed pertinent publications retrieved by a selective search of the PubMed and EU databases (www. zusatzstoffe-online.de, www.codexalimentarius.de). Much of this literature is of very recent date; in identifying important articles, we benefited from our experience working in international guideline committees (the "Kidney Disease—Improving Global Outcomes" (KDIGO) initiative [MK]) (4) and developing nutrition programs for dialysis patients (the Phosphate Unit Program [Phosphat-Einheiten-Programm, PEP] [MKK]). Our literature search also revealed the lack of prospective controlled trials on this topic.

THE MAIN SOURCES OF EASILY RESORBABLE PHOS-PHATE

Phosphate occurs naturally in the form of organic esters in many kinds of food, including meat, potatoes, bread, and other farinaceous products; the consumption of such foods cannot be restricted without incurring the risk of lowering protein intake. Naturally occurring phosphate in food is organically bound, and only 40% to 60% of it is absorbed in the gastrointestinal tract (e1).

On the other hand, an avoidable risk to health that has not attracted sufficient attention to date arises from the increased use of phosphate as a food additive and preservative. This "free" (not organically bound) phosphate is very effectively absorbed in the gastrointestinal tract. Typical foods with large amounts of added phosphate are processed meat, ham, sausages, canned fish, baked goods, cola drinks, and other soft drinks. Dietary counseling is all the more difficult because the phosphate content in food—and, in particular, the added phospate content—is not marked on the package.

It used to be thought that the only health risk posed by phosphate lay in the promotion of calcification in blood vessels and bodily organs. Recently, however, important discoveries have been made about the hormonal regulation of phosphate metabolism. It is now known that the serum phosphate concentration is controlled by two newly discovered factors called fibroblast growth factor 23 (FGF23) and klotho; that phosphate causes lasting damage to the cardiovascular system, either by a direct mechanism or by way of these hormonal factors; and that phosphate accelerates aging processes in animal models (e2, e3).

In particular, phosphate added to animal fodder accelerates age-related organ complications such as muscle and skin atrophy, the progression of chronic renal failure, and cardiovascular calcifications (e2). Phosphate added to human food probably has similar effects in man. Inexpensive food containing additives (processed food), and fast food in particular, are extraordinarily rich in phosphate additives. Such foods are consumed in greater amounts by the poor. In the USA, hyperphosphatemia has been found to be twice as common among persons of low income as among persons of high income. In the Chronic

Renal Insufficiency Cohort Study, which was carried out in the USA, patients with mildly to moderately impaired renal function were studied over time; the multivariate adjusted risk of hyperphosphatemia (defined as a serum phosphate concentration above 1.45 mmol/L) was higher in persons of the lowest income class than in persons of the highest income class, by a factor of 2.5 to 2.7 (9). As the total phosphate intake among persons in these two groups was roughly the same (1156 versus 1190 mg/day), the finding is presumably attributable to the different types of phosphate that were consumed.

The purpose of this article is to inform physicians of these new aspects of nutritional medicine and, more broadly, to acquaint laypersons interested in health policy with the problem of excessive phosphate intake.

HORMONAL PHOSPHATE REGULATION

Until recently, medical students were taught that phosphate was resorbed in the intestines proportionally to the amount consumed in food, and that the resorbed phosphate was then excreted without any further difficulty by the kidneys. Normally, up to 80% of dietary phosphate is resorbed in the intestines, but the transport rate varies depending on the source of phosphate (and on the individual's vitamin D status). About two-thirds of dietary phosphate is eliminated in the urine, and one-third in the feces (10). It has been discovered only in the last five years that phosphate homeostasis and the renal excretion of phosphate are regulated by a complex endocrine feedback system. The key hormone for phosphate homeostasis is FGF23 (11, e4). In genetically manipulated mice, the absence of FGF23 led to severe hyperphosphatemia and simultaneously to increased renal calcitriol synthesis by way of increased 1-alpha-hydroxylation (12). On the other hand, an elevated concentration of FGF23 leads to increased renal excretion of phosphate and diminishes the activation of vitamin D to calcitriol.

Two Hormone Systems Prevent Phosphate Accumulation

FGF23 is mainly produced in bone osteocytes (10, 11, e4). Its secretion is stimulated by high intestinal phosphate resorption and a high serum phosphate concentration; it is unclear whether this stimulation occurs directly or by way of as yet unidentified intestinal messenger molecules (10). The FGF23-induced increase in renal phosphate excretion can long delay the development of hyperphosphatemia even in patients with progressive renal failure. Definitely elevated serum phosphate concentrations are not seen until the glomerular filtration rate (GFR) drops below 30 mL/min (CKD stage 4). Normophosphatemia in the early stages of renal failure comes at a price, however: steadily increasing concentrations of FGF23 and parathormone (PTH). A high concentration of FGF23 indirectly stimulates the secretion of parathormone by suppressing renal calcitriol synthesis (secondary hyperparathyroidism). PTH has a phosphaturic effect, as does FGF23. Thus, the body uses two different hormone systems (FGF23, PTH) to stave off hyperphosphatemia as long as possible in the setting of progressive renal failure (13, e5).

FGF23 exerts its effects only in the kidneys and parathyroid glands, which are the only organs that express klotho. Klotho is a beta-glucuronidase located in the cell membrane that serves as a coreceptor for FGF23 (10, 11, e4). The absence of klotho leads to the same changes as a total lack of FGF23, namely, hyperphosphatemia and an elevated concentration of active vitamin D. Moreover, both klotho knockout mice and FGF23-deficient mice express a phenotype that resembles premature aging, with vascular calcification, osteoporosis, skin atrophy, pulmonary emphysema, infertility, and early death (14).

It follows from all of the above that high-normal phosphate concentrations are associated with elevated morbidity and mortality even in persons with normal renal function, particularly if they suffer from cardiovascular disease (6, 7) *(Figure 2)*.

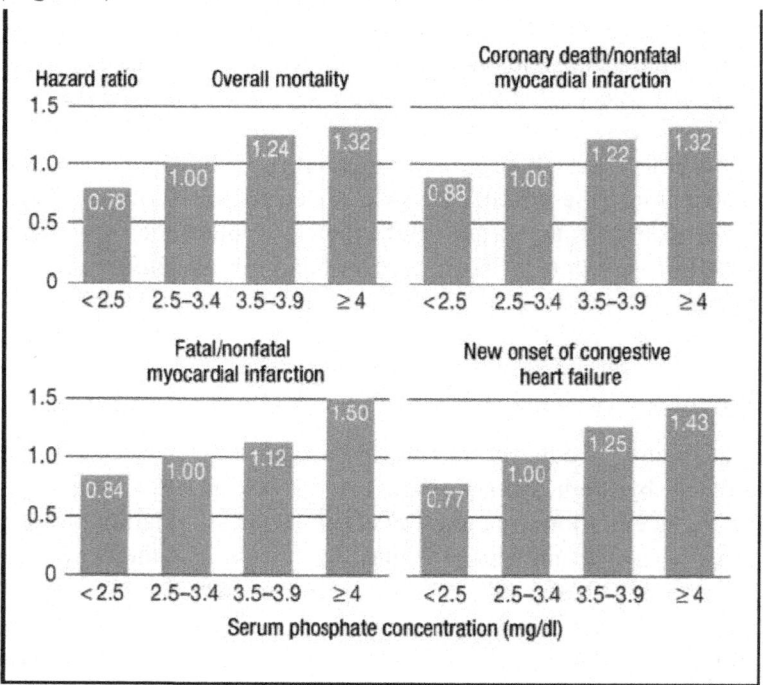

Figure 2: Post-hoc analysis in the Cholesterol and Recurrent Events (CARE) study (n

= 4127): A high-normal phosphate concentration is an independent predictor of mortality and further cardiovascular events in patients who have sustained a myocardial infarction

Kuro-o, who first described the klotho/FGF23 system, views the serum phosphate concentration as a surrogate marker for the activity of the phosphate/FGF23/klotho endocrine axis. Low klotho activity is associated with a high serum phosphate concentration. Kuro-o calls phosphate the "signal molecule of aging," but this interpretation is mainly based on data from animal experiments (15).

Phosphate and Blood Vessels

Phosphate induces vascular calcification both in vitro and in vivo (16, e6). What occurs is not merely the passive precipitation of calcium × phosphate, but rather an active cellular process in which smooth-muscle cells in blood vessels are reprogrammed to become osteoblast-like cells ("osteogenic transdifferentiation") (16). This process, originally identified in cell culture and in animal experiments, has since been demonstrated in human arteries as well (17, e7). Moreover, it has recently been shown that increased phosphate intake leads to a marked impairment of endothelial-cell function in the vascular system, both in experimental animals and in man (18). Phosphate-induced vascular changes may be the link connecting elevated serum phosphate concentrations to premature aging and death.

The Risks of Hypophosphatemia

Lastly, it should be mentioned that the very rare genetic or tumor-associated disturbances of the klotho/FGF23 system lead to severe hypophasphatemia that can be expressed as hypophosphatemic rickets in children and as osteomalacia in adults (10,11, e4). Aside from these conditions, clinically relevant hypophosphatemia is seen only in extremely malnourished persons, e.g., in septic patients in intensive care, who may have phosphate concentrations of 0.5 mmol/L or lower. In some cases, rhabdomyolysis may ensue as the result of depleted energy stores (intracellular ATP [adenosine triphosphate] deficiency).

PHOSPHATE AS A FOOD ADDITIVE

As mentioned above, organic phosphate esters are found mainly in protein-rich foods such as dairy products, fish, meat, sausages, and eggs. They are slowly hydrolyzed in the gastrointestinal tract and then slowly resorbed from the intestine. About 40% to 60% of the organic phosphate esters consumed in the diet are resorbed (e1, e8).

The phosphates found in grains, nuts, and legumes are mainly in the form of phytic acid (hexaphospho-inositol), which cannot be split in the human intestine because of the lack of the enzyme phytase (19). The bioavailability of vegetable phosphate esters is usually less than 50% (8, 20) and thus much lower than that of the phosphate esters in protein-rich foods. It follows that the phosphate content of food cannot be automatically equated with the phosphate load.

The phosphate content of industrially processed food is much higher than that of natural food, because polyphosphates are commonly used as an additive in industrial food production *(Table)*.

Table: The phosphate content of various food groups

TABLE

The phosphate content of various food groups

	Portion size	mg of phosphate	Presence of phosphate additives
Meat, sausages, fish, poultry			
Pork, veal, beef, lamb	150 g	200–300	–/+ (frozen products)
Cold cuts	50 g	50–100	+ (labeling)
Sausages (frankfurters, knockwurst, veal sausages, frying sausages, etc.)	150 g	200–300	+ (labeling)
Fish, seafood	150 g	300–400	–/+ (canned products)
Cheese, dairy products, eggs			
Soft cheese (camembert, gorgonzola, mozzarella, butterkäse, etc.)	50 g	100–200	–/+
Hard and sliced cheese (edam, gouda, emmentaler, raclette, etc.)	50 g	200–300	–/+
Processed/parmesan/American cheese	50 g	400–500	++
Milk, regardless of fat content	200 mL	100–200	–
Yoghurt, regardless of fat content	150 g	100–200	–/+
Quark cheese, regardless of fat content	150 g	200–300	–
Egg (hen's)	60 g	100–200	–
Vegetable spread for bread	100 g	100–200	+
Vegetables, fruit, baked goods, additives for baking			
Potatoes, rice, noodles, semolina	150 g	50–100	–
Salad, fruit	150 g	0–50	–
White bread	100 g	50–100	–/+ (baking mix)
Whole-wheat bread	100 g	100–200	–/+ (baking mix)
Peanuts, almonds, pistachios	100 g	400–500	–
Chocolate (whole milk)	50 g	100–200	–/+
Baker's yeast	cube	200–300	+
Baking powder	packet	1500	++
Beverages			
Cola, mixed drinks containing cola	200 mL	50–100	++
Beer	200 mL	50–100	–/+
Fruit juice (non-perishable)	200 mL	50–100	+
Coffee	150 mL	0–100	–/+ (instant products)
Cocoa powder	20 g	100–200	+

– no added phosphate;
–/+ added phosphate in some products;
+ added phosphate in most products;
++ large quantities of added phosphate in most products

In the European Union, sodium phosphate (E 339), potassium phosphate (E 340), calcium phosphate (E 341), and salts of orthophosphoric acid diphosphate (E 450), triphosphate (E 451), and polyphosphate (E 452) can legally be added to food as preservatives, acidifying agents, acidity buffers, and emulsifying agents. Phosphate salts are also added to many foods as stabilizers or taste intensifiers.

Fast food and ready-to-eat processed foods are the main contributors to today's rising dietary consumption of phosphate. Because of the increased use of food additives, the estimated daily intake of phosphate-containing food additives has more than doubled since the 1990s, from just under 500 mg/day to 1000 mg/day (21, e9). The findings of another recent study were on the same order of magnitude: In processed meat and poultry products, the phosphate content was nearly twice that of the natural product because of added phosphate (22). Thus, not only patients with impaired renal function need to be put on a low-phosphate diet; patients with cardiovascular diseases should also consume less phosphate, and indeed the general population should as well.

Phosphate additives play an especially important role in the meat industry, where they are used as preservatives. They are also used as a component of melting salts in the production of soft cheese. Phosphates loosen the structure of protein, enabling it to bind more water. Phosphate additives are also found in large quantities in flavored soft drinks and in sterilized, ultra-heat treated, thickened, and powdered milk. A further use of phosphates is to prevent the agglomeration of food powders such as powdered coffee and pudding. Cola drinks and flavored soft drinks often contain large quantities of phosphoric acid (E 338) as an acidifying agent. Such agents are given to lower the pH of food and thereby inhibit the growth of yeast, fungi, and bacteria.

Without Added Phosphate, Cola Drinks Would Be Black

In particular, the phosphate that is added to cola drinks interrupts a glycation reaction, which, if unhindered, would produce so-called advanced glycation end products (AGE) and color the beverage pitch-black. Thus, cola drinks owe their brown color to phosphate. European regulations allow up to 700 mg/L of phosphate in cola drinks; if this much phosphate were added, one liter of cola would already provide 50% to 75% of the recommended daily allowance of phosphate for adults. The actual amount of phosphate added to each liter of cola is somewhat less, however, about 520 mg.

More than 300 food additives have been approved for use in Europe and given a uniform designation with an E number. According to the guidelines of the European Union, the E numbers of all food additives in packaged foods must be marked on the package. The EU eco-regulation further restricts the use of food additives for organically grown foods; among additives containing phosphate, only calcium phosphate can be used in organic food. The labeling requirement is, unfortunately, only qualitative, and not quantitative. The consumer, or the patient, cannot determine how much phosphate is actually present in each item, as neither the overall phosphate content nor the quantity of added phosphate is indicated.

THE NEED TO TAKE ACTION

In view of the known connection between dietary phosphate and organ calcification in patients with renal failure, as well as the growing realization that phosphate can damage health even in persons with normal kidneys, one may ask whether concrete interventions in health policy ought to be taken now, even though such steps cannot yet be supported by any findings from prospective interventional trials.

Public Education and Food Labeling

One important step would be to inform physicians and the public thoroughly about the potential risks to cardiovascular and renal function arising from dietary phosphate consumption. Phosphate has long been known to elevate the cardiovascular risk in dialysis patients, but analogous effects have only recently been shown in persons with moderately impaired renal function (of whom the number is growing) and even in persons with normal renal function (6, 7, 23). The changing age structure of the population, with ever more elderly people, further deepens the implications of this problem for health policy, as does the high prevalence of "diseases of civilization," such as diabetes mellitus, hypertension, and coronary heart disease that damage the kidneys and accelerate the age-related decline of renal function. The link between phosphate and progressive renal failure was already suspected and investigated in the early 1980s (24, e10).

To raise public awareness of the health risks of phosphate, we recommend that relevant information should be provided through the mass media. It is important that the subject should be presented in a manner appropriate for laymen, yet without loss of scientific accuracy.

The Labeling of Phosphate Content

Comprehensive labeling of phosphate additives in food—ideally, with a "traffic-light" scheme—would also be desirable, as would a quantitative restriction of phosphate additives. The amount of added phosphate, whether low, medium, or high, should be indicated with a green, yellow, or red sign on the package, as is currently done in Finland and the United Kingdom to indicate sodium chloride content. In order for such measures to be implemented, support should be sought from the food industry, consumer protection organizations, medical societies, and governmental and quasi-governmental entities (the statutory health insurance carriers, the German Federal Ministry of Health, and the European Union). The public is already well informed about the damaging effects of excessive salt consumption as a result of thorough scientific research and the effective public education efforts of medical institutions (25).

It remains to be clarified whether the association of a high serum phosphate concentration with increased morbidity and mortality reflects a direct toxic effect of phosphate or is rather due to pathological concentrations of the phosphate-regulating hormones FGF23 and klotho. Phosphate might thus also be a surrogate parameter for the functioning of this hormone system. We believe that comprehensive public education, with a scientifically well-grounded explanation of the adverse effects of high phosphate intake along with easily understandable labeling of the phosphate content of food, could help considerably to limit the damage done by this newly recognized cardiovascular risk factor.

ACKNOWLEDGMENTS
Translated from the original German by Ethan Taub, M.D.

REFERENCES
1. Sullivan C, et al.: Effect of food additives on hyperphosphatemia among patients with end-stage renal disease: a randomized controlled trial. JAMA 2009; 301: 629–35.

2. Block GA, et al.: Mineral metabolism, mortality, and morbidity in maintenance hemodialysis. JASN 2004; 15: 2208–18.

3. Kestenbaum B, et al.: Serum phosphate levels and mortality risk among people with chronic kidney disease. JASN 2005; 16: 520–8.

4. KDIGO clinical practice guideline for the diagnosis, evaluation, prevention, and treatment of Chronic Kidney Disease-Mineral and Bone Disorder (CKD-MBD). Kidney Int Suppl 2009: S1–130.

5. Noori N, et al.: Association of dietary phosphorus intake and phosphorus to protein ratio with mortality in hemodialysis patients. CJASN 2010; 5: 683–92.

6. Foley RN, et al.: Serum phosphorus levels associate with coronary atherosclerosis in young adults. JASN 2009; 20: 397–404.

7. Dhingra R, et al.: Relations of serum phosphorus and calcium levels to the incidence of cardiovascular disease in the community. Arch Intern Med 2007; 167: 879–85.

8. Tonelli M, et al.: Relation between serum phosphate level and cardiovascular event rate in people with coronary disease. Circulation 2005; 112: 2627–33.

9. Gutierrez OM, et al.: Low socioeconomic status associates with higher serum phosphate irrespective of race. JASN 2010; 21: 1953–60.

10. Berndt T, Kumar R: Novel mechanisms in the regulation of phosphorus homeostasis. Physiology (Bethesda) 2009; 24: 17–25.

11. Liu S, Quarles LD: How fibroblast growth factor 23 works. JASN 2007; 18: 1637–47.

12. Shimada T, et al.: Targeted ablation of Fgf23 demonstrates an essential physiological role of FGF23 in phosphate and vitamin D metabolism. J Clin Invest 2004; 113: 561–8.

13. Wetmore JB, Quarles LD: Calcimimetics or vitamin D analogs for suppressing parathyroid hormone in end-stage renal disease: time for a paradigm shift? Nat Clin Pract Nephrol 2009; 5: 24–33.

14. Kuro-o M, et al.: Mutation of the mouse klotho gene leads to a syndrome resembling ageing. Nature 1997; 390: 45–51.

15. Kuro-o M: A potential link between phosphate and aging—lessons from Klotho-deficient mice. Mech Ageing Dev 2010; 131: 270–5.

16. Giachelli CM: The emerging role of phosphate in vascular calcification. Kidney Int 2009; 75: 890–7.

17. Shroff RC, et al.: Chronic mineral dysregulation promotes vascular smooth muscle cell adaptation and extracellular matrix calcification. JASN 2010; 21: 103–12.

18. Shuto E, et al.: Dietary phosphorus acutely impairs endothelial function. JASN 2009; 20: 1504–12. Bohn L, Meyer AS, Rasmussen SK: Phytate: impact on environment and human nutrition. A challenge for molecular breeding. J Zhejiang Univ Sci B 2008; 9: 165–91.

19. Lei XG, Porres JM: Phytase enzymology, applications, and biotechnology. Biotechnol Lett 2003; 25: 1787–94.

20. Kalantar-Zadeh K, et al.: Understanding sources of dietary phosphorus in the treatment of patients with chronic kidney disease. CJASN 2010; 5: 519–30.

21. Sherman RA, Mehta O: Phosphorus and potassium content of enhanced meat and poultry products: implications for patients who receive dialysis.

CJASN 2009; 4: 1370–3. Tonelli M, Pannu N, Manns B: Oral phosphate binders in patients with kidney failure. NEJM 2010; 362: 1312–24.

22. Haut LL, et al.: Renal toxicity of phosphate in rats. Kidney Int 1980; 17: 722–31.

23. Klaus D, Hoyer J, Middeke M: Salt restriction for the prevention of cardiovascular disease. Dtsch Arztebl Int 2010; 107(26): 457–62. VOLLTEXT

24. Uribarri J: Phosphorus homeostasis in normal health and in chronic kidney disease patients with special emphasis on dietary phosphorus intake. Semin Dial 2007; 20: 295–301.

25. Ohnishi M, Razzaque MS: Dietary and genetic evidence for phosphate toxicity accelerating mammalian aging. FASEB J 2010; 24: 3562–71.

26. Dhingra R, et al.: Relations of serum phosphorus levels to echocardiographic left ventricular mass and incidence of heart failure in the community. Eur J Heart Fail 2010; 12: 812–8.

27. Prie D, Urena Torres P, Friedlander G: Latest findings in phosphate homeostasis. Kidney Int 2009; 75: 882–9.

28. Martin KJ, Gonzalez EA: Prevention and control of phosphate retention/ hyperphosphatemia in CKD-MBD: what is normal, when to start, and how to treat? Clin J Am Soc Nephrol 2011; 6: 440–6.

29. Stubbs JR, et al.: Role of hyperphosphatemia and 1,25-dihydroxyvitamin D in vascular calcification and mortality in fibroblastic growth factor 23 null mice. J Am Soc Nephrol 2007; 18: 2116–24.

30. Moe SM, et al.: Medial artery calcification in ESRD patients is associated with deposition of bone matrix proteins. Kidney Int 2002; 61: 638–47.

31. Kayne LH, et al.: Analysis of segmental phosphate absorption in intact rats. A compartmental analysis approach. J Clin Invest 1993; 91: 915–22.

32. Calvo MS, Park YK: Changing phosphorus content of the U.S. diet: potential for adverse effects on bone. J Nutr 1996; 126(4 Suppl): 1168S–80S.

33. Alfrey AC: The role of abnormal phosphorus metabolism in the progression of chronic kidney disease and metastatic calcification. Kidney Int Suppl 2004: S13–7.

Chapter 8

EVALUATION OF FOOD PRESERVATIVES, LOW TOXICITY CHEMICALS, LIQUID FRACTIONS OF PLANT EXTRACTS AND THEIR COMBINATIONS AS ALTERNATIVE OPTIONS FOR CONTROLLING CITRUS POST-HARVEST GREEN AND BLUE MOULDS IN VITRO

Emad I. Hussein[1], Ghassan J. M. Kanan[1], Khalid M. Al- Batayneh[1], Khalaf Alhussaen[2], Wesam Al Khateeb[1], Janti Qar[1], Jacob H. Jacob[3], Riyadh Muhaidat[1] and Mohamed I. Hegazy[4]

[1]Department of Biological Sciences, Yarmouk University, Irbid, Jordan

[2]Department of Plant Production and Protection, Faculty of Agriculture, Jerash, University, Jerash, Jordan

[3]Department of Biological Sciences, Al al-Bayt University, Mafraq, Jordan

[4]Department of Microbiology, Faculty of Agriculture, Zagazig University, Egypt

ABSTRACT

The post-harvest moulds Penicillium digitatum and Penicillium italicum are important plant pathogens and spoilage-causing molds especially against citrus fruits. If not treated, post-harvest moulds can cause enormous economic losses during storage and marketing. Therefore, more investigations are needed to examine new antifungal agents against such fungi. In this work, we aimed to evaluate the antifungal activity of some plant extracts (namely, Harmal seeds (Peganum harmala L.), cinnamon bark (Cinnamomum cassia L.) and sticky fleabane leaves (Inula viscosa L.), food preservatives (namely, sodium benzoate, sodium molybdate, ammonium heptamolybdate tetrahydrate, potassium carbonate and sodium bicarbonate) and their mixtures, i.e., plant extracts and food preservatives against P. digitatum and P. italicum. Both disc agar diffusion method and broth dilution methods was used to evaluate the antifungal activity of the plant extracts and food preservatives. Results revealed that methanolic fractions of cinnamons' bark and sticky fleabane leaves showed the highest efficacy. MIC values of 150 and 37.5 µg mL^{-1} were obtained with cinnamons' fraction against P. italicum and P. digitatum, respectively. Sodium benzoate was the most effective against tested fungal species. The obtained MIC values against P. digitatum and P. italicum were 37.5 and 75 µg mL^{-1}, respectively. Mixtures of tested chemicals showed synergistic effects against both fungal species. Mixtures of sodium benzoate and fractions of either cinnamon or sticky fleabane reflected synergistic effects against P. italicum and antagonistic effects against P. digitatum. Inhibition zones against P. italicum ranged between 38-57 mm.

INTRODUCTION

The post-harvest green and blue moulds Penicillium digitatum (Pers: Fr) Sacc. and Penicillium italicum, respectively, are considered universal diseases that lead to the spoilage of almost all kinds of mature citrus fruits (Plaza et al., 2004; Prusky et al., 2004; Samson et al., 2004). Significant economic losses caused by post-harvest pathogens during storage and marketing are greater than what most people believe. These unavoidable losses between the farm gate and the consumer are currently of big concern (Soylu et al., 2005; Smilanick et al., 2005). Citrus industry relies heavily on the extensive use of chemical fungicides as a common practice for the control of post-harvest fungal decay of citrus fruits (McGrath, 2001; Bouzerda et al., 2003; Tripathi and Dubey, 2004). However, the consumer demands for fungicide-free products and the increasing fungal resistance for fungicides necessitates the search for alternative control options (Obagwu and Korsten, 2003; Soylu et al., 2005; Lee et al., 2004; Ikeura et al., 2011; Bhyan et al., 2007; Reddy et al., 2010).

Plant extracts and their essential oils are one of the non-synthetic chemical control options that have recently received attention for controlling plant diseases (Soylu et al., 2005; Abad et al., 2007; Nahunnaro, 2008; Zaker and Mosallanejad, 2010; Hasan et al., 2005). Seeds and roots extracts of Harmal (Peganum harmala) have been shown to contain a variety of active alkaloids, including harmaline which was reported as a very strong antifungal agent (Telezhenetskaya and D'yakonov, 2004). In addition, eugenol and cinnamaldehyde from cinnamon bark have consistently been reported to have antifungal activity (Delaquis et al., 2002). Other studies were carried out to elucidate the biological activity of sticky fleabane extracts (Wang et al., 2004). All types of Inula viscosa extracts have been proved to exert significant antifungal activity. The extracts from oily leaves paste showed in vitro antifungal activity against some dermatophytes, Candida spp. and Downy mildew (Cafarchia et al., 2002; Cohen et al., 2006). Furthermore, the sesquiterpene lactone isolated from I. viscosa flowers was found to possess an in vitro activity against Microsporum canis, Microsporum gypseum and Trichophyton mentagrophytes (Abu-Zarga et al., 1998; Cafarchia et al., 2002).

Recently, many chemical compounds were tested as alternative control options for citrus post-harvest diseases either alone or in combination with physical or biological treatments (Palou et al., 2001; Reddy et al., 2007; Bonjar, 2004; Bonjar and Nik, 2004). Over a wide range of commonly used chemicals, sodium bicarbonate, potassium sorbate, sodium benzoate, sodium carbonate and ammonium molybdate were the most promising fungicides alternatives to control penicillium species (Palou et al., 2002b; Smilanick et al., 2002). Soda ash (sodium carbonate) has been used for over 70 years to control post-harvest decay of citrus fruit in California (Palou et al., 2001). It has been reported that soda ash reduces the incidence of green mould by over 90% (Smilanick et al., 1997). In addition, the combination of sodium bicarbonate and fungicides was used to manage post-harvest green and blue moulds of citrus in California (Smilanick et al., 1999; Ismail and Zhang, 2004). However, sodium bicarbonate has partially controlled the green mould and other fungal diseases of citrus fruits (Larrigaudiere et al., 2002; Smilanick et al., 1999; Palou et al., 2001; Larrigaudiere et al., 2002). Although potassium sorbate was reported to delay rather than to stop green mould's infection, it has been reported to act (at low levels) synergistically along with heat and fungicides in controlling the disease. Therefore, sorbate has commercially substituted sodium bicarbonate for improving the performance of fungicides in retarding the development of fungal resistant isolates (Palou et al., 2002a; Smilanick et al., 2005). Moreover, Buazzi and Maeth (1992), indicated that sodium benzoate which is usually used as food preservative, has prevented the growth of many microorganisms including yeasts, bacteria and moulds. Sodium and ammonium molybdate that are applied as molybdenum sources for foliar and soil applications (i.e., as fertilizers) have effectively controlled the growth of green and blue moulds of citrus fruits (Palou et al., 2002a). In addition, the use of antifungal mixtures which selectively had different modes of action is considered a highly recommended practice in avoiding the risks of pathogens resistance. The purpose of this study is to investigate potentially novel treatments for combating pathogenic fungi to citrus fruits, namely, Penicillium digitatum and Penicillium italicum, the green and blue moulds, respectively. Our emphasis was placed on identifying combined treatments of novel and existing antifungal agents which may exhibit potent synergistic effect against fungal pathogens. Novel antifungal agents include mixtures of plant liquid fractions (natural molecules), combinations of food preservatives and mixtures of both plant fractions and food preservatives.

MATERIALS AND METHODS

Origin of fungal isolates: Two wild-type species of Penicillium, Penicillium digitatum and Penicillium italicum, were isolated from mature spoiled orange (Citrus sinensis L.) and lemon (Citrus limon L.) fruits and identified using morphological and physiological characters. Fruits were obtained from the local market Irbid-Jordan.

Growth media: The Aspergillus nidulans complete (CM) medium described previously by Cove (1966) and modified by Al-Najar (2007) was used to achieve optimal growth conditions for the tested fungal isolates.

Purification of fungal isolates: Conidiospores from each isolate were grown for 7 to 10 days at 20-25°C on CM plates (Al-Najar, 2007) to confirm their purity and identity.

A suspension of conidiospores was prepared in 5 mL physiological saline/ Tween 80 (0.05%) solution at a concentration of approximately 108 spores per milliliter. Aliquots of 100 μL from a dilution of 10-6 or 10-7 were plated again on complete media in order to get a single colony as a source of pure culture (Zhang et al., 2004).

Tested decay control chemicals: Five decay control chemicals were tested for their effect on mould growth: Sodium benzoate ($C_7H_5O_2Na$), sodium molybdate ($MONa_2O_4.2H_2O$), ammonium heptamolybdate tetrahydrate (($NH_4)_6Mo_7O_{24}.4H_2O$), potassium carbonate (KCO_3) and sodium bicarbonate ($NaHCO_3$). Various concentrations (20, 30, 40, 80, 120, 160, 200, 240, 280, 320 and 400 μg mL^{-1}) of the chemicals were used in order to determine their influence on the growth of P. digitatum and P. italicum isolates. Each concentration of the different chemicals was aseptically added to agar wells. Conidiospores suspension from each tested fungal isolate was prepared as described above. Petri dishes of complete media (chemical and plant extract free media) were inoculated with conidial suspension (200 μL) loaded onto the surface of the plate and spread over the whole plate. Three plates were used for each tested concentration per chemical. Control treatments were set up using sterile distilled water and/or dimethylsulphoxide (DMSO) and each treatment was repeated at least twice. The inoculated plates were incubated for 5 days at 20 or 25°C as required. The diameter of inhibition zone was measured in two directions at right angles to each other (Schroeder and Bullerman, 1985).

Plant extracts preparation: Extracts of three plant materials were tested: Harmal seeds (Peganum harmala L.), cinnamon bark (Cinnamomum cassia L.) and sticky fleabane leaves (Inula viscosa L.). Each plant material was dried in shade, ground to a fine powder using liquid nitrogen and extracted (48 h) with absolute ethanol in Soxhlet extractor (Ndukwe et al., 2006). The solvent was removed using rotary evaporator (Heidolph, VV2000) under reduced pressure at temperatures below 55°C. The resulting crude extracts were stored at -20°C until tested. Stock solutions and serial dilutions of each extract were prepared in dimethylsulphoxide (DMSO) (Ambrozin et al., 2004) and used as control.

Fractionation of plant crude extracts: Each crude extract was fractioned into three parts (aqueous, hexane and methanolic fraction). The emulsions that may form between layers were also tested. Each crude extract sample was fractionated with (1:1) ratio of water/dichloromethane (v/v). The resultant aqueous fraction was further extracted with dichloromethane and then concentrated to dryness using rotary evaporator. This dichloromethane fraction was subsequently partitioned with (1:1) n-hexane/90% methanol. The hexane and methanol fractions produced were concentrated to dryness using rotary evaporator and kept in sterile containers at 4°C until used. Each fraction type was prepared at the required concentration (μg mL^{-1}) by dissolving in dimethylsulphoxide then tested for its antifungal activity (Souza-Fagundes et al., 2002).

Antifungal assay of plant crude extracts, their fractions, decay control chemicals and their combinations: Aliquot of 100 μL spore suspension (ca. 1x108 spores mL^{-1}) of each tested isolate was streaked in radial patterns on the surface of complete media plates. Stock solutions of each crude extract or liquid fraction were filter sterilized through a 4 μm Millipore filter (Soylu et al., 2005). Agar wells (6 mm in diameter) were made in solidified complete growth media. Each tested concentration from the plants crude extracts or their liquid fractions (20, 50, 100, 200, 400 and 600 μg mL^{-1}) or the five tested chemicals (20, 30, 40, 80, 120, 160, 200, 240, 280, 320 and 400 μg mL^{-1}) were loaded into the agar wells. In addition, the following pair-wise combinations were tested using agar well diffusion method: combinations from each crude extract and each of its liquid fractions (200 μg mL^{-1} of each), combinations of crude extracts or fractions referring to different plant species (200 μg mL^{-1} of each), combinations made between tested chemical substances (160 μg mL^{-1} of each) and combinations (each of 200 μg mL^{-1}) between chemicals and each of the tested plant extracts or fractions. DMSO was used as control for the ethanolic extracts or fractions. The cultured plates were incubated for 3-5 days at 20 or 25°C as required. The radius of the inhibition zone was measured in two directions at right angles to each other. Experiments were carried out with three replicates per treatment and each treatment was repeated at least twice (Ndukwe et al., 2006). The Minimum Inhibitory Concentration (MIC) was defined as the lowest concentration of the extract, the fraction, or the tested chemical that was able to inhibit any visible fungal growth after 48 h in liquid cultures. The fungistatic or fungicidal effects of each tested material was determined by streaking on complete plates with 200 μL spore suspension taken from the liquid culture that specified the MIC value of that substance (Obagwu and Korsten, 2003).

Statistical analysis: The Minimum Inhibitory Concentration (MIC), the correlation coefficient values and the significance values at p≤0.05 or 0.01 levels (2-tailed) for tested chemical substances, plant crude extracts, their fractions and the mixtures made were calculated by regression analysis for the relationship between the size of fungal inhibition zone (mm) and the concentration (μg mL^{-1}) of the evaluated substances (Log value). Microsoft Excel 2003 software and the SPSS program version 15 were used for data analysis.

RESULTS

In vitro sensitivity of P. italicum and P. digitatum to different concentrations of plant extracts and their liquid fractions: Data of the regression analysis for the relationship between sizes of fungal growth inhibition zone (mm) and concentrations of plant (cinnamon bark; sticky fleabane leaves and Harmal seeds) crude extracts and their liquid fractions are shown in Table 1. There was a significant correlation (at either the 0.01 or the 0.05 level of significance) between the tested concentrations of plant crude extracts and mean inhibition zones of both fungal species (P. italicum and P. digitatum) (Table 1). An exception to this pattern was shown with Harmal's seed crude extract, where no significant correlation was found between tested concentrations and P. italicum inhibition zones (r = 0.806; p = 0.053). Cinnamon's bark crude extract was found to be the most effective against both fungal species (Table 1). The obtained MIC values against P. digitatum and P. italicum were 75 and 187.5 μg mL^{-1}, respectively. In contrast, Harmal's seed extract was the least effective against both fungal species (with inhibition zones of approximately 18 to 20 mm) and the obtained MIC values were within the range of 400 to 450 μg mL^{-1}. Regarding the inhibition effect of plant's liquid fractions, there was a significant correlation between tested concentrations of fractions and mean inhibition zones of both Penicillium species (Table 1). Exceptions to this pattern were hexane fraction (r = 0.797; p = 0.058) and water fraction (r =0.783; p = 0.066) of sticky fleabane against P. italicum and P. digitatum, respectively. Furthermore, no significant correlations were noticed with Harmal's hexane fraction against P. italicum (r = 0.798; p = 0.057) and P. digitatum (r = 0.803; p = 0.054). Additionally, the layer between Harmal's fractions reflected no significant correlation against P. italicum (r = 0.806; p = 0.053). On the other hand, methanolic fractions of cinnamon bark and sticky fleabane leaves showed the highest inhibitory efficacy against both fungal species where, the largest inhibition zones were obtained with these fractions (Table 1). In addition, MIC values of 150 and 37.5 µg mL^{-1} were obtained with cinnamon methanolic fraction against P. italicum and P. digitatum, respectively. Also, MIC values of 375 and 75 µg mL^{-1} were obtained with sticky fleabane methanolic fraction against the same species, respectively. In contrast, none of the tested concentrations (within a range of 20 to 600 µg mL^{-1}) of cinnamons' water fraction reflected inhibitory effect against both fungal species.

Table 1: In vitro sensitivity of two wild-type fungal isolates, Penicillium italicum and P. digitatum, to different concentrations (µg m^{-1}) of three plant crude extracts and their liquid fractions

Extract source[a]	Range of conc. (µg mL^{-1})[b]	Range of inhibition [c] zone (mm) Mean±SD	MIC [d]	Corr. value [e](r)	Sig. value[f]	Fungal isolate
Cinn/crude	20-100 200-600	10.5±2.12 to 18.5±3.53	187.5	0.981**	0.001	P. italicum
		29.5±2.12 to 35.5±3.54				
Cinn/crude	20-100 200-600	18.2±1.12 to 23.5±2.12	75	0.952**	0.003	P. digitatum
		39±1.41 to 45±2.85				
Cinn/met	20-100 200-600	16.5±1.71 to 21.25±3.2	150	0.941**	0.005	P. italicum
		23.5±3.53 to 33.5±2.4				
Cinn/met	20-100 200-600	23±1.4 to 29.5±2.12	37.5	0.968**	0.001	P. digitatum
		43.5±2.2 to 55.5±1.76				
Cinn/hex	20-100 200-600	0.0 to 0.0	300	0.903*	0.014	P. italicum
		14.5±0.71 to 18.5±2.1				
Cinn/hex	20-100 200-600	0.0 to 0.0	300	0.908*	0.012	P. digitatum
		17.5±0.71 to 24±1.10				
Cinn/wat	20-100 200-600	0.0 to 0.0 0.0 to 0.0				P. italicum
Cinn/wat	20-100 200-600	0.0 to 0.0 0.0 to 0.0				P. digitatum
Stick/crude	20-100 200-600	0.0 to 0.0	375	0.913*	0.011	P. italicum
		9.5±2.12 to 16±1.41				
Stick/crude	20-100 200-600	0.0 to 17±1.41	150	0.990**	0.000	P. digitatum
		29.5±3.54 to 38±2.14				
Stick/met	20-100 200-600	0.0 to 0.0	375	0.917*	0.010	P. italicum
		11.5±0.71 to 20.5±3.51				
Stick/met	20-100 200-600	19.5±2.12 to 22±5.7	75	0.940**	0.005	P. digitatum
		35±1.41 to 44±2.83				
Stick/hex	20-100 200-600	0.0 to 0.0 0.0 to 18±1.44	375	0.797	0.058	P. italicum
Stick/hex	20-100 200-600	0.0 to 0.0 18±2.16 to 27±2.38	375	0.912*	0.011	P. digitatum
Stick/wat	20-100 200-600	0.0 to 0.0 9.5±2.12 to 18.5±2.22	375	0.915*	0.011	P. italicum
Stick/wat	20-100 200-600	0.0 to 0.0 0.0 to 26.5±2.22	300	0.783	0.066	P. digitatum
Harm/crude	20-100 200-600	0.0 to 0.0 0.0 to 20.5±2.33	400	0.806	0.053	P. italicum
Harm/crude	20-100 200-600	0.0 to 0.0 13.5±1.4 to 18.5±1.14	450	0.908*	0.012	P. digitatum
Harm/met	20-100 200-600	0.0 to 0.0 12±1.41 to 24.5±3.45	450	0.908*	0.012	P. italicum
Harm/met	20-100 200-600	0.0 to 0.0 13.5±1.4 to 22.5±1.23	450	0.912*	0.011	P. digitatum
Harm/hex	20-100 200-600	0.0 to 0.0 0.0 to 13±1.11	600	0.798	0.057	P. italicum
Harm/hex	20-100 200-600	0.0 to 0.0 0.0 to 17.5±2.16	600	0.803	0.054	P. digitatum
Harm/wat	20-100 200-600	0.0 to 0.0 12±1.34 to 24±2.88	375	0.914	0.011	P. italicum
Harm/wat	20-100 200-600	0.0 to 17.25±0.5	375	0.908*	0.012	P. digitatum
		17.5±2.6 to 21.5±1.75				
Harm/bet	20-100 200-600	0.0 to 0.0 0.0 to 19.5±1.77	375	0.806	0.053	P. italicum
Harm/bet	20-100 200-600	0.0 to 15.75±0.5				
		17±0 to 22.5±2.25	375	0.930**	0.007	P. digitatum

[a]Extract source: Cinn: Cinnamon (Cinnamomum cassia) bark, stick: Sticky fleabane leaves (Inulla viscosa), Harm: Harmal (Peganum harmala) seeds, crude: Crude extract, met: Methanolic fraction, hex: Hexane fraction, wat:

Water fraction, bet: Between layers fraction, bA range of concentrations (20, 50 100, 200, 400 and 600 μg mL⁻¹) of extract was used against tested fungal isolates, cFungal inhibition zone was determined from the mean diameter (mm) zone±SD of three independent tests using agar well diffusion method, dMinimal inhibitory concentration of the extract, eCorrelation coefficient values, fLevel of significance at the 0.05 and 0.01 levels (2- tailed), **Correlation is significant at the 0.01 level (2-tailed), *Correlation is significant at the 0.05 level (2-tailed)

In vitro sensitivity of Penicillium species to combinations of cinnamon fractions, sticky fleabane fractions and fractions of both plant species: Since significant inhibitory effects were obtained with most of the tested plant extracts at concentrations of 200 μg mL⁻¹ or more, attention was turned to evaluate mixtures of extracts which may display potent synergistic effects against fungal pathogens. The investigated treatments included: combinations of 1:1 ratio (200 μg mL⁻¹ each) between fractions of the same plant material and combinations between fractions of different plants.

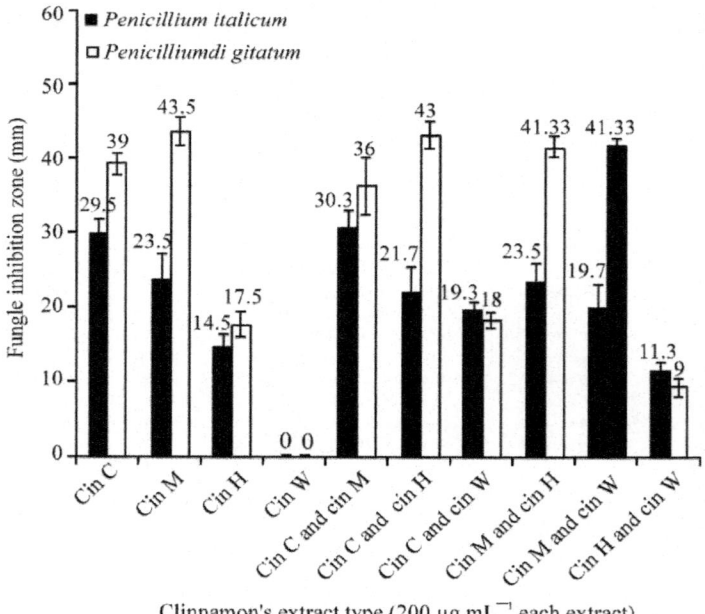

Figure 1: Inhibition zone of Penicillium italicum and P. digitatum wild-type isolate s obtained by 200 μg mL⁻¹ of cinnamon different fractions, Cin: Cinnamon, C: Crude extract, C: Crude extract, M: Methanolic fraction, H: Hexane fraction; W: Water fraction

The obtained results (Fig. 1) indicate that most of the combinations made between cinnamon's bark crude extract and its fractions resulted in antagonistic effects against both citrus-post harvest fungal pathogens. Exceptions to this pattern were the combinations made between cinnamon's crude extract and it's methanolic (against P. italicum) and hexane fractions (against P. digitatum) where inhibition zone diameters were approximately similar to those obtained by the crude extract itself (Fig. 1). However, the combinations between cinnamon's methanolic and hexane fractions or cinnamon's methanolic and water fractions have maintained or preserved the inhibitory effect of methanolic fraction against both fungal species (i.e., inhibition zone diameters similar to those obtained by methanolic fraction (around 42 mm) were obtained) (Fig. 1).

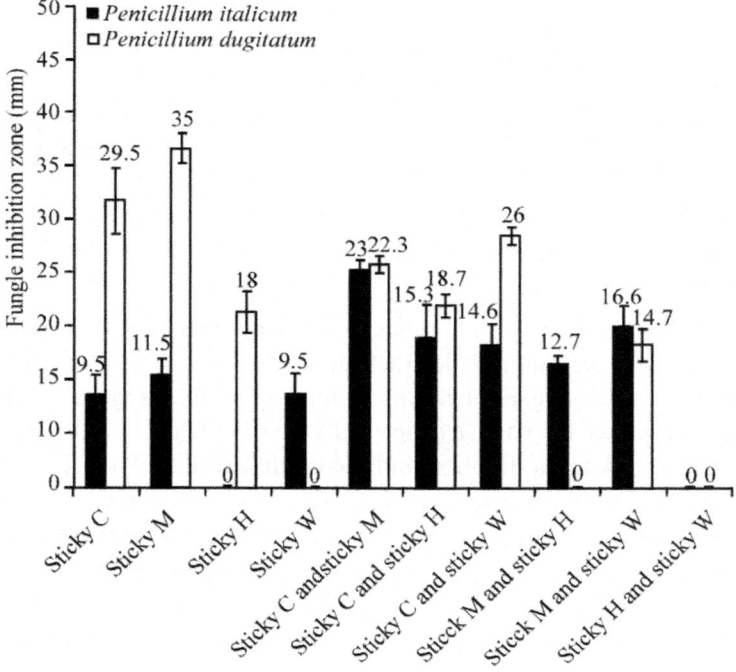

Sticky fleabane extract type (200 µg mL⁻¹ each extract)

Figure 2: Inhibition zone of Penicillium italicum and P. digitatum wild-type isolates obtained by 200 µg mL⁻¹ of sticky fleabane (Inulla viscosa) crude extract, its liquid fractions and combinations of these fractions. C: Crude extract, M: Methanolic fraction, H: Hexane fraction and W: Water fraction

The combined concentrations (200 µg mL⁻¹ each) of sticky fleabane crude extract and its hexane fraction resulted in synergistic effects against P.

italicum (Fig. 2). The mean inhibition zone obtained by the above mentioned combination was approximately 16 mm as compared to 0.0 and 9 mm obtained by the crude extract and its hexane fraction, respectively. However, the combination of sticky fleabane methanolic and hexane fractions has preserved the inhibitory effect obtained by the methanolic fraction against P. italicum (Fig. 2), where inhibition zones of approximately 12 mm were obtained. Mixtures between sticky fleabane crude extract and either it's methanolic or water fraction resulted in additive effects against P. italicum. Moreover, mixtures of sticky fleabane methanolic and water fractions had also generated additive effects against P. italicum. Regarding the effects of the remaining possible combinations made between sticky fleabane extract and its liquid fractions, all have generated antagonistic effects against both fungal species tested (Fig. 2). The mixture of methanolic and hexane fractions resulted in no inhibitory effect against P. digitatum, as compared to approximately 35 and 18 mm inhibition zones obtained by each fraction alone, respectively. Also, the combination made between sticky fleabane hexane and water fractions showed no inhibitory effect against both fungal species, although hexane fraction was relatively effective (approximately 18 mm) against P. digitatum but not P. italicum (no inhibitory effect). In addition, the water fraction generated 9 mm inhibition zone against P. italicum but no activity was detected against P. digitatum (Fig. 2).

Mixtures made of cinnamon's crude extract and sticky fleabane (each of 200 μg mL⁻¹) or its methanolic fraction generated inhibition zones against P. digitatum of similar sizes (i.e., approximately in the range of 32 to 39 mm) to these obtained by each extract individually (Fig. 3). Also, mixtures of cinnamon's methanolic fraction and the crude extract of sticky fleabane or its methanolic fraction have generated inhibition zones (in the range of 35 to 44 mm) against P. digitatum, with similar sizes to those obtained by each extract alone. The combinations made between cinnamon's methanolic fraction and that of sticky fleabane resulted in inhibition zones against P. italicum of similar sizes (approximately 24 mm) to those obtained by cinnamon's fractions when tested individually (Fig. 3).

In addition, the combinations made between cinnamon's hexane fraction and all fractions of sticky fleabane generated inhibition zones against P. italicum of approximately equal sizes to these obtained by cinnamon's hexane fraction (i.e., zones in the range of 12 to 15 mm).

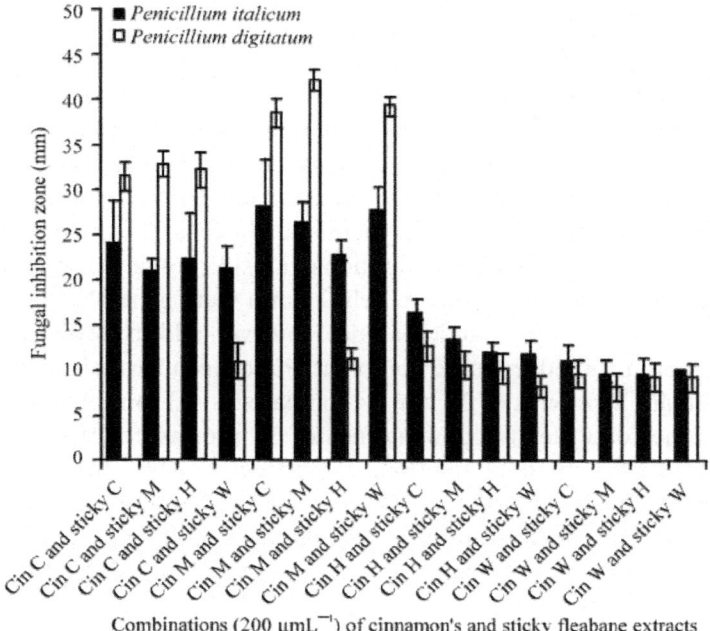

Figure 3: Inhibition zone of Penicillium italicum and P. digitatum wild-type isolates obtained by 200 μg mL^{-1} of all possible combinations between cinnamon and sticky fleabane extracts and fractions, Cin: Cinnamon (Cinnamomum cassia) bark, sticky: Sticky fleabane (Inulla viscosa) leaves, C: Crude extract, M: Methanolic fraction, H: Hexane fraction, W: Water fraction

In contrast, the combinations (each of 200 μg mL^{-1}) made between cinnamon's and sticky fleabane water fractions and between cinnamons water fraction and the crude extract of sticky fleabane have maintained the inhibition levels obtained by the sticky fleabane extract (i.e., inhibition zones of approximately 10 mm) against P. italicum. Similar sized zones to those obtained by cinnamons' methanolic fraction (approximately 25 mm) were generated against P. italicum by combined concentrations (each of 200 μg mL^{-1}) of cinnamons' methanolic fraction and sticky fleabane crude or water fraction, namely, 27 mm (Fig. 3). Surprisingly, synergistic effects were obtained by combined concentrations of cinnamons water and sticky fleabane hexane fraction against P. italicum where, inhibition zones of approximately 8 mm were obtained as compared to "no inhibition" by individual fractions. The same synergistic effect was obtained against P. digitatum by combinations of cinnamon and sticky fleabane water fractions (Fig. 3). The remaining combinations among cinnamons and sticky fleabane fractions all generated antagonistic effects towards both fungal species as compared to the effect of individually tested fraction (Fig. 3).

In vitro sensitivity of Penicillium species to combinations of cinnamon fractions and harmal fractions: There was a considerable synergistic effect from the combined concentrations of Harmals' crude extract and each of its methanolic, hexane and the layer between fractions against P. italicum (inhibition zones were almost 15 mm) as compared to "no inhibition" exerted by each fraction alone. In addition, synergistic effects of the mixture of harmal seeds crude extract and its hexane fraction against P. digitatum were observed, where inhibition zones of approximately 21 mm were generated as compared to no inhibition by hexane fraction alone. However, additive effects against P. digitatum were resulted from the combination between harmals' crude extract and its methanolic (zones of 25 mm as compared to 13 mm obtained by each fraction alone) or water fractions (zones of 33 mm as compared to 14 and 18 mm by their individual fraction, respectively). Also, the combinations made between the layer formed between harmal fractions and each of the crude extract or its methanolic fraction generated inhibition zones against P. digitatum of approximately similar sizes to those obtained by each fraction individually (Fig. 4).

Furthermore, the combination made between harmals' water fraction and the layer between fractions showed inhibition zones against P. digitatum with sizes approximately similar to those obtained by each fraction alone. The remaining possible combinations made between harmal's fractions resulted in antagonistic effects as compared to each fraction alone (Fig. 4).

Mixtures of cinnamons water fraction and each of the crude extract, hexane fraction and the layer between harmals fractions generated synergistic effects against P. italicum, where inhibition zones between 8^{-1} 5 mm were obtained as compared to no inhibition using individual fractions (Fig. 5). However, mixtures between cinnamon's methanolic fraction and each of harmal's fractions were found to reserve the same level of inhibition of cinnamon's fraction alone against P. italicum (in the range of 22 to 25 mm). Furthermore, the inhibition zones against Penicillium species obtained from the combinations between cinnamon's hexane fraction with harmal extract or any of its liquid fractions were of similar sizes (in the range of 13^{-1} 8 mm) compared to those obtained by cinnamon's hexane fraction alone.

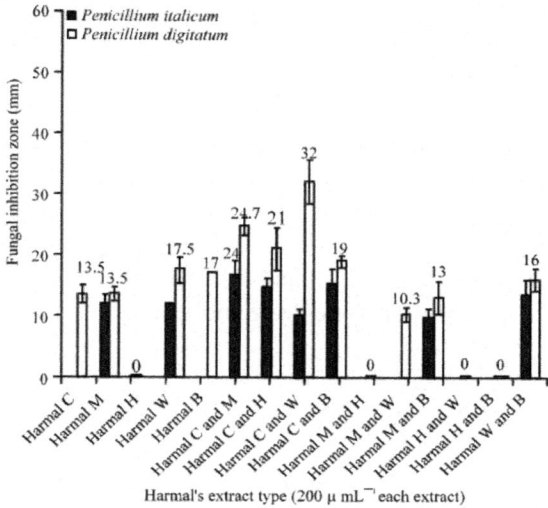

Figure 4: Inhibition zone of Penicillium italicum and P. digitatum wild-type isolates obtained by 200 μg mL^{-1} of harmal (Peganum harmala) seed crude extract, its liquid fractions and combinations of these fractions, C: Crude extract, M: Methanolic fraction, H: Hexane fraction, W: Water fraction, B: Between layers fraction

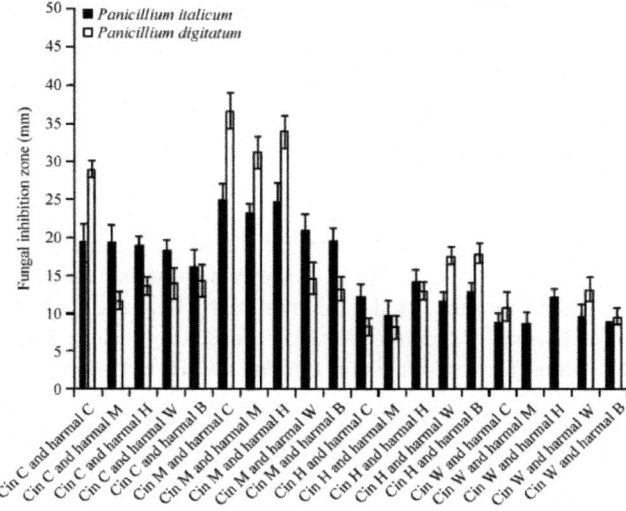

Figure 5: Inhibition zone of Penicillium italicum and P. digitatum wild-type isolates obtained by 200 μg mL^{-1} of all possible combinations between cinna-

mon and harmal extracts and fractions, Cin: Cinnamon (Cinnamomum cassia) bark, harmal: Harmal (Peganum harmala) seeds, C: Crude extract, M: Methanolic fraction, H: Hexane fraction, W: Water fraction and B: Between layers fraction

In contrast, the remaining possible combinations between fractions of cinnamon and harmal have reflected antagonistic effects, where reduced inhibition zones (in the range of 8-36 mm) were obtained from such combinations as compared to zones in the range of 18-44 mm resulted from any of the cinnamon's fractions alone (with the exception of cinnamon's water fraction, where no inhibition zones were noticed).

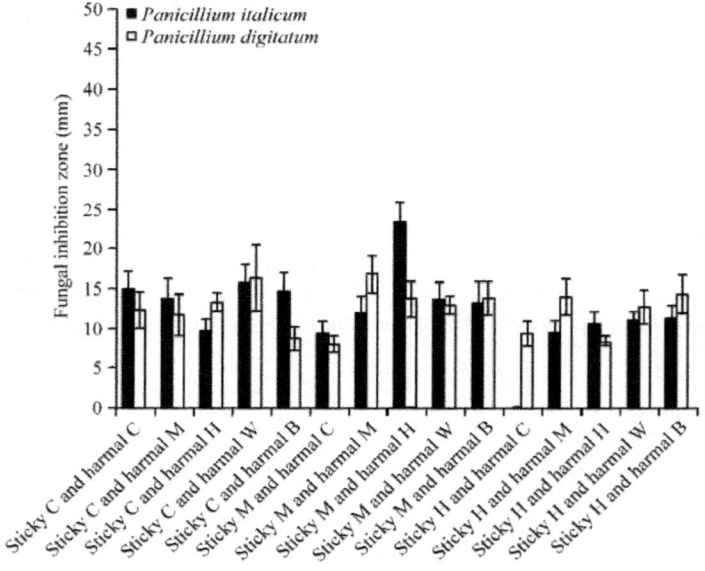

Figure 6: Inhibition zone of Penicillium italicum and P. digitatum wild-type isolates obtained by 200 µg mL^{-1} of all possible combinations between sticky fleabane and harmal extracts and fractions, sticky denotes sticky fleabane (Inulla viscosa) leaves, harmal denotes harmal (Peganum harmala) seeds, C: Crude extract, M: Methanolic fraction, H: Hexane fraction, W: Water fraction and B: Between layers fraction

In vitro sensitivity of Penicillium sp. to combinations of sticky fleabane fractions and harmal fractions: Results presented in Fig. 6, indicate that mixtures of sticky fleabane leaves and harmal seeds hexane fractions have generated synergistic effects against both Penicillium species. The obtained inhibition zones were approximately 11 mm as compared to no inhibition effect for single extracts (Fig. 2, 4). Also, mixtures of sticky fleabane methanolic fraction and harmal hexane fraction exerted synergistic effects against P. italicum. However, mixtures from sticky fleabane methanolic fraction and any of the harmals' fractions maintained the same inhibitory level of the sticky fleabane fraction (inhibition zone of 11-13 mm) compared to those obtained by the sticky fleabane fraction alone (Fig. 2). The remaining combinations between sticky fleabane and harmals' fractions showed antagonistic effects against P. digitatum, where smaller sized zones by such combinations were obtained (Fig. 6).

In vitro sensitivity of Penicillium isolates to various concentrations of low toxicity-food preservatives: A significant correlation (at the 0.01 level of significance) between tested concentrations (in the range of 20 to 400 $\mu g\ mL^{-1}$) of all examined food preservative chemicals and mean inhibition zones (mm) of both Penicillium species was obtained (Table 2). An exception to this trend was Na-molybdate, where the correlation between tested concentrations and inhibitory zones of P. italicum was not significant (r = 0.464; p = 0.150). Na-benzoate was the most effective against both fungal species. The MIC values against P. digitatum and P. italicum were 37.5 and 75 $\mu g\ mL^{-1}$, respectively. Furthermore, NH4-molybdate and K-carbonate showed the next level of strength after Na-benzoate against the tested fungal species (MIC values approximately 175 $\mu g\ mL^{-1}$). Na-bicarbonate and Na-molybdate were the least effective against the tested fungal species (Table 2), where no inhibition zones against P. digitatum were obtained with Na- molybdate at a range of concentrations from 20 to 400 $\mu g\ mL^{-1}$.

Table 2: In vitro sensitivity of two wild-type fungal isolates, Penicillium italicum and P. digitatum, to different concentrations ($\mu g\ mL^{-1}$) of five chemical substances used as food preservatives

Chemical substance[a]	Range of conc ($\mu g\ mL^{-1}$)[b]	Range of Inhibition[c] zone (mm) mean±SD	MIC[d]	Corr. value[e] (r)	Sig. value[f]	Fungal isolate
Na-benzoate	20-120 160-400	0.0-20±1.44	75	0.464	0.150	P. italicum
		20.0±2.12 to 28.5±1.6				
Na-benzoate	20-120 160-400	0.0-24±2.45	37.5	0.914**	0.000	P. digitatum
		24.5±2.34 to 62.5±2.12				
Na-molybdate	20-120 160-400	0.0-0.0 0.0-22.5±1.32	200	0.635*	0.036	P. italicum
Na-molybdate	20-120 160-400	0.0-0.0 0.0-0.0				P. digitatum
NH4-molybdate	20-120 160-400	0.0-14±1.67	275	0.925**	0.000	P. italicum
		14±1.67 to 17.5±1.12				
NH4-molybdate	20-120 160-400	0.0-15±2.77	175	0.973**	0.000	P. digitatum
		24.5±2.45 to 37±2.56				
K-carbonate	20-120 160-400	0.0-15.5±2.11	175	0.973**	0.000	P. italicum
		18±1.33 to 35.5±3.11				
K-carbonate	20-120 160-400	0.0-15±1.44	175	0.973**	0.000	P. digitatum
		17±2.11 to 26±2.18				
Na- bicarbonate	20-120 160-400	0.0-0.0 0.0 to 17±1.55	275	0.742**	0.009	P. italicum
Na- bicarbonate	20-120 160-400	0.0-0.0 0.0 to 13±1.23	350	0.808**	0.003	P. digitatum

a: Five chemical substance were tested against two wild-type fungal isolates representing two Penicillium species indicated above. b: a range of concentrations (20, 30, 40, 80, 120, 160, 200, 240, 280, 320 and 400 µg mL^{-1}) of each chemical substance was used against tested fungal isolates. c: Fungal inhibition zone was determined from the mean diameter (mm) zone±SD of three independent tests using agar well diffusion method. d: Minimal inhibitory concentration of the tested chemical substance. e: correlation coefficient values. f: level of significance at the 0.05 and 0.01 levels (2- tailed). ** Correlation is significant at the 0.01 level (2-tailed).*Correlation is significant at the 0.05 level (2-tailed)

Moreover, no inhibition zones were obtained with either Na-molybdate or Na-bicarbonate within a range of concentrations from 20 to 160 µg mL^{-1}.

In vitro sensitivity of Penicillium isolates to various combinations of low toxicity-food preservatives: Substantial synergistic killing effects against both fungal species from almost all possible combinations made between various tested chemicals (each of 160 µg mL^{-1}) were obtained (Fig. 7). The combinations between Na-benzoate and the remaining chemicals generated inhibition zones of approximately 55 mm against P. italicum, while the inhibition zones from each chemical alone were in the range of 0.0 to 18 mm. Furthermore, inhibition zones in the range of 37 to 63 mm against both fungal species were obtained from combinations made between Na-molybdate or NH4-molybdate and the remaining chemicals. Moreover, synergistic effects against both Penicillium species from the combined concentrations of carbonate and bicarbonate were observed which were in the range of 53 mm to 55 mm compared to zones of 0.0 to 18 mm obtained by each chemical, respectively.

In vitro sensitivity of Penicillium isolates to various combinations of cinnamon's extracts and low toxicity-food preservatives: Synergistic effects against P. italicum from all combinations made between Na- benzoate and cinnamons' extract or its liquid fractions were generated (Fig. 8). In contrast, antagonistic effects by these combinations against P. digitatum were noticed. Zones of inhibition against P. italicum by benzoate and cinnamon fractions were in the range of 38 to 57 mm, whereas, zones obtained by each substance alone were in the range of 0.0 to 24 mm (Fig. 1, 8).

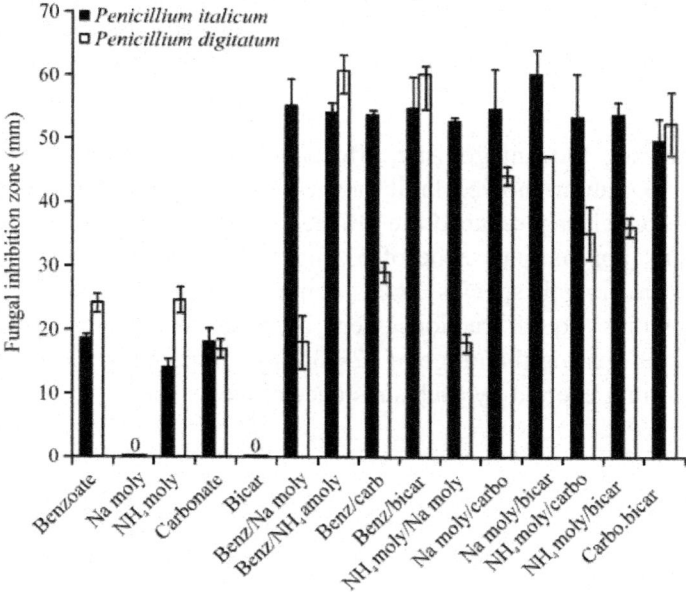

Figure 7: Inhibition zone of Penicillium italicum and P. digitatum wild-type isolates obtained by 160 µg mL^{-1} of each chemical type (as single) and combination of two substances (each of 160 µg mL^{-1}), tested chemical substances are: Na-benzoate, Na-molybdate, NH4-molybdate, K-carbonate and Na-bicarbonate tested as singles and pair-wise combinations

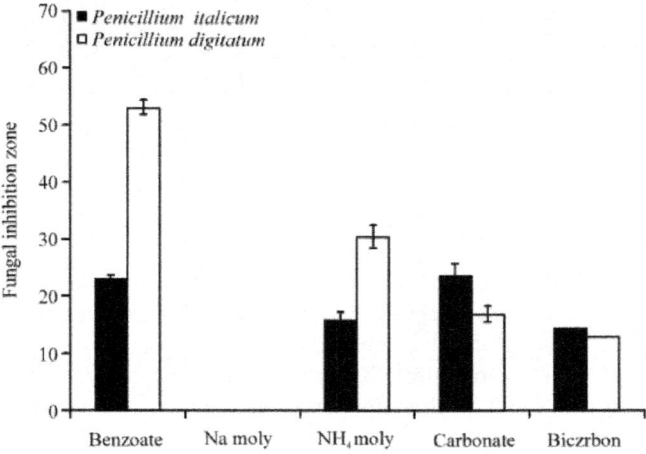

Figure 8: Inhibition zone of Penicillium italicum and P. digitatum wild-type

isolates obtained by five chemical substances (200 μg mL⁻¹ of each chemical
alone). Tested chemical substances are: Na-benzoate, Na-molybdate, NH4-
molybdate, K-carbonate and Na-bicarbonate, addition of Na-molybdate generated no inhibition

In contrast, the antagonistic effects generated by these combinations
against P. digitatum showed inhibition zones in the range of 18 to 22 mm.
Such mixtures clearly reduced the efficacy of benzoate alone (Fig. 9) against
P. digitatum (approximate zone of 53 mm). Furthermore, synergistic effects
against both fungal species from combinations between Na-molybdate
and cinnamon's liquid fractions were obtained. Inhibition zones by such
combinations were in the range of 42-53 mm compared to zones in the range
of 0.0 to 24 mm exerted by each substance alone.

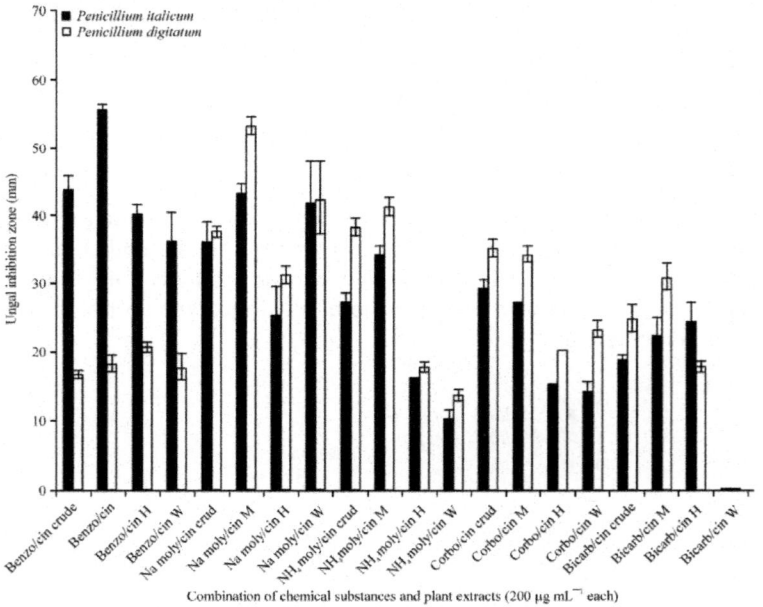

Combination of chemical substances and plant extracts (200 μg mL⁻¹ each)

Figure 9: Inhibition zone of Penicillium italicum and P. digitatum wild-type isolates
obtained by combinations of cinnamon crude extract or its liquid fractions and chemical substances (200 μg mL⁻¹ each) Tested chemical substances are: Na-benzoate,
Na-molybdate, NH4- molybdate, K-carbonate and Na-bicarbonate, each chemical
substance was tested in a combination with cinnamon crude extract and its liquid fractions: cin crude: Cinnamon crude extract, cin M: Cinnamon methanolic fraction, cin
H: Hexane fraction, cin W: Water fraction

Moreover, the combination made between Na- molybdate and cinnamons' water fraction has enhanced the inhibitory effect against both fungal species considerably, where zones of approximately 43 mm were obtained (Fig. 9) compared to no inhibitory effect by each individual substance (Fig. 1, 8). An exception to this synergistic pattern was in the activity obtained by the mixture of Na-molybdate and cinnamon's crude extract (approximate zone of 37 mm), where the addition of Na-molybdate (which generated no inhibition) did not negatively influence the activity of cinnamons extract (inhibition zones were in the range of 30-39 mm). The mixtures made between NH4-molybdate or Na-carbonate and cinnamon's fractions exerted inhibitory effects (against both fungal species) that were similar to those obtained by cinnamons' fractions when tested individually, i.e., the activity of cinnamon fraction was maintained and has not been influenced by the combined chemicals. An exception to this was the combination of Na-carbonate and cinnamon's water fraction (no inhibition zones when tested individually), where a synergistic effect against P. digitatum was obtained (i.e., double sized zones of 28 mm).

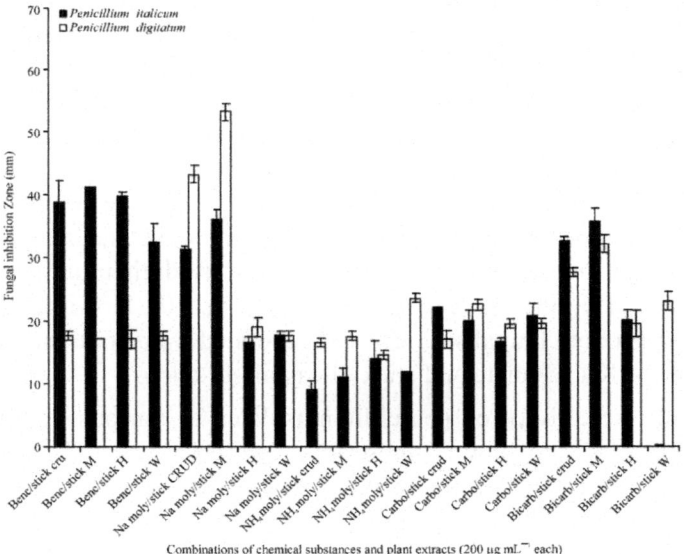

Combinations of chemical substances and plant extracts (200 μg mL^{-1} each)

Figure 10: Inhibition zone of Penicillium italicum and P. digitatum wild-type isolates obtained by combinations of chemical substances and plant extract or its liquid fractions (200 μg mL^{-1} each), tested chemical substances are: Na-benzoate, Na-molybdate, NH4- molybdate, K-carbonate and Na-bicarbonate, each chemical substance was tested in combination with sticky fleabane crude extract and its liquid fractions: Stick cru: Sticky fleabane crude extract, stick, M: Sticky fleabane methanolic fraction, stick H: Hexane fraction, stick W: Water fraction

In contrast, the addition of cinnamon's water fraction to Na- bicarbonate had completely abolished the effect of bicarbonate, where no inhibitions by this combination against both fungal species were obtained.

In vitro sensitivity of Penicillium isolates to various combinations of sticky fleabane extract and low toxicity-food preservatives: The combinations made between Na-benzoate and any of the sticky fleabane fractions had synergistic effects against P. italicum and antagonistic effects against P. digitatum (Fig. 10). The obtained inhibition zones against P. italicum by mixtures of benzoate and either methanolic or hexane fraction of sticky fleabane leaves were approximately of 42 mm in size compared to zones of 0.0 to 22 mm generated by each substance alone (Fig. 2 and 8). An exception to this was the activity of mixtures of benzoate and either the crude extract or water fraction of sticky fleabane, where additive effects were monitored against P. italicum (Fig. 10). The obtained antagonistic effects against P. digitatum by benzoate and sticky fleabane fractions indicated that the combined plant fractions have negatively influenced the activity of benzoate. The obtained inhibition zones by benzoate alone (approximately 53 mm) were greatly reduced to approximately 18 mm in all tested combinations. In contrast, all combinations of Na-molybdate and sticky fleabane fractions had synergistic effects against both fungal species. The obtained inhibition zones by these combinations ranged from 16 to 35 mm against P. italicum and from 18 to 54 mm against P. digitatum. However, the obtained zones by each substance alone against P. italicum and P. digitatum ranged from 0.0 to 11 and 0.0 to 35 mm, respectively. The combinations made between Na- bicarbonate and sticky fleabane fractions had synergistic effects against P. italicum, whereas activities against P. digitatum were similar to those obtained by the plant fractions alone (Fig. 10). However, antagonistic effects against both fungal species were generated from combinations of NH4-molybdate or K-carbonate and sticky fleabane fractions.

In vitro sensitivity of penicillium isolates to various combinations of harmal extracts and low toxicity-food preservatives: The combinations between benzoate and either harmal's crude extract or its hexane fraction exerted synergistic effects against P. italicum (inhibition zones were in the range of 35 to 38 mm) (Fig. 11), while zones in the range of 0.0 to 22 mm were obtained by each substance alone (Fig. 4, 8). The combinations of benzoate and harmal's methanolic fraction, also generated additive effects against P. italicum (approximately 32 mm inhibition zone).

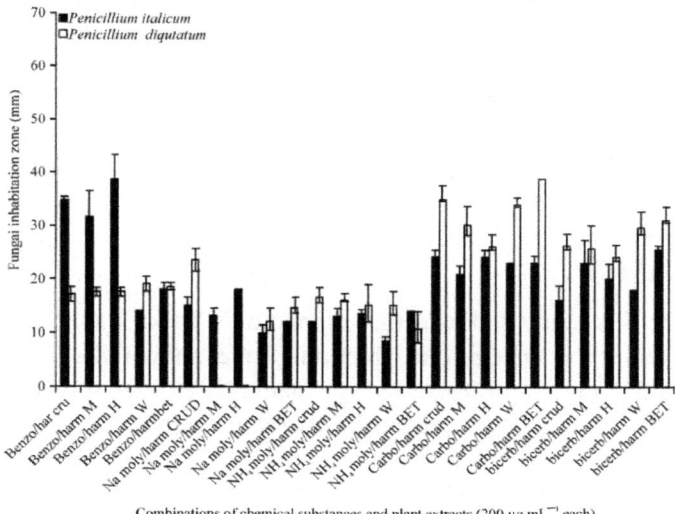

Combinations of chemical substances and plant extracts (200 μg mL⁻¹ each)

Figure 11: Inhibition zone of Penicillium italicum and P. digitatum wild-type isolates obtained by combinations between each chemical substance (200 μg mL⁻¹) and harmal extract or liquid fraction (200 μg mL⁻¹), tested chemical substances are: Na-benzoate, Na- molybdate, NH4-molybdate, K-carbonate and Na-bicarbonate, each chemical substance was tested in a combination with harmal crude extract and its liquid fractions: Harm cru denotes, harmal crude extract, harm M: Harmal methanolic fraction, harm H: Hexane fraction, harm W: Water fraction and harm bet: The layer between fractions

In contrast, the above mentioned mixture generated antagonistic effects against P. digitatum (approximately 17 mm inhibition zone) where the plant fraction greatly reduced the activity obtained by benzoate alone (approximately 53 mm inhibition zone). Furthermore, synergistic effects against P. italicum were generated by mixtures of Na-molybdate and each of harmals crude extract, hexane fraction or the layer between fractions (Fig. 11). Inhibition zones of approximately 14 mm were obtained by these combinations as compared to zero zones by any of the tested substances individually. In contrast, the combinations between Na-molybdate and either methanolic or hexane fraction of harmal generated no inhibitory effect against P. digitatum. In contrast, mixtures of carbonate or bicarbonate and a harmal's fraction generated additive effects against P. digitatum, where inhibition zones in the range of 28 mm to 38 mm were obtained. An exception to this pattern was the combination between carbonate and bicarbonate and harmal's hexane fraction, where synergistic effects against P. digitatum were obtained (inhibition zones in the range of 24 to 26 mm) compared to 0.0 to 18 mm by each substance alone (Fig. 4, 8). However, the combinations between carbonate and bicarbonate and harmal's fractions generated inhibitory zones (in the range of 22 to 24 mm) against P. italicum of similar sizes to those obtained by each chemical alone. Furthermore, the combinations of NH4-molybdate and harmal's fractions exerted antagonistic effects against P. digitatum (i.e., half-sized zones of 16 -18 mm), whereas inhibition zones of similar sizes (i.e., inhibition zones of 13-16 mm) to these obtained by NH4-molybdate alone against P. italicum were obtained.

DISCUSSION

Results of this study indicate that cinnamon's bark, sticky fleabane leaves extracts and their methanolic fractions have shown the highest efficacy against both fungal species tested. Cinnamon's extract showed MIC values of 75 and 187.5 μg mL^{-1} against P. digitatum and P. italicum, respectively. Whereas, sticky fleabane extract revealed MIC values of 150 and 375 μg mL^{-1} against the same tested species, respectively. These findings disagreed with that obtained by Al-Najar (2007), where none of the mentioned extract types has resulted in complete inhibition of fungal growth within a range of concentrations from 50 to 520 μg mL^{-1}, although such extracts were the most effective (among other tested extracts) against tested isolates (in vitro) of both fungal species. In contrast, previous results from the in vivo study against P. italicum isolates revealed that the same extract types have completely inhibited the growth of all P. italicum tested isolates that infect orange rather than lemon fruits (Al-Najar, 2007). Concerning the effect of cinnamon's and sticky fleabane methanolic fractions, the results indicate that MIC values of 37.5 and 75 μg mL^{-1} were obtained against P. digitatum with the above mentioned plant extracts, respectively. Also, these fractions have generated MIC values of 150 and 375 μg mL^{-1} against P. italicum, respectively. Clearly, the results obtained previously with the same fractions (Al-Najar, 2007; Kanan and Al-Najar, 2008, 2009) are consistent with our findings, where these fractions have completely inhibited the growth of both fungal species tested. The IC50 values obtained with cinnamons fraction against several P. digitatum isolates were within the range of 5-23 μg mL^{-1}, whereas, IC50 values within the range of 27.25- 31.75 μg m^{-1} were obtained with sticky fleabane fraction (Kanan and Al-Najar, 2008). In addition, the same liquid fractions have completely inhibited the growth of several P. italicum isolates where, IC50 values within the range of 11.2- 24 and 25-36 μg mL^{-1} were obtained with cinnamons' and sticky fleabane fractions, respectively (Kanan and Al-Najar, 2009). The combinations (200 μg mL^{-1} each) made between cinnamon's methanolic fraction and that of sticky fleabane or harmal's fractions have maintained the inhibitory effect obtained by cinnamon's fraction, towards tested fungal species and here, similar sized inhibition zones to those obtained by cinnamon's fraction were generated. The obtained results suggest that the high efficacy of cinnamon's extract is related to the presence of certain bioactive agents which might include cinnamaldehyde, eugenol, cinnamic acid as well as flavonoids, alkaloids, tannins, anthraquinones and phenolic compounds. These active components have been shown previously to have a strong antifungal activity (Inouye et al., 2000; Gill and Holley, 2004). However, the detected efficacy may be traced mainly to cinnamaldehyde that works as an inhibitor for enzymes such as β-(1,3)-glucansynthase which is involved in chitin and β- glucans cell wall components biosynthesis (Cowan, 1999). This possibility coincides with the findings of Rojas et al. (1992), who identified such components as active antifungal agents. Furthermore, eugenol and cinnamaldehyde have consistently been reported to be the main components of cinnamon exhibiting high fungitoxic activity (Jham et al., 2005). As indicated above, the sticky fleabane (Inula viscosa) crude extract and its methanolic fraction possess high efficacy against the tested fungi which is in agreement with previous findings by Kanan and Al-Najar (2008, 2009). Wang et al. (2004) proposed that sticky fleabane extract exhibit significant activity against several pathogenic fungal species of various crop plants including dermatophytes and downy mildew. However, the obtained results disagreed with some findings of (Muller-Riebau et al., 1995), who found small amounts of antifungal essential oils or phenolics. The antifungal oils and phenolics seem to affects chitin biosynthesis due to high content of flavonoids, phenolic compounds and anthraquinones in methanolic and aqueous extracts of sticky fleabane (Cohen et al., 2006). These observations are in agreement with those reported previously by other researchers (Ali-Shtayeh and Abu Ghdeib, 1999; Cafarchia et al., 2002; Cohen et al., 2006; Al-Najar, 2007). The mixtures of sticky fleabane methanolic and hexane fractions have abolished the inhibitory effect of both fractions against P. digitatum. In addition, combining both fractions (sticky fleabane hexane and water) did not show any inhibition effect against the tested fungal species. It can thus be suggested that active components in fractions had antagonistic effect on each other when combined, since their availability in a pair-wise form or in the crude extract has reduced their efficacy. Our results indicate that among tested plant materials, harmal's extract and its liquid fractions were the least effective against both fungal species tested. These findings support previous findings by Kanan and Al-Najar (2009) which showed that none of the harmal's extracts has led to complete inhibition of any of four tested P. italicum isolates. However, P. digitatum isolates were found to be more susceptible to the plant extracts tested (Kanan and Al-Najar, 2008). The inhibitory activity of the crude extract may be related to high content of alkaloids (harmine, harmaline and tetrahydroharmine) and phenolic compounds, where these compounds may alter fungal cell membrane permeability and thus allow the deficit of macromolecules (Rasooli and Razzaghi-Abyaneh, 2004). This could be due to the ability of phenols to inactivate essential enzymes that react with cell membrane proteins or by disrupt functions of the genetic material (Kartal et al., 2003; Telezhenetskaya and D'yakonov, 2004). The in vitro treatment of fungal isolates with sodium benzoate was the most effective of all tested chemicals in controlling both P. digitatum and P. italicum.

These findings coincided with those obtained previously by Al-Najar (2007) where, MIC values from 4 to 50 mM was shown to be effective for complete fungal growth inhibition. Also, our findings agreed with that proposed by Buazzi and Maeth (1992), who stated that benzoate prevented the growth of nearly all examined microorganisms. However, the effectiveness of sodium benzoate was shown to be pH dependent because its undissociated form is mostly accountable for the antimicrobial activity, since most organic acid salts are effectively inhibitory at pH 5-5.5 and lower (the PKa value of potassium benzoate is 4.2) (Palou et al., 2002a). Ammonium molybdate and potassium carbonate were ranked as the second most effective (after Na-benzoate) against tested fungal species (MIC values of approximately 175 µg mL^{-1}). Results obtained with the molybdate agreed with findings of Palou et al. (2002a) who declared that 5 mM NH4-molybdate totally controlled both blue and green moulds on all cultivars of orange fruits. Furthermore, Al-Najar (2007) proposed that potassium molybdate has significantly inhibited the growth of four tested strains of P. digitatum, where the toxicity of molybdate has resulted in the formation of dark bluish-petroleum, color in the inoculation site, when a concentration of 400 mM was used. As compared to the activity of NH4-molybdate, Na-molybdate showed no inhibition effect against both fungal species tested within a concentration range of 20-400 µg mL^{-1}. However, these findings disagreed with findings of Al-Najar (2007) who proposed that MIC values in the range of 250-500 mM (IC50 values from 47.5-350 mM) were required to inhibit the growth of P. italicum isolates. Also, these results disagreed with findings of Palou et al. (2002a) who stated that both sodium and ammonium molybdate have effectively controlled the growth of green and blue moulds of citrus fruits. However, results of this study indicate that the synergistic effects against tested pathogenic fungi were obtained by activity of the following mixtures (200 µg mL^{-1} each): sodium and ammonium molybdate mixture; mixtures of either sodium or ammonium molybdate and any of the remaining tested chemicals, mixture of sodium molybdate and cinnamons' fractions and mixtures of carbonate and bicarbonate.

The generated synergistic effects by mixtures of substances having different origins and modes of actions may propose that these mixtures leave the fungus unable to disrupt binding of the mixture to their protein at the binding site. In addition, the used mixtures may render the efflux mechanisms inefficient to pump the toxic substances out of the cell. Furthermore, the tested mixtures may inhibit certain metabolic pathways that usually aid in converting the toxic materials into non-toxic forms, i.e., the mixture might inhibit specific proteins that usually can detoxify the substance.

CONCLUSION

Results of this study indicate that cinnamon's bark extract was the most efficient against tested fungal species (MIC values of approximately 75 and 187.5 μg mL^{-1} were obtained against P. digitatum and P. italicum, respectively). Methanolic fractions of cinnamon's bark and gluey fleabane leaves extract exhibited the highest value against both fungal species tested. However, cinnamon's fraction was more efficient at lower concentrations (MIC values of approximately 37.5 μg mL^{-1}). Mixtures of cinnamon's methanolic fraction and either hexane or water fraction of cinnamon have maintained the efficiency of the methanolic fraction against both fungal species. The synergistic effects against P. italicum were obtained from mixtures of sticky fleabane extract and its hexane fraction, whereas additive effects against the same species were obtained by mixtures of sticky fleabane extract and its methanolic or water fraction. Also, a synergistic effect against P. italicum were obtained by mixtures of harmal extract and each of its methanolic, hexane or layer between fractions compared to no inhibitory effect by each extract type alone. The same synergistic effect was generated against P. digitatum by mixtures of harmal's extract and its hexane fraction, although no inhibition was achieved by hexane fraction alone. However, additive effects were obtained against P. digitatum by mixtures of harmal's extract and either it's methanolic (zones of approximately 25 mm) or water fraction (zones of approximately 33 mm). Na-benzoate was the most effective (MIC values in the range of 75 μg mL^{-1} - 37.5) against fungal species tested. NH4-molybdate and K-carbonate were ranked the second in terms of efficacy against tested fungal species (MIC of approximately 175 μg mL^{-1}). All combinations between tested chemicals have generated synergistic inhibitory effects against both fungal species. Combinations made between Na-benzoate and either cinnamons or sticky fleabane fractions generated synergistic effects against P. italicum and antagonistic effects against P. digitatum. The combinations made between Na-molybdate and cinnamon's liquid fractions resulted in synergistic effects against both fungal species. The mixture of Na-molybdate and cinnamon's water fraction has enhanced the inhibitory effect of both substances against fungal species tested, where zones of approximately 43 mm were obtained as compared to no inhibition effect of each individual substance. All combinations of Na-molybdate and sticky fleabane fractions have reflected synergistic effects against both fungal species. The mixture of Na-benzoate and harmal's methanolic fraction generated additive effects against P. italicum and antagonistic effects against P. digitatum where, the plant fraction has significantly reduced the activity of benzoate. Synergistic effects against P. italicum were also obtained by the combination of either Na-benzoate or Na-molybdate with harmal's extract or its hexane fraction.

ACKNOWLEDGMENTS

The Authors would like to thank Yarmouk University (Jordan), for providing the required facilities and the suitable environment for this research work.

REFERENCES

1. Abad, M.J., M. Ansuategui and P. Bermejo, 2007. Active antifungal substances from natural sources. ARKIVOC, 7: 116-145.

2. Abu-Zarga, M.H., E.M. Hamed, S.S. Sabri, W. Voelter and K.P. Zeller, 1998. New sesquiterpenoids from the Jordanian medicinal plant Inula viscose. J. Nat. Prod., 61: 798-800.

3. Al-Najar, R.A., 2007. Selection and evaluation of alternatives to synthetic fungicides for the control of post-harvest citrus fruits rot caused by Penicillium italicum (blue mould) in Jordan. M.Sc. Thesis, Mutah University, Jordan.

4. Ali-Shtayeh, M.S. and S.I. Abu Ghdeib, 1999. Antifungal activity of plant extracts against dermatophytes. Mycoses, 42: 665-672.

5. Ambrozin, A.R., P.C. Vieira, J.B. Fernandes, M.F. da Silva, S. Albuquerque, 2004. Trypanocidal activity of Meliaceae and Rutaceae plant extracts. Mem. Inst. Oswaldo Cruz., 99: 227-231.

6. Bhyan, S.B., M.M. Alam and M.S. Ali, 2007. Effect of plant extracts on Okra mosaic virus incidence and yield related parameters of Okra. Asian J. Agric. Res., 1: 112-118.

7. Bonjar, G.H.S. and A.K. Nik, 2004. Antibacterial activity of some medicinal plants of Iran against Pseudomonas aeruginosa and P. fluorescens. Asian J. Plant Sci., 3: 61-64.

8. Bonjar, G.H.S., 2004. Screening for antibacterial properties of some iranian plants against two strains of Escherichia coli. Asian J. Plant Sci., 3: 310-314.

9. Bouzerda, L., H. Boubaker, E.H. Boudyach, O. Akhayat and A.A. Bin Aoumar, 2003. Selection of antagonistic yeasts to green mold disease of

citrus in Morocco. Food Agric. Environ., 1: 215-218.

10. Buazzi, M.M. and E.H. Maeth, 1992. Characteristics of sodium benzoate injury of Listeria monocytogenes. Microbios, 70: 199-207.

11. Cafarchia, C., N. de Laurentis, M.A. Milillo, V. Losacco and V. Puccini, 2002. Antifungal activity of essential oils from leaves and flowers of Inula viscosa (Asteraceae) by Apulian region. Parassitologia., 44: 153-156.

12. Cohen, Y., W. Wang, B.H.B. Daniel and Y. Ben-Daniel, 2006. Extracts of Inula viscosa control downy mildew of grapes caused by Plasmopara viticola. Phytopathology, 96: 417-424.

13. Cove, D.J., 1966. The induction and repression of nitrate reductase in the fungus Aspergillus nidulans. Biochem. Biophys. Acta, 113: 51-56.

14. Cowan, M.M., 1999. Plant products as antimicrobial agents. Clin. Microbiol. Rev., 12: 564-582.

15. Delaquis, P.J., K. Stanich, B. Girard and G. Mazza, 2002. Antimicrobial activity of individual and mixed fractions of dill, cilantro, coriander and eucalyptus essential oils. Int. J. Food Microbiol., 74: 101-109.

16. Gill, A.O. and R.A. Holley, 2004. Mechanisms of bactericidal action of cinnamaldehyde against Listeria monocytogenes and of eugenol against L. monocytogenes and Lactobacillus sakei. Applied Environ. Microbiol., 70: 5750-5755.

17. Hasan, M.M., S.P. Chowdhury, Shahidul Alam, B. Hossain and M.S. Alam, 2005. Antifungal effects of plant extracts on seed-borne fungi of wheat seed regarding seed germination, Seedling health and vigour index. Pak. J. Biol. Sci., 8: 1284-1289.

18. Ikeura, H., N. Somsak, F. Kobayashi, S. Kanlayanarat and Y. Hayata, 2011. Application of selected plant extracts to inhibit growth of Penicillium expansum on Apple fruits. Plant Pathol. J., 10: 79-84.

19. Inouye, S., T. Tsuruoka, M. Watanabe, K. Takeo, M. Akao, Y. Nishiyama and H. Yamaguchi, 2000. Inhibitory effect of essential oils on apical growth of Aspergillus fumigatus by vapour contact. Mycoses, 43: 17-23.

20. Ismail, M. and J. Zhang, 2004. Postharvest citrus diseases and their control. Outlooks Pest Manage., 15: 29-35.

21. Jham, G.N., O.D. Dhingra, C.M. Jardim and V.M.M. Valente, 2005. Identification of major fungitoxic component of cinnamon bark oil. Fitopatologia, 30: 404-408.

22. Kanan, G.J. and R.A. Al-Najar, 2008. In vitro antifungal activities of various plant crude extracts and fractions against citrus post-harvest

disease agent Penicillium digitatum. Jor. J. Biol. Sci., 1: 89-99.

23. Kanan, G.J. and R.A. Al-Najar, 2009. In vitro and In vivo Antifungal Activities of Various plant extracts and their fractions against citrus post-harvest disease agent Penicillium italicum. J. plant. Prot. Res., 49: 341-352.

24. Kartal, M., M.L. Altun and S. Kurucu, 2003. HPLC method for the analysis of harmol, harmalol, harmine and harmaline in the seeds of Peganum harmala. J. Pharm. Biomed. Anal., 31: 263-269.

25. Larrigaudiere, C., J. Pons , R. Torres and J. Usall, 2002. Storage performance of elementines treated with hot water, sodium carbonate and bicarbonate dips. J. Horit. Sci. Biotech., 77: 314-319.

26. Lee, K.W., H. Everts and A.C. Beynen, 2004. Essential oils in broiler nutrition. Int. J. Poult. Sci., 3: 738-752.

27. McGrath, M.T., 2001. Fungicide resistance in Cucurbita Powdery Mildew: Experiences and challenges. Plant Dis., 85: 236-245.

28. Muller-Riebau, F., B. Berger and O. Yegen, 1995. Chemical composition and fungitoxic properties to phytopathogenic fungi of essential ois of selected aromatic plants growing wild in Turkey. J. Agric. Food Chem., 43: 2262-2266.

29. Nahunnaro, H., 2008. Effects of different plant extracts in the control of yam rot induced by Rhizopus stolonifer on stored yam (Dioscorea sp.) in Yola, Adamawa State Nigeria. Agric. J., 3: 382-387.

30. Ndukwe, I.G., J.D. Habila, I.A. Bello and E.O. Adeleye, 2006. Phytochemical analysis and antimicrobial screening of crude extracts from the leaves, stem bark and root bark of Ekebergia senegalensis A. Juss. Afr. J. Biotechnol., 5: 1792-1794.

31. Obagwu, J. and L. Korsten, 2003. Control of citrus green and blue molds with garlic extracts. Eur. J. Plant Pathol., 109: 221-225.

32. Palou, L., J. Usall, J.A. Munoz, J.L. Smilanick and I. Vinas, 2002. Hot water, sodium carbonate and sodium bicarbonate for the control of postharvest green and blue molds of clementine mandarins. Postharvest Biol. Technol., 24: 93-96.

33. Palou, L., J. Usall, J.L. Smilanick, M.J. Aguilar and I. Vinas, 2002. Evaluation of food additives and low-toxicity compounds as alternative chemicals for the control of Penicillium digitatum and Penicillium italicum on citrus fruit. Pest Manage. Sci., 58: 459-466.

34. Palou, L., J.L. Smilanick, J. Usall and I. Vinas, 2001. Control of postharvest blue and green molds of oranges by hot water, sodium

carbonate and sodium bicarbonate. Plant Dis., 85: 371-376.

35. Plaza, P., A. Sanbruno, J. Usall, N. Lamarca, R. Torres, J. Pons and I. Vinas, 2004. Integration of curing treatments with degreening to control the main postharvest diseases of Calementine mandarins. Postharv. Biol. Technol., 34: 29-37.

36. Prusky, D., J.L. McEvoy, R. Saftner, W.S. Conway and R. Jones, 2004. Relationship between host acidification and virulence of Penicillium spp. on apple and citrus fruit. Phytopathology, 94: 44-51.

37. Rasooli, I. and M.R. Abyaneh, 2004. Inhibitory effects of Thyme oils on growth and aflatoxin production by Aspergillus parasiticus. Food Control, 15: 479-483.

38. Reddy, K.R.N., C.S. Reddy and K. Muralidharan, 2007. Exploration of ochratoxin a contamination and its management in rice. Am. J. Plant Physiol., 2: 206-213.

39. Reddy, K.R.N., S.B. Nurdijati and B. Salleh, 2010. An overview of plant-derived products on control of mycotoxigenic fungi and mycotoxins. Asian J. Plant Sci., 9: 126-133.

40. Rojas, A., L. Hernandez, R. Pereda-Miranda and R. Mata, 1992. Screening for antimicrobial activity of crude drug extracts and pure natural products from Mexican medicinal plants. J. Ethnopharmacol., 35: 275-285.

41. Samson, R.A., K.A. Seifert, A.F.A. Kuijpers, J.A.M.P. Houbraken and J.C. Frisvad, 2004. Phylogenetic analysis of Penicillium subgenus Penicillium using partial β-tubulin sequences. Stud. Mycol., 49: 175-200.

42. Schroeder, L.L.. and L.B. Bullerman, 1985. Potential for development of tolerance by Penicillium digitatum and Penicillium italicum after repeated exposure to potassium sorbate. Appl. Environ. Microbiol., 50: 919-923.

43. Smilanick, J.I., B.E. Mackey, R. Reese, J. Usall and D.A. Margosan 1997. Influence of the concentration of soda ash, temperature and immersion period on the control of post-harvest green mold of oranges Plant Dis., 81: 379-382.

44. Smilanick, J.L., D.A. Margosan, F. Mlikota, J. Usall and I.F. Michael, 1999. Control of citrus green mold by carbonate and bicarbonate salts and the influence of commercial postharvest practices on their efficacy. Plant Dis., 83: 139-145.

45. Smilanick, J.L., J. Aiyabei, F. Mlikota, J. Gabler, D. Sorenson and B. Mackey, 2002. Quantification of the toxicity of aqueous chlorine to spores of Penicillium digitatum and Geotrichum citri-aurantii. Plant Dis.,

86: 509-514.

46. Smilanick, J.L., M.F. Mansour, D.A. Margosan, F. Mlikota-Gabler and W.R. Goodwine, 2005. Influence of pH and NaHCO3 on the effectiveness of imazalil to inhibit germination of spores of Penicillium digitatum and to control green mold on citrus fruit. Plant Dis., 89: 640-648.

47. Souza-Fagundes, E.M., A.B.R. Queiroz, O.A.M. Filho, G. Gazzinelli, R. Corra-Oliveira, T.M.A. Alves and C.L. Zani, 2002. Screening and fractionation of plant extracts with antiproliferative activity on human peripheral blood mononuclear cells. Mem. Inst. Oswaldo. Cruz., 97: 1207-1212.

48. Soylu, E.M., F.M. Tok, S. Soylu, A.D. Kaya and G.A. Evrendilek, 2005. Antifungal activities of the essential oils on post-harvest disease agent Penicillium digitatum. Pak. J. Biol. Sci., 8: 25-29.

49. Telezhenetskaya, M.V. and A.L. D'yakonov, 2004. Alkaloids of Peganum harmala. Unusual reaction of peganine and vasicinone. Chem. Nat. Comp., 27: 471-474.

50. Tripathi, P. and N.K. Dubey, 2004. Exploitation of natural products as an alternative strategy to control postharvest fungal rotting of fruit and vegetables. Postharvest Biol. Technol., 32: 235-245.

51. Wang, W., B.H. Ben-Daniel and Y. Cohen, 2004. Extracts of Inula viscose control downy milew caused by Plasmopara viticola in grape-vines. Phytoparasitica, 32: 208-211.

52. Zaker, M. and H. Mosallanejad, 2010. Antifungal activity of some plant extracts on Alternaria alternata, the causal agent of alternaria leaf spot of potato. Pak. J. Biol. Sci., 13: 1023-1029.

53. Zhang, H.Y., C. X. Fu, X. D. Zheng, D. He, L. J. Shan and X. Zhan, 2004. Effect of Cryptococcus laurentii (Kufferath) skinner in combination with sodium bicarbonate on biocontrol of post-harvest green mold decay of citrus fruit. Bot. Bull. Acad. Sin., 45: 159-164.

Chapter 9

PROLIFERATION OF ILLEGAL AND POTEN-TIALLY HAZARDOUS FOOD ADDITIVES IN PROCESSED AND PACKAGED FOODS IN AFRICA: A CASE STUDY AND HAZARD IDENTIFICATION IN GHANA

Courage Kosi Setsoafia Saba

Department of Biotechnology, Faculty of Agriculture, University for Development Studies, Tamale Ghana

ABSTRACT

In the developed countries, a lot of researches have been carried out on the effects of food additives on consumers but few studies have been reported in Africa and particularly Ghana. The objective of the study was to survey labels of processed and packaged foods in Ghana and document all the potentially harmful food additives in processed and packaged foods in some Ghanaian food products. We purchased 63 processed and packaged food products from the Ghanaian market and documented food additives on their labels as well as trade name, type of food, company name, and country of origin and whether the Ghana Food and Drugs Authority (GFDA) certified them. Thirty seven percent of all the products sampled on the market were not registered with the GFDA. Seventy one percent of all the products sampled contained one or more additives that are likely to cause adverse reactions when consumed. The general public is at risk of consuming potentially hazardous food additives. There is the need for a research to determine whether the levels of additives are above the recommended acceptable daily intake proposed by the Joint FAO/WHO Expert Committee on Food Additives (JECFA).

INTRODUCTION

Food additives are normally added to food or feed to enhance their palatability, shelf life and attractiveness. There are several categories of food additives: acids, acidity regulators, antioxidants, anticaking agents, antifoaming agents, bulking agents, food colouring, preservative sweeteners, etc. The Joint FAO/WHO Expert Committee on Food Additives (JECFA) regulates food additives globally but individual countries have specific regulations, which are enforced by their food regulating bodies. There are several controversies surrounding the use of certain food additives and the acceptable daily intake (McCann et al., 2007; Bosetti et al., 2009; El-Wahab and Moram, 2013; Ceyhan et al., 2013; Abhilash et al., 2014; Masone and Chanforan, 2015).

In most African countries, regulations and enforcement of the recommended food additives is weak due to inadequate experts, inadequate equipment for testing and lack of knowledge on the part of most consumers to read labels on especially processed and packaged foods (Van der Merwe et al., 2013). This situation is worsened by the proliferation of the African markets with products from especially countries that are noted to be food violators: India, China, Mexico, France, USA, Vietnam, Brazil etc. (Scott and Zak, 2014). There is little information about food additives especially illegal ones in Africa as well Ghana. Most people do not know about the designated numbers for the various food additives. This makes it easier for other countries to dump processed and packaged foods with illegal food additives in African countries. The objective of this study was to survey processed and packaged foods in Ghana and document all the illegal or potentially harmful food additives that are used to prepare them and their possible adverse effects on Ghanaians.

MATERIALS AND METHODS

Sample collection and documentation: Wide ranges of samples were randomly purchased on the Ghanaian Market from September 2013 to June 2014. In all, 63 items were purchased and include: Fruit juices, soft drinks, milks, ice creams, energy drinks, spices, noodles and margarine. Information documented on labels or packaging materials include: trade name, type of food, company name, and country of origin, additives/preservatives and whether they were certified by the GFDA.

Verification of the products for Food and Drugs Authority Certification: All the products sampled were checked for certification from the Ghana Food and Drugs Authority's registered products site (http://www.fdaghana.gov.gh/index.php?option=com_fdasearch&Itemid=36) to verify whether they were duly registered in the country.

Determination of food additives that are likely to cause adverse effects: The food additives in the various products purchased were first checked in the food intolerance network website (http://fedup.com.au/information/ information/complete-lists-of-additives-6#code) to determine whether they were listed among the food additives that may cause adverse effects. Based on this list, we calculated in percentages the products that had the highest and lowest number of food additives likely to cause adverse effects. We did not, however, perform any chemical analysis on the products to check the levels of those potentially harmful additives.

Data analysis: The results were calculated using Microsoft excel and presented with graphs and tables with descriptive statistics.

RESULTS

Countries of origin of products sampled: Figure 1 shows the country of origin of all the processed or packaged food sampled in this research. Majority of the products were manufactured here in Africa (46%) followed by Europe (30%), Asia (21%) and USA (3%). Out of the percentage manufactured in Africa (46%), 62% was manufactured in Ghana by either foreign or local companies followed by Nigeria (15%), Algeria (10%), Togo (7%), Ivory Coast (3%) and South Africa (3%). Among the European countries, UK recorded the highest number of products (26%) followed by Bulgaria (21%), Germany (17%), Netherlands, Austria (11%) and Romania, Spain (7%). Among the Asian countries the highest number of products was from China (24%), followed by Dubai (19%), Lebanon (17%) while Bangladesh, Turkey, Thailand, Taiwan and Vietnam had 8%.

Categories of foods sampled: The highest category of food sampled was soft drinks (38%), followed by energy drinks (18%), fruit juice (13%), ice cream (11%), milk (5%), noodles (6%), spices (3%) and biscuit and margarine (1.6%) each. Among the soft drinks, 46% originated from Europe (Bulgaria, Germany, Netherlands, Northern Ireland and UK), 38% were produced in Africa (Algeria, Ghana Nigeria and Togo) and 17% originated from Asia (China, Turkey and Dubai).

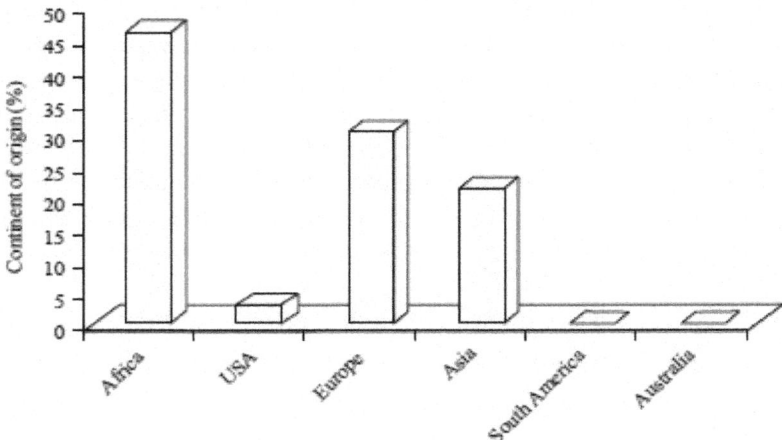

Figure 1: Percentage of countries of origin represented by the various continents from which the sampled products originated

Only 8% of the soft drinks sampled originated from Ghana. Most of the energy drinks originated from Europe (46%) followed by Asia (27%), USA (18%) with only 9% coming from Africa. The only energy drink produced from Africa in this study was from Ghana. Most of the fruit juices were produced in Africa (50%) followed by Europe (25%) and Asia (25%). All the ice creams sampled originated from Ghana (100%). Forty percent (40%) of the milk originated from Africa (all from Ghana), 40% from Europe and 20% from Asia. Most of the noodles were from Asia (75%) and Africa 25%. All the spices sampled originated from Africa (Ghana and Cote de Ivoire) while the biscuit and margarine originated from Ghana.

Verification of the products for Food and Drugs Authority Certification: Upon verification from the Ghana Food and Drugs Authority's product register, 37% of all the products sampled on the market were not registered with the GFDA at the time of collecting samples for this research. The Ghana Food and Drugs Authority certified only 63% of the products. Of all the uncertified products, majority of the products originated from Europe (52%), of which Bulgaria represented 34%, followed by UK (17%), Germany (17%), Austria, Netherlands, Northern Ireland and Romania represented 8% each. Twenty six percent of the non-certified products originated from Asia, of which China represented 50% followed by Thailand, Turkey and Bangladesh that represented 17% each. The uncertified products from Africa represented 17% and originated from Algeria, Ghana, Nigeria and Togo representing 25% each. Only 5% of the illegal products came from USA. The results show that most of the illegal products sampled came from European countries followed by Asia,

Africa and USA. Only 1 Ghanaian product out of the total of 17 from all the 63 products sampled was not registered. Majority of the uncertified products were soft drink (50%), noodles (18%), energy drinks (14%), Fruit juice and Milk represented 9% each. Of all the uncertified products, 82% of them contained 1 or more food additives that are potentially hazardous.

Food additives in products sampled: Out of the total number of 63 products sampled, 6 (10%) indicated there were no additives added on their labels. Majority (50%) of products with "No additives" labels were fruit juices followed by milk (33%) and a soft drink (17%). Ninety percent (90%) of all products sampled contained 1 or more food additives. All the food additives found in this survey are presented in Table 1. Seventy-one percent of all the products sampled contained one or more additives that are likely to cause adverse reactions when consumed.

Table 1: Categories of potential harmful food additives in sampled products

Category	Food additive
Artificial colours	*Tartrazine (E102), sunset yellow (E110), Amaranth (E123), brilliant scarlet or Ponceau (E124)* and *Allura red (E129)*
Natural colours	*Annatto extracts (E160b)*
Preservatives	*Sodium benzoate (E211), potassium sorbate (E202), tert-Butylhydroquinone, tBHQ (E319), butylated hydroxytoluene, BHT (E321), butylated hydroxyanisole, BHA (E320)* and *sodium metabisulphite (E223)*
Flavour enhancers	*Monosodium L-glutamate (E621), disodium guanylate (E627)* and *disodium inosinate (E631)*
Artificial sweeteners	*Aspartame (E951), sucralose (E955), cyclamates (E952), saccharin (E954)* and *acesulphame-K (E950)*
Miscellaneous (1)	*Carrageenan (E407), anthocyanins (E163), paprika oleoresins, (E160c), cochineal or carmines red (E120)* and *disodium 5'-ribonucleotides (E635)*
Miscellaneous (2)	Sodium phosphates (E339), potassium phosphates (E501), lecithin (E322), guar gum (E412), sodium alginate (E401), locust bean gum (E410), Mono and di-glycerides of fatty acids (E471), sodium carboxymethylcellulose (E466), glycerin (E422), starch sodium octenylsuccinate (E1450), sucrose acetate isobutyrate (E444), phosphoric acid (E338), citric acid (E330), malic acid (E296), caramel i (E150a), caramel iii (E150c), caramel iv (E150d), magnesium carbonates (E504), acetylated distarch adipate (E1422), calcium disodium EDTA (E385) and gum arabic (E414)

Italicized words are food additives that are likely to cause adverse reactions

Potentially harmful artificial and natural colours found in the products sampled: With the artificial colours: Tartrazine (E102) was found in 43% of the ice creams, 13% of soft drinks, 25% of fruit juice, 25% of noodles sampled, sunset yellow (E110) was found in 43% of ice creams, 18% of energy drinks and 13% of soft drinks sampled, amaranth (E123) was found in only 29% of the ice creams sampled, brilliant scarlet (E124) was found in 14% of the ice creams, 9% of the energy drinks and 50% of the spices sampled and allura red (E129) was found in only 9% of the energy drinks.

Potentially harmful preservatives in products sampled: Sodium benzoate (E211) was found in 46% of the soft drinks, 27% of the energy drinks and 38% of the fruit juices. Potassium sorbate (E202) was found in 29% of the soft drinks, 36% of the energy drinks, 25% of the fruit juices and 50% of the spices. Sodium metabisulphite (E223) was found in only 1 of the fruit juices (13%). tert-butylhydroquinone, tBHQ (E319) was found in both spices (50%)

and noodle (25%) only. Butylated hydroxytoluene, BHT (E321) was found in only one noodle (25%) while Butylated hydroxyanisole, BHA (E320) was also found in only one noodle (25%).

Potentially harmful artificial sweeteners in products sampled: Aspartame (E951) was found in 21% of the soft drinks, sucralose (E955) was found in 8% of the soft drinks, cyclamates (E952) was found in 21% of the soft drinks, saccharin (E954) was found in 13% of the soft drinks, acesulphame potassium (E950) was found in 17% of the soft drinks while saccharin was found in 4% of the energy drink.

Potentially harmful flavour enhancers in products sampled: Monosodium L-glutamate (E621) was found in 50% of the spices and 100% in the noodles, disodium guanylate (E627) was found in 75% of the noodles while disodium inosinate (E631) was also found in 75% of the noodles.

DISCUSSION

Countries of origin of products sampled: Most of the products sampled originated from Africa (46%) and most of the products from Africa originated from Ghana (62%). However, 54% of the products came from outside Africa. This results shows that what is happening in other countries in terms of regulation of food additives must be of concern to us in Ghana and Africa since we import most of our products from outside. What this study designed but was constrained by funding, was to verify whether these additives with potentially adverse effects are consumed in the countries of origin. This is to ascertain whether some of the products that contain potentially hazardous food additives were intentionally dumped into Africa. There is a tendency of dumping of products that contained illegal or banned additives in Africa because most of the people in our countries in Africa are either ignorant of the E numbers or do not know the adverse effects of some food additives.

Categories of foods sampled: Out of the top 4 most common categories of products sampled, 3 categories of products (soft drink, fruit juice and ice cream) are mostly consumed by children and women. This means any illegal or adverse food additives found in those products are likely to have a deleterious effects on children and women.

Verification of the products for Food and Drugs Authority Certification: It was clear that, some products entered the Ghanaian market without the approval of the Food and Drugs Authority. The number of potentially hazardous food additives (82%) that is present in the uncertified products is really a cause for concern. The influx of these products in the country may pose potentially hazardous adverse effect to our people if not properly regulated. The situation is more alarming considering the fact that 50% of the uncertified product was soft drinks, which are mostly consumed by the vulnerable (children and women) in the population. This clearly shows the loophole in our food regulatory body and the risk of contracting any disease if all trust is put in our regulatory bodies alone. The government must, therefore, encourage other researcher especially in our universities to carry out research to compliment the efforts of the regulatory bodies.

Food additives in products sampled: The result of this study is really a course for concern. Some of these products especially the soft drinks, fruit juice and ice creams are taken daily. With the little knowledge that exists in the population about the Acceptable Daily Intake (ADI) of these additives, most people are likely to take them above the acceptable limit. This may lead to toxic effect in consumers and may lead to severe complications. The worrying aspect of this study is that artificial colours or a sodium benzoate preservative (Table 1), either mixed or individuals are known to increase hyperactivity in 3-year-old children in the general population (McCann et al., 2007; Turner and Kemp, 2012). The additive, sodium benzoate (E211) was found in 46% of the soft drinks, 27% of the fruit juice and 27% of the energy drinks sampled. Some of these products are given to children in Ghana and this may lead to hyperactivity in Ghanaian children. It is worth mentioning that 2 energy drinks from Europe (Germany and UK) provided cautions on the labels of their products that sunset yellow (E110) and ponceau 4 R (E124) may have an adverse effect on the activity and attention in children. It was therefore, surprising and scary that most of the products sampled on the Ghanaian market are not having such labels warning consumers about the health effect of consuming such additives.

Potentially harmful artificial and natural colours found in the products sampled: Patients with allergic disorders have been found to have high leukocyte hypersensitivity to tartrazine in 10.8%, sunset yellow in 4.8% and ponceau or brilliant scarlet in 13.2%, (Titova, 2011). In a similar study, some of these food colorants (amaranth and tartrazine) had a toxic potential to human lymphocytes and have been found to bind directly to DNA (Mpountoukas et al., 2010). This means that either children or adults who take these products regularly or in excess are at risk of allergic disorders and possible DNA damage, which may result in cancer (Axon et al., 2012). The alarming revelation in this study is that, tartrazine and other artificial colours are prohibited in several countries including Iran (Hajimahmoodi et al., 2013) and strictly regulated in the UK by the Food and Standard Agency because of their possible effects on children (http://www.food. govuk/science/additives/foodcolours/). Tartrazine (E102) has been reported to produce several harmful effects in humans and mice of which hyperactivity, asthma and urticaria are most common (El-Wahab and Moram, 2013; Saxena and Sharma, 2014). However, some studies have also debunked the claim that there are no adverse effects of these additives (Pestana et al., 2010; Shimada et al., 2010). Sunset yellow (E110) has also been reported to produce adverse effects in people especially children (Yadav et al., 2013) and has been regulated in most countries. The other artificial colours Allura red (129), amaranth (E123) and ponceau (E124) have also been reported to have adverse effects in several studies (Ceyhan et al., 2013). The only natural colour found to be potentially harmful among the sampled products was annatto extracts (E160 b), which was found in one of the spices. Annatto extracts (E160 b) has been implicated to cause allergy reactions in certain products (Ebo et al., 2009). Even though annatto extracts (E160 b) causes allergy reactions, the risk of its allergenicity in Ghana is very low since it was found in one out of the 63 products sampled. But it effect cannot be overemphasized since there is proliferation of spices in the Ghanaian market. In order to be on the safer side in the light of these findings and counter findings, consumers need to be careful when consuming these additives.

Potentially harmful preservatives in products sampled: Sodium benzoate has been found to cause allergic reactions and other medical conditions when included in food and drugs (Schnuch et al., 2011; Mori et al., 2012). The cytotoxicity and genotoxicity of tert-butylhydroquinone (E319) has been demonstrated (Eskandani et al., 2014). Butylated hydroxyanisole, BHA (E320) has also been reported to have carcinogenic effects on epithelial cells (Schilderman et al., 1995). Liver toxicity has been linked to the over consumption of butylated hydroxytoluene, BHT (E321) (Engin et al., 2011). Even though there seem to be very low amount of these food additives in the products sampled, they are present in one of the spices that is widely used in the Ghanaian homes and restaurants, so it is a cause for concern in the general public. It worth to note that one of the spices (popular known in Ghana as cube) in our research contained the highest number of additives that are likely to cause adverse effect. It contains 8 potentially harmful additives [Potassium sorbate, (E202), ponceau 4R (124), monosodium L-glutamate (E621), disodium 5'-ribonucleotides (E635), potassium metabisulphite (E224) butylated hydroxytoluene (E321) butylated hydroxyanisole (E320) and tert-butylhydroquinone, (E319)] out of the 11 additives used in its preparation.

Potentially harmful artificial sweeteners in products sampled: Artificial sweeteners have generated a lot of controversies. Long-term aspartame exposure could alter the brain antioxidant status and can induce apoptotic changes in brain (Ashok and Sheeladevi, 2014). Aspartame has also been reported to have a possible effect on cognitive functions (Abhilash et al.,

2014). Artificial sweeteners are also implicated in certain types of cancers (Soffritti et al., 2010). Studies have also reported that there is no adverse effect for the consumption of artificial sweeteners with regards to its carcinogenic potentials (Bosetti et al., 2009). Long term use of artificial sweeteners have also been linked with significant increase in body weight (Polyak et al., 2010). Acesulphame potassium has also been reported to trigger allergy reactions (Stohs and Miller, 2014) as well as affect cognitive functions (Cong et al., 2013). There are several controversies on the consumption of artificial sweeteners. Considering the long-term exposure to artificial sweeteners, consumers must be educated in Ghana to know the possible side effects of products that contain them. There is more cause for concern especially in children who take soft drinks in Ghana indiscriminately.

Potentially harmful flavour enhancers in products sampled: A study has reported that high doses of monosodium L-glutamate (E621) may lead to partial infertility in male (Iamsaard et al., 2014). Monosodium L-glutamate treated rats have also been reported to be more susceptible to develop anxiogenic and depressive like behaviour (Quines et al., 2014). Oral consumption Monosodium L-glutamate appears to cause alkaline urine and may increase the risks of kidney stones with hydronephrosis in rats (Sharma et al., 2013). With the percentage of monosodium L-glutamate that is found in noodles products sampled (100%), there is a serious cause for concern due to the high craze for noodles especially by children in Ghana. In recent times these noodles have flooded the Ghanaian markets. This means that consumers in Ghana are at risks of all the harmful effects of these products.

CONCLUSION

This study is to survey processed and packaged foods in Ghana and document all the illegal or potentially harmful food additives that are used to prepare them. The study shows that most of the products sampled (71%) contain one or more food additives that could potentially cause adverse effects in the Ghanaian populations. The potentially adverse effects that these food additives could trigger in the Ghanaian populace range from allergy, hyperactivity, hypersensitivity, asthma, urticaria and several types of cancer effect on cognitive functions to partial infertility in male. Most of the products that contained the potentially harmful additives (soft drinks, fruit juice and ice creams) are mostly consumed by children, who have been reported in several studies to be the most vulnerable to the consumption of potentially harmful additives. The serious concerns is the long-term effects these additives have on it consumers. The long-term effect of these additives, if not properly monitored will have a serious repercussion on the future generations in our country and Africa. In the course of this study, most people seem to have little knowledge about food additives and cared less about the labels to look for any potentially hazardous additives. This means that there is generally low public knowledge about food additives in Ghana. The Food and Drugs Authority must be resourced very well to carry out extensive test on products for some of these harmful additives whether they are local or foreign. There is the need for a research into the levels of inclusion of these additives to establish if they meet the acceptable daily intake recommended by JECFA.

ACKNOWLEDGMENT

This study was financed by Courage Kosi Setsoafia Saba.

REFERENCES

1. Abhilash, M., M. Alex, W. Mathews and R.H. Nair, 2014. Chronic effect of aspartame on ionic homeostasis and monoamine neurotransmitters in the rat brain. Int. J. Toxicol., 33: 332-341.

2. Ashok, I. and R. Sheeladevi, 2014. Biochemical responses and mitochondrial mediated activation of apoptosis on long-term effect of aspartame in rat brain. Redox Biol., 2: 820-831.

3. Axon, A., F.E.B. May, L.E. Gaughan, F.M Williams, P.G. Blain and M.C. Wright, 2012. Tartrazine and sunset yellow are xenoestrogens in a new screening assay to identify modulators of human oestrogen receptor transcriptional activity. Toxicology, 298: 40-51.

4. Bosetti, C., S. Gallus, R. Talamini, M. Montella, S. Franceschi and C. La Vecchia, 2009. Artificial sweeteners and the risk of gastric, pancreatic and endometrial cancers in Italy. Cancer Epidemiol Biomarkers Prev., 18: 2235-2238.

5. Ceyhan, B.M., F. Gultekin, D.K. Doguc and E. Kulac, 2013. Effects of maternally exposed coloring food additives on receptor expressions related to learning and memory in rats. Food Chem. Toxicol., 56: 145-148.

6. Cong, W.N., R. Wang, H. Cai, C.M. Daimon and M. Scheibye-Knudsen et al., 2013. Long-term artificial sweetener acesulfame potassium treatment alters neurometabolic functions in C57BL/6J mice. PLoS ONE, Vol. 8. 10.1371/journal.pone.0070257

7. Ebo, D.G., S. Ingelbrecht, C.H. Bridts and W.J. Stevens, 2009. Allergy for cheese: Evidence for an IgE-mediated reaction from the natural dye annatto. Allergy, 64: 1558-1560.

8. El-Wahab, H.M. and G.S. Moram, 2013. Toxic effects of some synthetic food colorants and/or flavor additives on male rats. Toxicol. Ind. Health, 29: 224-232.

9. Engin, A.B., N. Bukan, O. Kurukahvecioglu, L. Memis and A. Engin, 2011. Effect of butylated hydroxytoluene (E321) pretreatment versus L-arginine on liver injury after sub-lethal dose of endotoxin administration. Environ. Toxicol. Pharmacol., 32: 457-464.

10. Eskandani, M., H. Hamishehkar and J.E.N. Dolatabadi, 2014. Cytotoxicity and DNA damage properties of tert-butylhydroquinone (TBHQ) food additive. Food Chem., 153: 315-320.

11. Hajimahmoodi, M., M. Afsharimanesh, G. Moghaddam, N. Sadeghi and M.R. Oveisi et al., 2013. Determination of eight synthetic dyes in foodstuffs by green liquid chromatography. Food Addit. Contam.: Part A, 30: 780-785.

12. Iamsaard, S., W. Sukhorum, R. Samrid, J. Yimdee and P. Kanla et al., 2014. The sensitivity of male rat reproductive organs to monosodium glutamate. Acta Medica Academica, 43: 3-9.

13. Masone, D. and C. Chanforan, 2015. Study on the interaction of artificial and natural food colorants with human serum albumin: A computational point of view. Comput. Biol. Chem., 56: 152-158.

14. McCann, D., A. Barrett, A. Cooper, D. Crumpler and L. Dalen et al., 2007. Food additives and hyperactive behaviour in 3-year-old and 8/9-year-old children in the community: A randomised, double-blinded, placebo-controlled trial. Lancet, 370: 1560-1567.

15. Mori, F., S. Barni, N. Pucci, M.E. Rossi, M. de Martino and E. Novembre, 2012. Cutaneous adverse reactions to amoxicillin-clavulanic acid suspension in children: The role of sodium benzoate. Curr. Drug Saf., 7: 87-91.

16. Mpountoukas, P., A. Pantazaki, E. Kostareli, P. Christodoulou and D. Kareli et al., 2010. Cytogenetic evaluation and DNA interaction studies of the food colorants amaranth, erythrosine and tartrazine. Food Chem. Toxicol., 48: 2934-2944.

17. Pestana, S., M. Moreira and B. Olej, 2010. Safety of ingestion of yellow tartrazine by double-blind placebo controlled challenge in 26 atopic adults. Allergologia Immunopathologia, 38: 142-146.

18. Polyak, E., K. Gombos, B. Hajnal, K. Bonyar-Muller and S Szabo et al., 2010. Effects of artificial sweeteners on body weight, food and drink intake. Acta Physiologica Hungarica, 97: 401-407.

19. Quines, C.B., S.G. Rosa, J.T. Da Rocha, B.M. Gai, C.F. Bortolatto, M.M.M.F. Duarte and C.W. Nogueira, 2014. Monosodium glutamate, a food additive, induces depressive-like and anxiogenic-like behaviors in young Rats. Life Sci., 107: 27-31.

20. Saxena, B. and S. Sharma, 2014. Serological changes induced by blend of sunset yellow, Metanil yellow and tartrazine in swiss albino rat, Rattus norvegicus. Toxicol. Int., 21: 65-68.

21. Schilderman, P.A.E.L., F.J. ten Vaarwerk, J.T. Lutgerink, A. Van Der Wurff, F. ten Hoor and J.C.S. Kleinjans, 1995. Induction of oxidative DNA damage and early lesions in rat gastro-intestinal epithelium in relation to prostaglandin H synthase-mediated metabolism of butylated hydroxyanisole. Food Chem. Toxicol., 33: 99-109.

22. Schnuch, A., H. Lessmann, J. Geier and W. Uter, 2011. Contact allergy to preservatives. Analysis of IVDK data 1996-2009. Br. J. Dermatol., 164: 1316-1325.

23. Scott, W. and S. Zak, 2014. Analysis of international food safety violations-2013. Food Sentry Organization http://www.foodsentry.org/analysis-international-food-safety-violations-2013/. Accesed on 16 August 2014.

24. Sharma, A., V. Prasongwattana, U. Cha'on, C. Selmi and W. Hipkaeo et al., 2013. Monosodium glutamate (MSG) consumption is associated with urolithiasis and urinary tract obstruction in rats. PLoS One, 10.1371/journal.pone.0075546

25. Shimada, C., K. Kano, Y.F. Sasaki, I. Sato, S. Tsudua, 2010. Differential colon DNA damage induced by azo food additives between rats and mice. J. Toxicol. Sci., 35: 547-554.

26. Soffritti, M., F. Belpoggi, M. Manservigi, E. Tibaldi, M. Lauriola, L. Falcioni and L. Bua, 2010. Aspartame administered in feed, beginning prenatally through life span, induces cancers of the liver and lung in male Swiss mice. Am. J. Ind. Med., 53: 1197-1206.

27. Stohs, S.J. and M.J.S. Miller, 2014. A case study involving allergic reactions to sulfur-containing compounds including, sulfite, taurine, acesulfame potassium and sulfonamides. Food Chem. Toxicol., 63: 240-243.

28. Titova, N.D., 2011. [Use of the granulocytic myeloperoxidase release reaction to diagnose food additive allergies]. Klinicheskaia Laboratornaia Diagnostika, 3: 42-44, (In Russian).

29. Turner, P.J. and A.S. Kemp, 2012. Intolerance to food additives-does it exist ? J. Paediatr. Child Health 48: E10-E14.

30. Van der Merwe, D., M. Bosman, S. Ellis , H. de Beer, A. Mielmann, 2013. Consumers' knowledge of food label information: an exploratory investigation in Potchefstroom, South Africa. Public Health Nutr., 16: 403-408.

31. Yadav, A., A. Kumar, A. Tripathi and M. Das, 2013. Sunset yellow FCF, a permitted food dye, alters functional responses of splenocytes at non-cytotoxic dose. Toxicol. Lett., 217: 197-204.

Chapter 10

PRODUCTION OF OLIGOSACCHARIDES AS PROMISING NEW FOOD ADDITIVE GENERATION

Hélène Barreteau[1], Cédric Delattre[2] and Philippe Michaud[3]

[1] Laboratoire des Enveloppes Bactériennes et Antibiotiques, IBBMC, UMR 8619 CNRS, Bâtiment 430, Université de Paris-Sud, FR-91405 Orsay, France

[2] Vellore Institut of Technology – Deemed University (VIT), Vellore 632014, Tamilnadu, India

[3] Laboratoire de Génie Chimique et Biochimique, Université Blaise Pascal – CUST, 24 avenue des Landais, BP206, FR-63174 Aubière cedex, France

SUMMARY

Recent research in the area of carbohydrate food ingredients has shown the efficiency of oligosaccharides when they are used as prebiotics or biopreservatives. Considering the former, they have various origins and structures, whereas the latter are described mostly as oligochitosans or as low molecular mass chitosans. If new manufacturing biotechnologies have significantly increased the development of these functional food ingredients, the main drawback limiting their applications is the difficulty to engender specific glycosidic structures. The present review focuses on the knowledge in the area of food bioactive oligosaccharides and catalogues the processes employed to generate them.

INTRODUCTION

In food industries, as chemical additives are becoming less and less welcome by consumers, there has been an increasing interest in the use of saccharidic natural substances known as prebiotic and biopreservative oligosaccharides. Traditionally, oligosaccharides are defined as polymers of monosaccharides with degrees of polymerization (DP) between 2 and 10 (3 and 10 according to the IUB-IUPAC nomenclature) but DPs up to 20–25 are often assimilated with them. Prebiotic oligosaccharides are noncariogenic, nondigestible (NDO) and low calorific compounds stimulating the growth and development of gastrointestinal microflora described as probiotic bacteria. It is claimed that these bacteria belonging to Bifidobacteria and Lactobacilli have several health-promoting effects (1, 2). Moreover, the recent development of commercial prebiotic oligosaccharides and probiotic bacteria has led naturally to a new concept, that of symbiotic one, combining probiotics and prebiotics (3). Paradoxically, other oligosaccharides and more specifically chitosan oligosaccharides (COS) or low molecular mass chitosans (LMMC) are described as food additives for their antimicrobial effects against pathogenic bacteria or fungi (4–6). Additionally, data suggest that specific COS or LMMC could also have beneficial effect on the growth of Bifidobacteria and Lactobacilli (7,8). Structural features of these oligomers appear as modulators for their biological activities.

In the current context of functional foods generating a global market of 33 billion US dollars (9), oligosaccharides could play a major role as functional ingredients compared to dietary fibers, sugar alcohols, peptides, probiotics, polyunsaturated fatty acids and antioxidants. Nonetheless, they will have a very large development in the future, depending on the viability of their large scale production. Oligosaccharides have currently two origins: they can be synthesized by chemical glycosylation and de novo using glycosidase and glycosyltransferase activities, or they can derive from chemical, physical or biological degradation of polysaccharides. As a consequence, this review focuses on the present uses of oligosaccharides as nutraceuticals, but also on the recent developments in the area of their production.

OLIGOSACCHARIDES AS PREBIOTICS

Presently, standard prebiotics are largely used depending on their putative positive action on the host's health. For these reasons, this new class of food ingredients has been added to human and domestic animals' foods. Concerning carbohydrates, the term prebiotic may be ambiguous because a lot of saccharidic compounds are present in feeding with or without prebiotic action (dietary fibers for example). In this context, Gibson et al. (10) established clear criteria for classifying a food ingredient as a prebiotic. Accordingly, a prebiotic oligosaccharide firstly needs to be resistant to gastric acidity, hydrolysis by mammalian enzymes and gastrointestinal absorption. Secondly, this oligomer has to be fermented by the intestinal microflora. Thirdly, it stimulates selectively the growth and/or activity of intestinal bacteria associated with health and wellbeing such as Bifidobacteria and Lactobacilli.

We have noticed that the majority of studies focuses on the in vitro metabolism of prebiotic oligosaccharides, and the mechanisms operating in vivo need to be elucidated. Uses of simulators of the human intestinal microbial ecosystem (SHIME) could lead to the design of more effective forms of prebiotics in the future (11). Furthermore, quantification strategies of prebiotic effects are currently assayed in vitro on faecal batch cultures (12) and a prebiotic index (PI) has been created (13).

FOS and GOS prebiotics

If some NDOs with prebiotic activities occur naturally in human milk (14) and plants (15), most of them are synthesised or isolated from plant polysaccharides such as fructooligosaccharides (FOS), galactooligosaccharides (GOS) or trans-galactooligosaccharides (TGOS), isomaltooligosaccharides (IMO), xylooligosaccharides (XOS), plant cell wall derived polysaccharides and other. At this time, FOS and GOS are leaders on the world market. There is little information about the structure-function relationships of these oligosaccharides apart from the studies comparing the fermentation properties of commercial products. The prebiotic effects of FOS or inulin (a mixture of FOS and polysaccharides) have been investigated by studying the metabolism of this mixture with DP from 3 up to 40–50 using Bifidobacterium sp. or Lactobacillus sp. (9, 16). We noted a high degree of variability in DP distribution depending on industrial preparations. All FOS consist of a glucose monomer á − 1,2)

linked to two (GF_2) or more (GF_n) â − (2,1) fructosyl units. Generally, in vitro data support the selective stimulation of bacterial growth by FOS and inulin using pure bacterial or/and faecal batch culture. However, metabolism of pure oligomers with controlled DP by colonic microflora has not been studied much even if the presence of fermentable mono- or disaccharides is well known in commercial preparations obtained from natural sources (e.g. inulin) or naturally synthesized from sucrose. In this context, the use of pure FOS mixtures containing three FOS species $(GF_2, GF_3$ and $GF_4)$ led to the identification of only 2 oligosaccharides $(GF_2$ and $GF_3)$ consumed by Lactobacillus strains. None of the examined strains was able to metabolise the GF_4 species, which suggests an intracellular metabolism after the FOS transport (17). This transfer has been recognised as mediated by an ATP-dependent transport system having specificity for a narrow range of substrates (18). Nonetheless, another paradigm is the FOS degradation by probiotic cell-associated exoglycosidases

and notably â − fructofuranosidases. With this mechanism, identified in a Bifidobacterium infantis strain, the monosaccharides generated are taken up by the bacteria (19).

Using animal models or volunteers fed with aliments containing inulin or FOS, in vivo experiments support the bifidogenic effect of FOS with large variations depending on the subjects, faecal microflora composition, doses and categories of FOS and/or inulin (9, 10). Authors noted the end of the prebiotic effect and the decrease of colonic microflora when the addition of FOS in food was stopped (20).

The GOS and TGOS fermentations are also well documented. TGOS are GOS produced by transgalactosylation of lactose using a â – galactosidase. In the final product, different linkages between the galactose and the reducing terminal glucose have been identified [(1, 2); (1, 3) and (1, 4)] and branched glucose residues occur. The galactan fragment is (1, 4) or (1, 6) linked as for GOS (21, 22). This high degree of variability in glycosidic linkage could implicate an incomplete resistance to gastric acidity and mammalian enzymes, as suggested by Tomomatsu (23). Generally, studies of the impact of TGOS on colic microflora species have shown that if many strains of enteric bacteria are unable to metabolise TGOS, Bifidobacteria and in a minor rate Lactobacilli will metabolise them (24). Bifidobacterium adolescentis, one of the predominant human faecal bacterium, can degrade and metabolise TGOS with DP 3 or higher, contrary to Bifidobacterium infantis and Lactobacillus acidophilus, which can use only TGOS with DP 3. This particularity is related to a â – galactosidase probably attached to the membrane (25). GOS metabolism was also investigated with fractionated GOS used as substrate for Bacillus lactis and for Lactobacillus rhamnosus by Gopal et al. (26), who noted that B. lactis, contrary to L. rhamnosus, was able to metabolise tri- and tetrasaccharidic fractions, suggesting a specific transport system. Data from in vivo experiments confirmed the increase of Bifidobacteria and Lactobacilli when TGOS were added in foods (27,28)

XOS, IMO and SOS prebiotics

Compared to FOS and GOS/TGOS, other prebiotic oligosaccharides are less documented, except for xylooligosaccharides (XOS), isomaltooligosaccharides (IMO), soybean oligosaccharides (SOS) and lactulose. However, even if the lactulose, resulting from the isomerisation of lactose to form galactosyl â – (1,4) fructose, is well known as a prebiotic oligosaccharide, its status in the oligosaccharide nomenclature (IUB-IUPAC) is not well established. Moreover, considering the possible lactose isomerisation during food engineering and notably during heat treatment of milk, this disaccharide may be naturally present in significant concentrations in food products. Nevertheless, in vitro comparative data showed that lactulose is one of the most efficient prebiotics in Bifidobacteria strains (13). Comparative results were found with in vivo experiments (29–31).

Considering XOS, IMO and SOS, their first commercial uses as prebiotics are presently being developed in Japan. Like the xylan, XOS are very resistant to acids and mammalian enzymes. They are manufactured by xylanase degradation of xylans and lead to an oligomeric mix where the xylobiose is the most representative compound (32). Data relating to XOS metabolism by intestinal microflora are ambiguous and Gibson et al. (10) concluded their recent review without the classification of XOS, as their fermentation does not seem to be selective (33,34).

Like XOS, IMO have a real positive effect, resulting in higher populations of Bifidobacteria (10,11,13,35). However, these á − (1,4)(1,6) oligoglucans produced from starch hydrolysis by á − amylases and pullulanases are potentially digestible by mammalians and can be metabolised by a wide range of bacteria. Consequently, their belonging to prebiotic oligosaccharides is not actually really defined (10).

The soybean oligosaccharides (raffinose and stachyose) are well known á -galactosyl sucrose derivatives extracted from soybeans (11,12). They are present in soygerm powder, whose fermentation properties have been successfully tested on Lactobacilli in the SHIME with faecal bacteria inoculum (11). A comparative in vitro evaluation of SOS on predominant gut bacterial groups showed that SOS have comparable effects with other galactooligosaccharides (12). On pure cultures, similar results have been obtained with individual purified compounds or mixture of oligosaccharides (36).

New prebiotic oligosaccharides

At present, the advancement of knowledge about polysaccharides from plant cell wall and plant cell wall polysaccharide cleavage enzymes allows the development of novel prebiotics. Effectively, these polysaccharides are available in large amounts notably from food industry by-products. Therefore, the use of specific hydrolysis conditions leads to processes for oligosaccharide productions. These oligomers have a large variety of structures and could become an interesting way to increase the value of plant by-products in the future. We noted that some of these oligomers are naturally produced during processing of food where glycanases are used for technological benefit.

In this way, arabinogalactooligosaccharides, arabinoxylooligosaccharides, arabinooligosaccharides, galacturonan oligosaccharides, rhamnogalacturonan oligosaccharides and pectic oligosaccharides have been successfully experimented with (25,37–39). These oligomers have been fermented in pure cultures by intestinal bacteria such as Bifidobacteria, Lactobacilli, Bacteroides sp., Clostridium sp., Escherchia coli and Klebsiella sp

In addition, recent literature has detailed numerous other oligosaccharidic structures as glucooligosaccharides and oligosaccharides from melibiose, mannan oligosaccharides, oligodextrans and gentiooligosaccharides with prebiotic activities (13,38,40,41). We also noted that some probiotic bacteria could produce by themselves polysaccharides (but no oligosaccharides) having prebiotic effects (42).

Use of prebiotics for additional beneficial effects

As classical prebiotic oligosaccharides added in food, human milk oligosaccharides stimulate the proliferation of bifidogenic microflora in breastfed children (43), but have also other important roles in the local intestine immune system (44). They play a role of additional defence mechanism as receptors (45) or block the progress of inflammatory responses (46). All these functions detected for sialylated and fucosylated oligosaccharides from human milk have not yet been tested with commercial FOS or GOS, but it is possible that these compounds have these effects as well.

Moreover, in addition to the increase of Bifidobacteria and Lactobacilli, prebiotic oligosaccharides have other identified effects that could enhance their use for therapeutic actions. One of them is the detection of short chain fatty acid (as propionate or butyrate) production as end fermentative products. These compounds have been recognized for their role in the prevention of colon cancer (47). It is also reported that FOS significantly increase the effects of different cytotoxic drugs used in human cancer treatment (48). The proliferation of beneficial bacteria under the influence of prebiotic oligosaccharides has also a significant impact on the prevention of the proliferation of pathogenic bacteria. This has been attributed to the low pH environment created during the fermentation of FOS in the colon (49).

Other data described a role of FOS in mineral absorption (mainly magnesium and calcium) because of the pH decrease in colon during their fermentation (50). The role of FOS in the control of diabetes has also been suggested (51). However, the important rate of residual monosaccharides in commercial FOS limits their uses in diabetic food products. FOS have also been implicated in the lipid metabolism and a lot of data suggest that FOS in foods modify the hepatic metabolism of lipids (52), inhibit secretion of triacylglycerol (TAG)-rich very low density lipoproteins (VLDL) (9) and reduce blood levels of TAG (53). FOS are also known to decrease the cholesterol in insulin-independent diabetic patients (54).

OLIGOSACCHARIDES AS NATURAL FOOD PRESERVATIVES

The term biopreservative includes a wide range of natural products from both plants and microorganisms, able to extend shelf life of foods, reduce or eliminate survival of pathogenic bacteria and increase overall quality of food products (55). These natural occurring antimicrobials can be, for example, peptides such as bacteriocins (56,57) or lipophilic substances such as essential oils (58). Compared to these two kinds of antimicrobial molecules, sugar molecules seem to be less investigated as potential food preservatives. In this context, one of the currently most studied polysaccharides is indisputably chitin. This linear homopolymer of â — (1,4)-linked-N-acetyl-D-glucosamine residues (Fig. 1) is one of the most abundant renewable natural polymers, second to cellulose. Chitin is commonly found in the exoskeletons or cuticles of many invertebrates like crustaceans and arthropods, in the cell walls of most fungi and is extracted commercially from shellfish wastes (59). As it is estimated to be synthesised in nature at a level of up to 10^9–10^{10} tonnes a year, the potential of chitin is evident in various industrial fields. Because of its limited solubility in aqueous solutions and organic solvents, many studies were realised on its low acetylated form, called chitosan (Fig. 1c). This biopolymer is easily obtained by alkali N-deacetylation of chitin. Polycationic at pH=6, biodegradable, nontoxic, soluble in acetic acid solutions, chitosan offers properties with great potential for many industrial applications. Accordingly, chitosan attracted considerable attention since it has been reported to exhibit interesting activities, notably to improve human health (60) and food quality with its antioxidative (61) and antimicrobial (62,63) properties. Moreover, concerning this last point and with respect to antimicrobial activity, chitosan seems superior to chitin since it contains amino groups which could interact with the negatively charged bacterial cell membranes and then inhibit the bacterial growth (64–67). Other mechanisms for antimicrobial activity of chitosan have also been suggested, as the blockage of RNA transcription by adsorption of penetrated chitosan to bacterial DNA (68) or the chelating action of chitosan with metal trace elements or essential nutrients, leading to microbial growth inhibition (69).

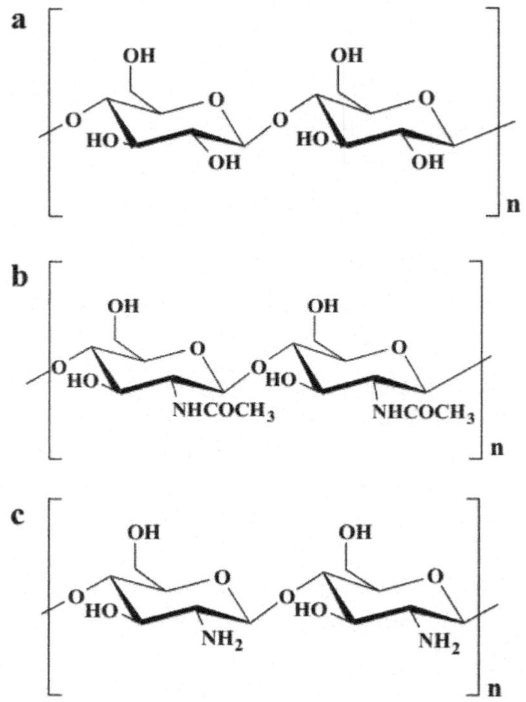

Figure 1: Structures of (a) cellulose, (b) chitin and (c) chitosan

Use of chitosan as potential food preservative

Most commercial native chitosans have a degree of deacetylation greater than 70 % and a molecular mass ranging between 100 and 1200 kDa. The legislation about their uses as food additives varies according to the country. Chitosan is sold in the European market in the form of dietary capsules to assist mass loss; it is reportedly used in Japan as a preservative in many food products (6,70), whereas the United States Food and Drug Administration (USFDA) approved its use in 1983 only as a feed additive (71) and has recently recognized it as a GRAS (Generally Recognised As Safe) component (72)

Chitosan antioxidative activities

Several studies reported antioxidative activities exhibited by chitosan. As the use of molecules with such properties is one way to extend the shelf life of food products, this biopolymer was tested on muscle foods, such as meat or seafoods, which contain highly unsaturated fatty acids particularly sensitive to oxidative change during storage (61). St. Angelo (73) reported that iron bound to proteins such as myoglobin or haemoglobin can be released during postharvest storage and cooking and then activate oxygen and initiate lipid oxidation. The mechanism involved in chitosan antioxidative activity is thought to be related to chelation of free iron. Effectiveness of chitosan treatment on oxidative stability of beef was also studied by Darmadji and Izumimoto (74) who observed that the addition of chitosan at 1 % concentration decreased the 2-thiobarbituric acid value of meat for 70 % after three days of storage at 4 °C.

Chitosan antimicrobial activities

Antimicrobial activities of chitosan were also demonstrated against many different kinds of microorganisms. Accordingly, chitosan was shown to inhibit food spoilage microorganisms, such as Candida sp., Escherichia coli and Staphylococcus aureus (74, 75). However, as the culture media employed poorly represent what really happens in complex food systems, this polysaccharide has also been tested in food products. Several studies were realized in fruit juices and emulsified sauces, but also in solid foods such as meat (74, 76), mayonnaise (66, 77), tofu (78), houmous and chilled salads (75). Finally, chitosan was also studied as an edible antimicrobial film to cover fresh fruits and vegetables (79), pizza (80) and meat (81).

Properties of chitosan oligosaccharides

If all the investigated studies recognize the antimicrobial activities of chitosan, those seem to depend on many factors, such as molecular mass, degree of acetylation, type of screened microorganisms and tested environmental conditions (82). Accordingly, DP is one of the most investigated factors. Finally, chitosan oligomers have received considerable attention since they were reported to be able to exhibit biological activities as interesting as those of their corresponding polymers even if the results about it are still controversial. In this way, No et al. (78) examined the antibacterial activities against several spoilage and food-borne bacteria of six chitosans and chitosan oligomers with widely different molecular mass. Their results led them to conclude that chitosans have higher antibacterial activities and markedly inhibited growth of most tested bacteria at a 0.1 % concentration. On the other hand, Tsai et al. (83) evaluated the antibacterial activity of a mixture of chitooligosaccharides prepared by digestion of shrimp chitosan against food-borne bacteria in nutrient broth. They showed that the required chitooligosaccharide mixture concentration to reach bactericidal effect on the tested bacteria was lower than that required with native chitosan. As to Kittur et al. (84), they showed that the growth inhibitory effect of a chitooligosaccharide preparation with DP ranging from 2 to 6 was better than that of native chitosan. For their part, Rhoades and Roller (75) investigated the antimicrobial actions in laboratory media of degraded and native chitosans against spoilage microorganisms. They proved that the inhibitory activity of chitosan against spoilage yeasts such as Zygosaccharomyces bailii and Saccharomyces cerevisiae was enhanced with chitosan degradation products. Tsai et al. (85) showed that LMM chitosan (12 kDa) has much higher antimicrobial activity than the mixture of chitooligosacharides (DP 1 to 8) against pathogenic microorganisms. In the same way Kittur et al. (84) demonstrated that the inhibition percentage of Bacillus cereus growth doubled from chitooligosaccharides trimer to hexamer at 10 % concentration. All these results were in agreement with those of Kendra and Hadwinger (86), who demonstrated that, if chitooligosaccharides can exhibit strong antibacterial effect, it is nevertheless greatly dependant on their DP and requires glucosamine oligomers with DP 7 or greater

In addition, environmental conditions and the nature of food products seem to influence the antimicrobial activity of chitosan and chitosan oligomers. Accordingly, Savard et al. (65) reported different inhibition patterns in liquid and solid media for chitosan oligomers tested against yeasts. Unlike previous results, the inhibitory activity of chitosan oligomers in agar medium was lost with the increase of the DP. As in liquid medium, the intermediate molecular mass chitosans were more toxic than the low molecular mass ones; the authors concluded that medium composition influenced the inhibitory activity of chitosan oligomers. Rhoades and Roller (75) explained the differences observed by the nature of solid medium which could restrict the mass transfer of macromolecules. Then, the choice of the food product to preserve from spoilage with chitosan or chitosan oligomers has to be judicious in order to keep a maximal antimicrobial activity. In the case of food products that need the use of microorganisms, such as lactic or vegetable fermentation, the antimicrobial action of chitosan or chitosan oligosaccharides may not permit their uses. Nonetheless, it can be conceivable to add them before packaging, in order to fight against spoilage yeasts (65).

Concerning the pH effects on various molecular mass chitosans, it affected their antibacterial activity independently of their molecular mass (78). This result clearly indicates that, whatever their molecular mass, applications of chitosans or chitosan oligomers in acidic food products would enhance their effectiveness.

OLIGOSACCHARIDE ENGINEERING

Even if the potentialities of oligosaccharides are real, in most cases the lack of oligosaccharide production processes is the main drawback limiting their applications. That is why the development of oligosaccharide engineering strategies actually represents a challenge. Bioactive oligosaccharides come from oligomer engineering with either (i) synthesis (using enzymatic or chemical engineering) or (ii) polysaccharide depolymerization (using physical, chemical or enzymatic methods)

Chemical and biochemical synthesis of oligosaccharides

The chemical or biochemical synthesis of oligosaccharidic structures is much more difficult than it was observed in the synthesis processes of other biopolymers such as peptides and nucleic acids because of the higher number of possibilities in monomeric unit combinations. So, the stereospecific introduction of glycosidic linkages appears as the challenge of oligosaccharide synthesis. Nevertheless, recent advances in enzymatic and chemical synthesis allow envisaging the preparation of a wide range of oligomers.

Chemical glycosylation in the production of oligosaccharides

Different strategies have been published for the chemical synthesis of oligosaccharides (87,88). Generally, the preparation of oligomers by chemical processes takes place as illustrated in Fig. 2. The glycosylation reaction is achieved by an inter-glycosidic condensation between a completely protected glycosyl donor (R-OH) that has an excellent leaving group (halogenides) at its anomeric position and a glycosyl acceptor that possesses only one free hydroxyl group. Among these glycosyl donors, the anomeric fluorides, trichloroacetimidates and thioglycosides (Fig. 3) are currently being employed (87). These compounds can be prepared under mild conditions and are sufficiently stable to be purified and stored for a considerable period of time. The preparation of specific glycosyl donor and acceptor implies many protection and deprotection steps to combine high yields with regioand stereoselectivity. With these strategies oligosaccharides of interest such as oligogalacturonides (OGAs) have been synthetised (89,90). These compounds used as glycosylation intermediates should permit the production of higher oligomers of D-galacturonates. Likewise, other oligomers as for example, xylooligosaccharides (DP 4 up to 10) have also been generated by a complex blockwise synthesis approach (91). These syntheses show that the formation of oligosaccharides is only possible when each step in the assembly of the glycosyl (donor and acceptor) is high yielding. Therefore, in spite of recent developments, synthesis of oligosaccharides by chemical glycosylations seems actually nonrealistic for industrial processes. Conversely, other chemical alternative methods such as chemomechanical synthesis of bifidogenic glucooligosaccharides have been successfully tested by extrusion process (92).

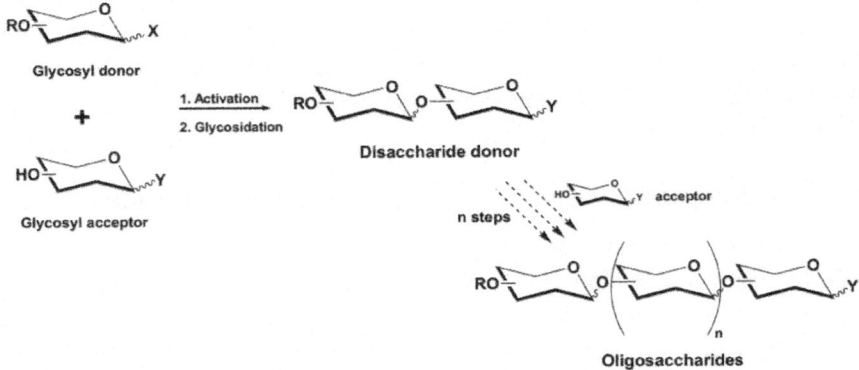

Figure 2: Current strategy of the chemical synthesis of oligosaccharides (R: H or saccharides)

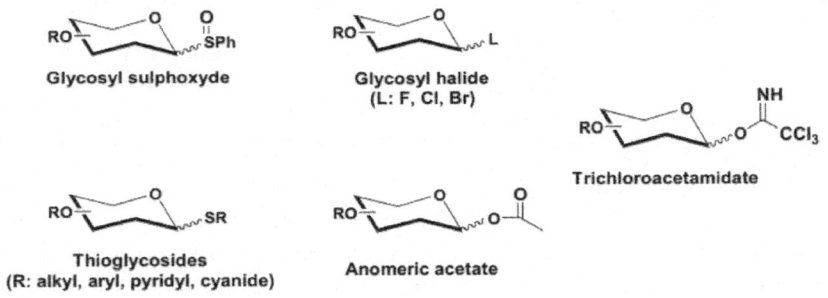

Figure 3: Currently used glycosyl donors in the chemical synthesis of oligosaccharides (R: H or saccharides)

The large-scale productions of oligosaccharides thanks to enzymatic synthesis have largely been developed these last decades. Effectively, specific studies of biosynthesis pathways of oligosaccharides and the use of glycosidases and glycosyltransferases permit to avoid the disadvantages of chemical methods, since the enzymes control both regio- and stereoselectivity of glycosylation.

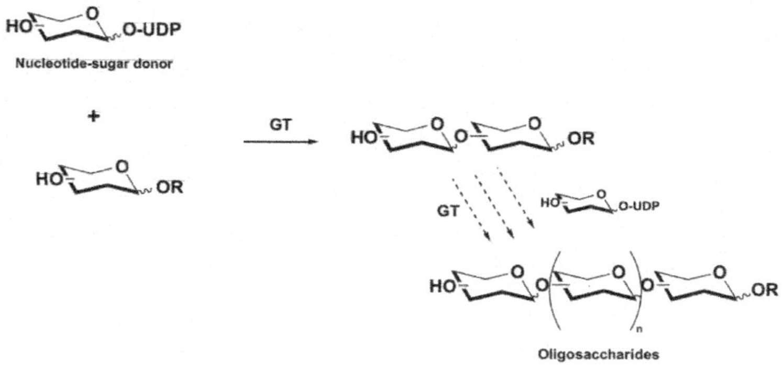

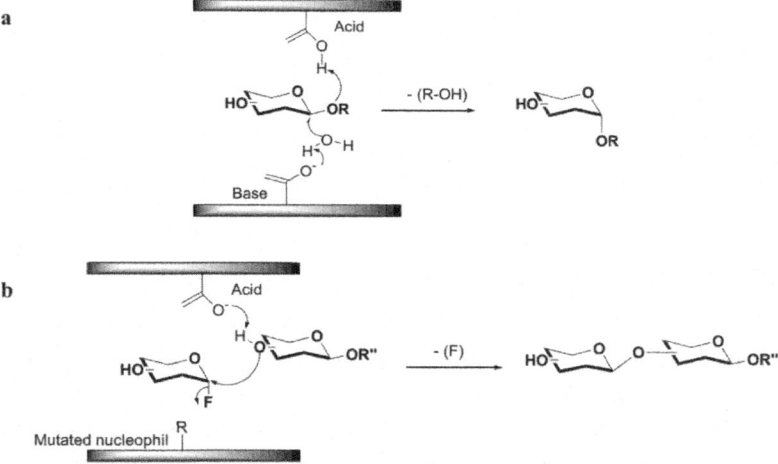

Figure 4: Current strategy of the enzymatic synthesis of oligosaccharides using glycosyltransferase (GT) (R: H or saccharides)

Glycosyltransferases (E.C. 2.4.xy) catalyse the transfer of sugar moieties from activated donor molecules to specific acceptor molecules, forming glycosidic bonds (Fig. 4). These enzymes are highly regio- and stereoselective and have been obtained after purification or after cloning and overexpression.

Figure 5: Reaction mechanism of (a) glycosidase and (b) glycosynthase (R, R'': saccharides)

Glycosylhydrolases (E.C. 3.2.1.-) are much more readily available than glycosyltransferases but are generally less stereoselective. These specific biocatalysts were traditionally used for the degradation of poly- and oligosaccharides. Nevertheless, the reverse hydrolytic activities under suitable reaction conditions and the utilizations of nucleotide-activated sugars permit the formation of glycosidic bond (Fig. 5a). Yields were generally lower with glycosidases except with mutated derivative named glycosynthase that operates more efficiently with appropriated activated glycosyl donors such as fluorides (Fig. 5b). Many bioactive oligosaccharides are produced by the enzymatic engineering such as FOS from sucrose using fructosyltransferase (9), TOS or

GOS from lactose using â − galactosidase (93) and gentiooligosaccharides from glucose by transglucosylation (94)

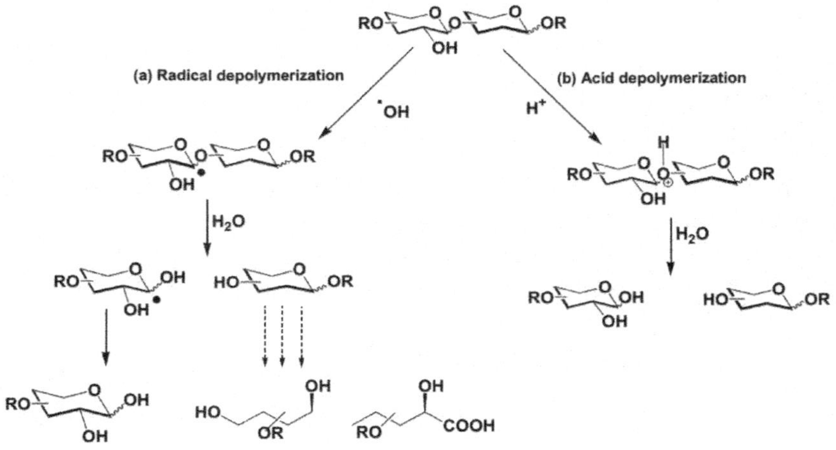

Figure 6: Reactional mechanism of (a) radical and (b) acid polysaccharide depolymerization (R: H or saccharide)

Polysaccharide depolymerization

Except for chemical treatments with acid and radical hydrolysis (Fig. 6) or physical treatments using thermal, microwave, ã -irradiation and ultrasonication degradations, the enzymatic depolymerization is the main approach to prepare large amounts of oligomers. The largely found strategies are microbiotechnology procedures that degrade various polysaccharides using regio- and stereospecifically microbial enzymes such as polysaccharide hydrolases and polysaccharide lyases. The polysaccharide hydrolases (E.C. 3.2.1.-) catalyze the hydrolysis of the glycosidic bonds via a general acid catalysis requiring two critical residues: a proton donor and a nucleophile/base. This hydrolysis occurs via two mechanisms giving rise to either an overall retention or an inversion of anomeric configuration. Concerning polysaccharide lyases (E.C. 4.2.2.-), they constitute a specific group of enzymes which generate the cleavage of polysaccharide chains via a â − elimination resulting in the formation of a double bond at the newly formed non-reducing end (Fig. 7).

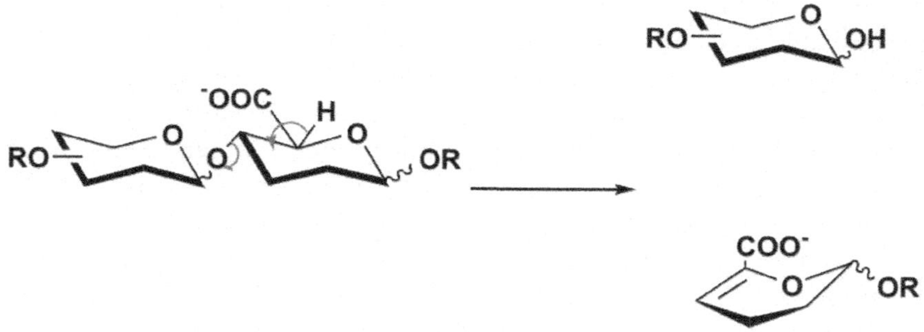

Figure7: Reactional mechanism of polysaccharide depolymerization by â − elimination (R: H or saccharide)

Actually, these selective biocatalysts are largely exploited and developed to prepare a majority of bioactive oligosaccharides. In this manner, several oligosaccharides with bifidogenic application were commercially prepared (94); for example, maltooligosaccharides from starch using á -amylases, isomaltooligosaccharides with the action of á , â − amylases and á -glucosidases, fructooligosaccharides from enzymatic hydrolysis of inulin, and xylooligosaccharides from xylan cleavage with â − (1,4) xylanases. In addition, pectinases, pectate lyases and other polygalacturonases (98) were used to generate oligogalacturonides from pectin cleavage.

CONCLUSION

The expansion of the knowledge of new oligosaccharidic structures, associated with the advances of glycobiotechnologies, could authorize a larger employment of these compounds as food ingredients in the future. The expansion of this market implies strategies for their low cost manufacturing. Even if the recent advances in polysaccharide cleavage enzymes and in the microorganisms are applied to generate some oligosaccharides in large scale, the improvement of glycosylation strategies (chemical or biochemical) seems to be much more promising and reliable in the manufacture of oligosaccharides and derivatives. Many di-, tri- and tetrasaccharides can be now synthesized routinely involving a minimum number of glycosylation steps and solid phase oligosaccharide synthesis is developed. It consequently opens the way to original oligomer preparations notably in regard to advances of automated systems. Nonetheless, as these methods are limited, chemists and biochemists focus on the preparation of large amounts of more oligosaccharides using enzymatic polysaccharide depolymerization. In this way, the chitinous biooligomers offer a wide range of unique applications to prevent foods from microbial deterioration. As they are able to exhibit interesting antioxidative and antimicrobial activities, they may be of advantageous use in the field of natural food additives. Considering the prebiotic oligosaccharides, the understanding of structure-function relationships related to probiotic bacteria targets could move the research towards specific structures with a higher level of functionality. Additionally, the screening of new oligosaccharidic structures and notably those obtained after bacterial polysaccharide degradations could enhance the prebiotic oligosaccharide families. Effectively, bacterial polysaccharide as exopolysaccharides offer a large range of structures and are produced in controlled conditions (reactors). So their degradations by specific enzymes will offer, in the future, numerous oligosaccharides with potentialities as prebiotics. Moreover, the next generations of prebiotic oligosaccharides could be supplied by probiotics themselves. Effectively, a part of Lactobacilli are exopolysaccharide-producing strains and no data have been reported related to the prebiotic activity of oligomers that come from them even if the polymeric patterns as levans are known as having prebiotic effects.

REFERENCES

1.   R.A. Rastall, G.R. Gibson, H.S. Gill, F. Guarner, T.R. Klaenhammer, B. Pot, G. Reid, I.R. Rowland, M.E. Sanders, Modulation of the microbial ecology of the human colon by probiotics, prebiotics and synbiotics to enhance human health: An overview of enabling science and potential applications, FEMS Microbiol. Ecol. 52 (2005) 145–152.

2.   D. Zopf, S. Roth, Oligosaccharide anti-infective agents, Lancet, 347 (1996) 1017–1021.

3.   C.J. Ziemer, G.R. Gibson, An overview of probiotics, prebiotics and synbiotics in the functional food concept: Perspectives and future strategies, Int. Dairy J. 8 (1998) 473– 479.

4.   B.K. Choi, K.Y. Kim, Y.J. Yoo, S.J. Oh, J.H. Choi, C.Y. Kim, In vitro antimicrobial activity of a chitooligosaccharide mixture against Actinobacillus actinomycetemcomitans and Streptococcus mutans, Int. J. Antimicrob. Agents, 18 (2001) 553– 557.

5.   Y.J. Jeon, P.J. Park, S.W. Kim, Antimicrobial effect of chitooligosaccharides

produced by bioreactor, Carbohydr. Polym. 44 (2001) 71–76.

6. S. Roller, N. Covill, The antifungal properties of chitosan in laboratory media and apple juice, Int. J. Food Microbiol. 47 (1999) 67–77.

7. H.W. Lee, Y.S. Park, J.S. Jung, W.S. Shin, Chitosan oligosaccharides, dp 2–8 have prebiotic effect on the Bifidobacterium bifidum and Lactobacillus sp., Anaerobe, 8 (2002) 319–324.

8. C.L. Vernazza, G.R. Gibson, R.A. Rastall, In vitro fermentation of chitosan derivates by mixed culture of human faecal bacteria, Carbohydr. Polym. 60 (2005) 539–545.

9. P.T. Sangeetha, M.N. Ramesh, S.G. Prapulla, Recent trends in the microbial production, analysis and application of fructooligosaccharides, Trends Food Sci. Technol. 16 (2005) 442–457.

10. G.R. Gibson, H.M. Probert, J. Van Loo, R.A. Rastall, M.B. Roberfroid, Dietary modulation of the human colonic mi- crobiota: Updating the concept of prebiotics, Nutr. Res. Rev. 17 (2004) 259–275.

11. P. De Boever, B. Deplancke, W. Verstraete, Fermentation by gut microbiota cultured in a simulator of the human intestinal microbiol ecosystem is improved by supplementing a soygerm powder, J. Nutr. 130 (2000) 2599–2606.

12. C.E. Rycroft, M.R. Jones, G.R. Gibson, R.A. Rastall, A comparative in vitro evaluation of the fermentation properties of prebiotic oligosaccharides, J. Appl. Microbiol. 91 (2001) 878–887.

13. R. Palframan, G.R. Gibson, R.A. Rastall, Development of a quantitative tool for the comparison of the prebiotic effect of dietary oligosaccharides, Lett. Appl. Microbiol. 37 (2003) 281–284.

14. S. Thurl, B. Muller-Werner, G. Sawatzki, Quantification of individual oligosaccharide compounds from human milk using high-pH anion-exchange chromatography, Anal. Biochem. 235 (1996) 202–206.

15. J.M. Campbell, G.C. Fahey Jr., B.W. Wolf, Selected indigestible oligosaccharides affect large bowel mass, cecal and fecal short-chain fatty acids, pH and microflora in rats, J. Nutr. 127 (1997) 130–136.

16. R.A. Rastall, V. Maitin, Prebiotics and synbiotics: Towards the next generation, Curr. Opin. Biotechnol. 13 (2002) 490–496.

17. H. Kaplan, R.W. Hutkins, Fermentation of fructooligosaccharides by lactic acid bacteria and Bifidobacteria, Appl. Environ. Microbiol. 66 (2000) 2682–2684.

18. H. Kaplan, R.W. Hutkins, Metabolism of fructooligosaccharides by Lactobacillus paracasei 1195, Appl. Environ. Microbiol. 69 (2003)

2217–2222.

19. S. Perrin, M. Warchol, J.P. Grill, F. Schneider, Fermentations of fructo-oligosaccharides and their components by Bifidobacterium infantis ATCC 15697 on batch culture in semi- -synthetic medium, J. Appl. Microbiol. 90 (2001) 859–865.

20. V.A. Rao, The prebiotic properties of FOS at low intake level, Nutr. Res. 21 (2001) 843–848.

21. J.B. Smart, Transferase reactions of beta-galactosidase – New products opportunities, Bull. Int. Dairy Fed. 289 (1993) 16– 22.

22. S. Zarate, M.H. Lopez-Leiva, Oligosaccharide formation during enzymatic lactose hydrolysis: A literature review, J. Food Prot. 53 (1990) 262–268.

23. H. Tomomatsu, Health effects of oligosaccharides, Food Technol. 48 (1994) 61–65.

24. R. Tanaka, H. Takayama, M. Morotomi, T. Kuroshima, S. Ueyama, K. Matsumoto, A. Kuroda, M. Mutai, Effects of administration of TOS and Bifidobacterium breve 4006 on the human fecal flora, Bifidobacteria Microflora, 2 (1983) 17– 24.

25. K.M.J. van Laere, R. Hartemink, M. Bosveld, H.A. Schols, A.G.J. Voragen, Fermentation of plant cell wall derived polysaccharides and their corresponding oligosaccharides by intestinal bacteria, J. Agric. Food Chem. 48 (2000) 1644– 1652.

26. P.K. Gopal, P.A. Sullivan, J. B. Smart, Utilisation of galacto-oligosaccharides as selective substrates for growth by lactic acid bacteria including Bifidobacterium lactis DR10 and Lactobacillus rhamnosus DR20, Int. Dairy J. 11 (2001) 19–25.

27. Y. Bouhnik, B. Flourié, L. D'Agay-Abensour, P. Pochart, G. Gramet, M. Duran, J.C. Rambaud, Administration of trans- -galacto-oligosaccharides increases fecal bifidobacteria and modifies colonic fermentation metabolism in healthy humans, J. Nutr. 127 (1997) 444–448.

28. I.R. Rowland, R. Tanaka, The effects of transgalactosylated oligosaccharides on gut flora metabolism in rats associated with a human fecal microflora, J. Appl. Bacteriol. 74 (1993) 667–674.

29. T. Tomoda, Y. Nalano, T. Kageyama, Effect of yogurt and yogurt supplemented with Bifidobacterium and/or lactulose in healthy persons: A comparative study, Bifidobacteria Microflora, 10 (1991) 123–130.

30. A. Terada, H. Hara, S. Kato, T. Kimura, I. Fujimori, K. Hara, T. Maruyama, T. Mitsuoka, Effect of lactosucrose (4G-beta-D-galactosylsucrose) on

fecal putrefactive products of cats, J. Vet. Med. Sci. 55 (1993) 291–295.

31. J. Ballongue, C. Schumann, P. Quignon, Effects of lactulose and lactitol on colonic microflora and enzymatic activity, Scand. J. Gastroenterol. 222 (1997) 41–44.

32. H. Yamada, Structure and properties of oligosaccharides from wheat bran, Cereal Food World, 38 (1993) 490–492.

33. R.G. Crittenden, M.J. Playne, Purification of food-grade oligosaccharides using immobilised cells of Zymomonas mobilis, Appl. Microbiol. Biotechnol. 58 (2002) 297–302.

34. M. Okazaki, S. Fujikawa, N. Matsumoto, Effects of xylooligosaccharides on growth of bifidobacteria, J. Jpn. Soc. Nutr. Food Sci. 43 (1990) 395–401.

35. T. Kohmoto, F. Fukui, H. Takaku, T. Mitsuoka, Dose-response test of isomaltooligosaccharides for increasing fecal bifidobacteria, Agr. Biol. Chem. (Tokyo), 55 (1991) 2157–2159.

36. J. Jaskari, P. Kontula, A. Siitonen, H. Jousimies-Somer, T. Mattila-Sandholm, K. Poutanen, Oat â − glucan and xylan hydrolyzates as selective substrates for Bifidobacterium and Lactobacillus strains, Appl. Microbiol. Biotechnol. 49 (1998) 175–181.

37. E. Olano-Martin, K.C. Mountzouris, G.R. Gibson, R.A. Rastall, Continuous production of oligosaccharides from pectin in an enzyme membrane reactor, J. Food Sci. 66 (2001) 966–971.

38. E. Olano-Martin, G.R. Gibson, R.A. Rastall, Comparison of the in vitro bifidogenic properties of pectins and pectic-oligosaccharides, J. Appl. Microbiol. 93 (2002) 505–511.

39. A. Oosterveld, G. Beldman, A.G.J. Voragen, Enzymatic modification of pectic polysaccharides obtained from sugar beet pulp, Carbohydr. Polym. 48 (2002) 73–81.

40. K.M.J. van Laere, R. Hartemink, G. Beldman, S. Pitson, C. Dijkema, H.A. Schols, A.G.J. Voragen, Transglycosidase activity of Bifidobacterium adolescentis DSM 20083 alpha-galactosidase, Appl. Microbiol. Biotechnol. 52 (1999) 681–688.

41. L.A. White, M.C. Newman, G.L. Comwell, M.D. Lindemann, Brewers dried yeast as a source of mannan oligosaccharides for weaning pigs, J. Anim. Sci. 80 (2002) 2619– 2628.

42. F. Dal Bello, J. Walter, C. Hertel, W.P. Hammes, In vitro study of prebiotic properties of levan-type exopolysaccharides from Lactobacilli and non-digestible carbohydrates using denaturing gels electrophoresis, Syst.

Appl. Microbiol. 24 (2001) 232–237.

43. C. Kunz, S. Rudloff, W. Baier, N. Klein, S. Strobel, Oligosaccharides in human milk: Structural, functional, and metabolic aspects, Annu. Rev. Nutr. 20 (2000) 699–722.

44. M.J. Gnoth, C. Kunz, S. Rudloff, Endotoxin-reduced milk oligosaccharide fractions suitable for cell biological studies, Eur. J. Med. Res. 5 (2000) 468–472.

45. N. Klein, A. Schwertmann, M. Peters, C. Kunz, S. Strobel, Immunomodulatory effects of breast milk oligosaccharides, Adv. Exp. Med. Biol. 478 (2000) 251–259.

46. C. Kunz, M. Rodriguez-Palmero, B. Koletzko, R. Jensen, Nutritional and biochemical properties of human milk, Part I, General aspects, proteins, and carbohydrates, Clin. Perinatal. 26 (1999) 307–333.

47. J.H. Cummings, Short chain fatty acids in the human colon, Gut, 22 (1981) 763–779.

48. H.S. Taper, M.B. Roberfroid, Inulin/oligofructose and anticancer therapy, Br. J. Nutr. (Suppl.), 87 (2002) 283–286.

49. A. Letllier, S. Messier, L. Lessard, S. Quessy, Assessment of various treatments to reduce carriage of Salmonella in swine, Can. J. Vet. Res. 64 (2000) 27–31.

50. S.K.E. Ahrens, J. Schrezenmeir, Inulin, oligofructose and mineral metabolism-experimental data and mechanism, Br. J. Nutr. (Suppl.), 87 (2002) 179–186.

51. J. Luo, M. Van Yperselle, S.W. Rizkalla, F. Rossi, F.R.J. Bornet, G. Slama, Chronic consumption of short-chain fructooligosaccharides does not affect basal hepatic glucose production or insulin resistance in type 2 diabetics, J. Nutr. 130 (2000) 1572–1577.

52. N.M. Delzenne, C. Daubioul, A. Neyrinck, M. Lasa, H.S. Taper, Inulin and oligofructose modulate lipid metabolism in animals: Review of biochemical events and future prospects, Br. J. Nutr. (Suppl.), 87 (2002) 255–259.

53. C.M. Williams, K.G. Jackson, Inulin and oligofructose: Effects on lipid metabolism from human studies, Br. J. Nutr. (Suppl.), 87 (2002) 261–264.

54. M.B. Roberfroid, Prebiotics and probiotics: Are they functional foods?, Am. J. Clin. Nutr. (Suppl.), 71 (2000) 1682–1687.

55. F.A. Draughon, Use of botanicals as biopreservatives in foods, Food Technol. 58 (2004) 20–28.

56. R.W. Jack, J.R. Tagg B. Ray, Bacteriocins of Gram-positive bacteria,

Microbiol. Rev. 59 (1995) 171–200.

57. J. Cleveland, T.J. Montville, I.F. Nes, M.L. Chikindas, Bacteriocins: Safe, natural antimicrobials for food protection. Int. J. Food Microbiol. 71 (2001) 1–20.

58. S.G. Deans, G. Ritchie, Antibacterial properties of plant essential oils, Int. J. Food Microbiol. 5 (1987) 165–180.

59. D. Knorr, Use of chitinous polymers in food, Food Technol. 38 (1984) 85–96.

60. K. Suzuki, T. Mikami, Y. Okawa, A. Tokoro, S. Suzuki, M. Suzuki, Antitumor effect of a hexa-N-acetyl-chitohexaose and chitohexaose, Carbohydr. Res. 151 (1986) 403–408.

61. J.Y.V.A. Kamil, Y.J. Jeon, F. Shahidi, Antioxidative activity of chitosans of different viscosity in cooked comminuted flesh of herring (Clupea harengus), Food Chem. 79 (2002) 69–77.

62. F. Shahidi, J.V.A. Kamil, Y.J. Jeon, Food applications of chitinand chitosans, Trends Food Sci. 10 (1999) 37–51.

63. S.K. Kim, N. Rajapakse, Enzymatic production and biological activities of chitosan oligosaccharides (COS): A review, Carbohydr. Polym. 62 (2005) 357–368.

64. A.B. Vishu Kamar, M.C. Varadaraj, L.R. Gowda, R.N. Tharanathan, Characterization of chito-oligosaccharides prepared by chitosanolysis with the aid of papain and Pronase, and their bactericidal action against Bacillus cereus and Escherichia coli, Biochem. J. 391 (2005) 167–175.

65. T. Savard, C. Beaulieu, I. Boucher, C.P. Champagne, Antimicrobial action of hydrolysed chitosan against spoilage yeasts and lactic acid bacteria of fermented vegetables, J. Food Prot. 65 (2002) 828–833.

66. H.I. Oh, Y.J. Kim, E.J. Chang, J.Y. Kim, Antimicrobial characteristics of chitosans against food spoilage microorganisms in liquid media and mayonnaise, Biosci. Biotechnol. Biochem. 65 (2001) 2378–2383.

67. D.H. Young, H. Kohle, H. Kauss, Effects of chitosan on membrane permeability of suspension-cultured Glycine max and Phaseolus vulgaris cells, Plant Physiol. 70 (1982) 1449– 1454.

68. J.Y. Kim, J.K. Lee, T. S. Lee, W.H. Park, Synthesis of chitooligosaccharide derivative with quaternary ammonium group and its antimicrobial activity against Streptococcus mutans, Int. J. Biol. Mol. 32 (2003) 23–27.

69. R.G. Cuero, G. Osuji, A. Washington, N-carboxymethyl chitosan inhibition of aflatoxin production: Role of zinc, Biotechnol. Lett. 13 (1991) 441–444.

70. A.J. Borderias, I. Sanchez-Alonso, M. Perez-Mateos, New applications of fibers in foods: Addition to fishery products, Trends Food Sci. Technol. 16 (2005) 458–465.

71. D. Knorr, Nutritional quality, food processing and biotechnology aspects of chitin and chitosan: A review, Process Biochem. 6 (1986) 90–92.

72. US Food and Drug Administration. Center for Food Safety and Applied Nutrition, Office of Premarket Approval, GRAS Notices (2001) (http://vm.cfsan.fda.gov).

73. A.J. St. Angelo, Lipid oxidation on foods, Crit. Rev. Food Sci. 36 (1996) 175–224.

74. P. Darmadji, M. Izumimoto, Effect of chitosan in meat preservation, Meat Sci. 38 (1994) 243–254.

75. J. Rhoades, S. Roller, Antimicrobial actions of degraded and native chitosan against spoilage organisms in laboratory media and foods, Appl. Environ. Microbiol. 66 (2000) 80–86.

76. S. Sagoo, R. Board, S. Roller, Chitosan inhibits growth of spoilage microorganisms in chilled pork products, Food Microbiol. 19 (2002) 175–182.

77. S. Roller, N. Covill, The antimicrobial properties of chitosan in mayonnaise and mayonnaise-based shrimp salads, J. Food Prot. 63 (2000) 202–209.

78. H.K. No, N.Y. Park, S.H. Lee, H.J. Hwang, S.P. Meyers, Antimicrobial activities of chitosans and chitosan oligomers with different molecular mass on spoilage bacteria isolated from tofu, J. Food Sci. 67 (2002) 1511–1514.

79. F. Devlieghere, A. Vermeulen, J. Debevere, Chitosan: Antimicrobial activity, interactions with food components and applicability as a coating on fruit and vegetables, Food Microbiol. 21 (2004) 703–714.

80. M.S. Rodriguez, V. Ramos, E. Agullo, Antimicrobial action of chitosan against spoilage organisms in precooked pizza, J. Food Sci. 68 (2003) 271–274.

81. B. Ouattar, R.E. Simard, G. Piett, A. Begin, R.A. Holley, Inhibition of surface spoilage bacteria in processed meats by application of antimicrobial films prepared with chitosan, Int. J. Food Microbiol. 62 (2000) 139–148.

82. R.G. Cuero, Antimicrobial action of exogenous chitosan, EXS, 87 (1999) 315–333.

83. G.J. Tsai, Z.Y. Wu, W.H. Su, Antibacterial activity of a chitooligosaccharides

mixture prepared by cellulase digestion of shrimp chitosan and its application to milk preservation, J. Food Prot. 63 (2000) 747–752.

84. F.S. Kittur, A.B. Vishu Kumar, M.C. Varadaraj, R.N. Tharanathan, Chitooligosaccharides – Preparation with the aid of pectinase isozyme from Aspergillus niger and their antibacterial activity, Carbohydr. Res. 340 (2005) 1239–1245.

85. G.J. Tsai, S.L. Zhang, P.L. Shieh, Antimicrobial activity of a low-molecular-weight chitosan obtained from cellulase digestion of chitosan, J. Food Prot. 67 (2004) 396–398.

86. D.F. Kendra, L.A. Hadwiger, Characterisation of the smallest chitosan oligomer that is maximally antifungal to Fusarium solani and elicits pisatin formation in Pisum sativum, Exp. Mycol. 8 (1984) 276–281.

87. G.J. Boons, Synthetic oligosaccharides: Recent advances, Drug Discov. Today, 8 (1996) 331–342.

88. G.J. Boons, Strategies in oligosaccharides synthesis, Tetrahedron, 52 (1996) 1095–1121.

89. D. Magaud, C. Grandjean, A. Doutheau, D. Anker, V. Shevchik, N. Cotte-Pattat, J. Robert-Baudouy, An efficient and highly stereoselective a(1®4) glycosylation between two D-galacturonic acid ester derivatives, Tetrahedron Lett. 38 (1997) 241–244.

90. D. Magaud, C. Grandjean, A. Doutheau, D. Anker, V. Shevchik, N. Cotte-Pattat, J. Robert-Baudouy, Synthesis of the two monomethyl esters of the disaccharide 4-O-a-D-galacturonosyl-D-galacturonic acid and of precursors for the preparation of higher oligomers methyl uronated in definite sequences, Carbohydr. Res. 314 (1998) 189–199.

91. K. Takeo, Y. Ohguchi, R. Hasegawa, S. Kitamura, Synthesis of (1,4)-â – D-xylo-oligosaccharides of dp 4–10 by a blockwise approach, Carbohydr. Res. 278 (1995) 301–313.

92. J.K. Hwang, Production of functional carbohydrates by the extrusion reactor, Food Sci. Biotechnol. 10 (2001) 455–459.

93. P. Czermak, M. Ebrahimi, K. Grau, S. Netz, G. Sawatzki, P.H. Pfromm, Membrane-assisted enzymatic production of galactosyl-oligosaccharides from lactose in a continuous process, J. Membr. Sci. 232 (2004) 85–91.

94. M.J. Playne, R. Crittenden, Commercially available oligosaccharides. Bulletin of IDF, 313 (1996) 10–22.

95. A. Tanriseven, S. Dogan, Production of isomalto-oligosaccharides using dextransucrase immobilized in alginate fibres, Process Biochem. 37 (2002) 1111–1115.

96. I. Ghazi, A.G. De Segura, L. Fernandez-Arrojo, M. Alcaldea, M. Yates, M.L. Rojas-Cervantes, F.J. Plou, A. Ballesteros, Immobilisation of fructosyltransferase from Aspergillus aculeatus on epoxy-activated Sepabeads EC for the synthesis of fructo-oligosaccharides, J. Mol. Catal. 35 (2005) 19–27.

97. T. Maugard, D. Gaunt, M.D. Legoy, T. Besson, Microwave- -assisted synthesis of galacto-oligosaccharides from lactose with immobilized a-galactosidase from Kluyveromyces lactis, Biotechnol. Lett. 25 (2003) 623–629.

98. S. Aldington, S.C. Fry: Oligosaccharins. In: Advances in Botanical Research, Vol. 19 (1993) pp. 1–101.

Chapter 11

THE FOOD ADDITIVE POLYGLYCEROL POLYRICINOLEATE (E-476): STRUCTURE, APPLICATIONS, AND PRODUCTION METHODS

Josefa Bastida-Rodríguez

Department of Chemical Engineering, University of Murcia, Campus de Espinardo, 30100 Murcia, Spain

ABSTRACT

The food additive named polyglycerol polyricinoleate (PGPR) and identified with the code E-476 (PGPR) is used as emulsifier in tin-greasing emulsions for the baking trade and for the production of low-fat spreads. However, the main application of PGPR is in the chocolate industry, where, besides its action as an emulsifier, it also has important properties as a viscosity modifier and thus improves the moulding properties of the molten chocolate. An additional property of PGPR in chocolate is its ability to limit fat bloom. Known chemical methods for preparing this emulsifier involve long reaction times and high operating temperatures, which adversely affect the quality of the final product leading to problems of coloration and odors that could make it inadvisable for the food industry. As an alternative, the enzymatic synthesis of PGPR by the catalytic action of two lipases has been developed. The enzymes act in mild reaction conditions of temperature and pressure, neutral pH, and in a solvent-free system, which makes the process environmentally friendly and avoids side reaction, so that the product has a higher purity and quality.

FOOD ADDITIVES

Food additives have been developed over the years to meet the needs of food production, as making foods on a large scale is a very different task to making them in the kitchen at home. Additives are needed to ensure processed food remains in a good condition throughout its journey from the factory to the shop and to the consumer at home. Some are so essential that they are even used in organic foods [1–3]. In the broadest of terms, food additives are substances intentionally added to food either directly or indirectly with one or more of the following purposes [4]:(1)to maintain or improve nutritional quality;(2)to maintain product quality and freshness;(3)to aid in the processing or preparation of food;(4)to make food more appealing.

On the other hand, food additives may only be authorized if [4](1)there is a technological need for their use;(2)they do not mislead the consumer;(3)they present no hazard to the health of the consumer.

The use of food additives must always be labelled on the packaging of food products by their category (antioxidant, preservative, colour, etc.) with either their name or E number. In the United States, food additives are regulated by the Food and Drug Administration. Two sections of the regulations govern their use: substances affirmed as GRAS, that is, Generally Recognized as Safe, (21CFR184) and Direct Food Additives (21CFR172). Substances that have been affirmed as GRAS usually have less stringent regulations attached to their use. However, Food and Drug Administration Standards of Identity may preclude their use in certain standardized foods. In comparison, Direct Food Additives may be allowed only in certain specific foods at low maximum allowable levels. The method of manufacture and analytical constants may also be defined.

The European Community (EC) regulates food emulsifiers in an analogous fashion to United States regulations, identifying the additives with E numbers. Specific regulations, however, must be consulted before food products are designed for international markets. For example, polyglycerol esters up to a degree of polymerization of 10 are widely accepted in the United States. For the EC, this value may not exceed 4. Standards of Identity may also differ significantly. Other countries, which have not formed trading communities, may have regulations, which are unique [5–7].

The food additive functional classes are based on the Codex Class Names and the International Numbering System (INS) for Food Additives (CAC/ GL 36-1989). There are four general categories of food additives: nutritional additives, processing agents, preservatives, and sensory agents. These are not strict classifications, as many additives fall into more than one category [8].(1) Nutritional additives. They are utilized for the purpose of restoring nutrients lost or degraded during production, fortifying or enriching certain foods in order to correct dietary deficiencies or adding nutrients to food substitutes.(2)Processing agents. These additives are added to foods in order to aid in processing or to maintain the desired consistency of the product.(3)Preservatives. They are classified into two main groups: antioxidants (compounds that delay or prevent the deterioration of foods by oxidative mechanisms) and antimicrobials (inhibit the growth of spoilage and pathogenic microorganisms in food).(4)Sensory agents. Such as colorants (natural or synthetics), flavorings, and sweeteners.

EMULSIFIERS

Emulsifiers are essential components of many industrial food recipes (see Table 1), whether they are added for the purpose of water/oil (W/O) emulsification in its simplest form, for textural and organoleptic modification, for shelf life enhancement, or as complexing or stabilizing agents for other components such as starch or protein [9–13].

Table 1: List of foodstuff that contain emulsifiers (http://www.understandingfoodadditives.org/).

Foods that commonly contain emulsifiers		
Biscuits	Toffees	Bread
Extruded snacks	Chewing gum	Margarine/low-fat spreads
Breakfast cereals	Frozen desserts	Coffee whiteners
Cakes	Ice cream	Topping powders
Desserts/mousses	Dried potato	Peanut butter
Soft drinks	Chocolate coatings	Caramels

The main applications of the emulsifiers are based on their ability to interact at the interface between phases. Multiphase systems consist of two or more distinctive phases; systems typical encountered in foods are water-in-oil (W/O), oil-in-water (O/W), solid-in-oil, gas-in-liquid, gas-, solid- or oil-in-water, and so on. These systems are often unstable due to the immiscibility and thus repulsion between the phases. Emulsifiers consist of molecules with ambiphilic properties; that is, part of the structure is of hydrophilic and another of lipophilic nature. In the multiphase systems these emulsifiers will orientate themselves in the position which is favorable with respect to energy, and the emulsifiers will reduce the interfacial tension between the phases in the multiphase system, typically an O/W or W/O system (Figure 1).

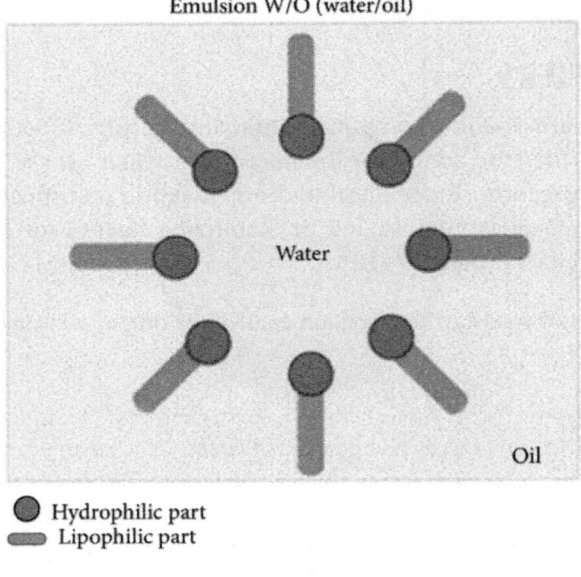

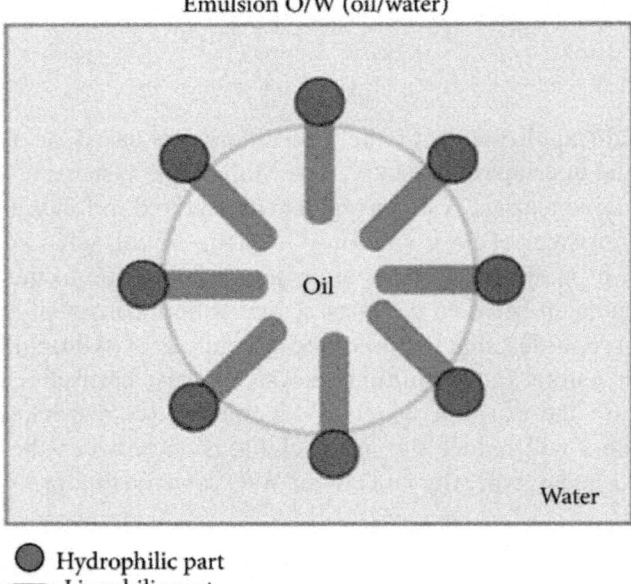

Figure 1: Types of emulsion. http://www.detergentsandsoaps.com/emulsifiers.html.

The lipophilic moiety of the emulsifier often consists of hydrocarbon chains of fatty acids, and the hydrophilic part originates from more polar molecules such as glycerol, lactic acid, citric acid, and polyglycerol. The types and sizes of the lipophilic and hydrophilic moieties determine the functional behavior in multiphase systems. An estimate for this relationship is the hydrophilic to lipophilic balance (HLB value), which can be calculated or determined experimentally. A high HLB value (the hydrophilic moiety dominates the emulsifier) typically stabilizes oil-in-water emulsions, whereas low HLB values will lead to water-in-oil type emulsion [14, 15].

Emulsifiers include compounds with a completely different chemical structure, and therefore with diverse mechanisms of action, and in turn different effects in dough and bread [16]. Emulsifiers are therefore classified either as ionic or nonionic. The potential for ionization is based on the electrochemical charge of the emulsifiers in aqueous systems. Nonionic emulsifiers (monoglycerides, distilled monoglycerides, epoxylated monoglycerides, and sucrose esters of fatty acids) do not dissociate in water due to their covalent bonds. Ionic emulsifiers may be anionic (diacetyl tartaric acid esters of monoglycerides, sodium stearoyl-2- lactylate) or cationic, but cationic emulsifiers are not used in foods. Amphoteric emulsifiers (lecithin) contain both anionic and cationic groups, and their surface-active properties are pH dependent [17, 18].

POLYGLYCEROL ESTERS OF FATTY ACIDS

Polyglycerol esters are nonionic emulsifiers that are allowed for food use in many countries. In addition to the stabilization of emulsions, foams, and dispersions, polyglycerol esters can act as aerating agents, dough strengtheners, rheology modifiers, crystal modifiers, antispattering agents, beverage clouding agents, humectants, solubilizers, or fat substitutes [19]. They have been used as food additives for many years in Europe and America since the 1940s and were approved for food use in the USA, in the 1960s. Due to the innocuous nature of these products, formulators continue to develop technologies that incorporate polyglycerol esters in a widening array of applications. [20–23]. From the legal point of view, food grade polyglycerol esters are divided in two classes: polyglycerol esters of edible fatty acids (E number: E475, also known as "PGFA") and polyglycerol polyricinoleate (E number: E476, also known as "PGPR").

The HLB balance of PGFA depends on the length of the polyglycerol chain (the number of hydrophilic hydroxyl groups present) and the degree of esterification. For example, decaglycerol monostearate has an HLB of 14.5, while triglycerol tristearate has an HLB of 3.6. Intermediate species have intermediate HLB values, and any desired value may be obtained by appropriate blending [15]. Depending upon their HLB, PGFA can act as water-in-oil (W/O) or oil-in-water (O/W) emulsifiers [23].

PGFAs allow a strong interfacial tension reduction between water and a wide variety of oils. In many systems, they have an even better surface activity than glycerol or homologous polyol esters [24, 25]. These esters also form highly stable α-gels in water, while gels produced from glycerol monostearate transform with time into aβ-crystal structure called coagel [26, 27]. The high stability of α-gels is a key advantage of PGFA in food applications. Indeed, it leads to better emulsification properties and a higher viscosity of the external water phase, resulting in enhanced stabilization of O/W emulsions and foams.

The polyglycerol portion can be prepared by three routes [15, 28]. (1)By polymerization of glycerin using strong base as a catalyst. This route leads to several different outcomes.(a)The polymers of interest, namely, polyglycerin, but with a broad distribution of homologues and significant levels of free glycerin.(b)The production of cyclic compounds similar to dioxane, which undergo further polymerization.(c)Dehydration by-products, namely, acrolein, which must be removed by distillation.(2)By polymerization of glycidol, which leads to linear polyglycerin.(3)By polymerization of epichlorohydrin, followed by hydrolysis, which also leads to linear polyglycerin. The by-product, sodium chloride, is removed by filtration.

Of these three routes used to produce polyglycerol, the polymerization of epichlorohydrin is the preferred method of manufacture, because polyglycerols obtained by the classical procedures may widely differ in composition. This process is designed to produce high quality products with reduced amounts of cyclic components which have reduced functionality (less hydroxyl groups available). Moreover, this process also allows an excellent control of the reaction, resulting in a narrow oligomer distribution and a minimum batch-to-batch variability [29]. Polymers of epichlorohydrin (or glycidol) are prepared by methods similar to those of other oxirane monomers. The process used is a catalytic ring opening of the oxirane group to form the polymer through an ether linkage between the monomers. Epichlorohydrin is expected to be completely consumed in this polymerization reaction. The final step in any oxirane polymerization is a steam distillation or deodorization step. This deodorization step employs the injection of steam into the reactor at temperatures between 150°C and 200°C with strong vacuum. Under these conditions, any residual monomer (epichlorohydrin or glycidol) is typically reduced to levels below 1 ppm. Specifically for the case of epichlorohydrin polymers, the material must then be hydrolyzed to produce polyglycerol. This is accomplished by base catalyzed reaction with water and excess base, either sodium or potassium hydroxide. If any residual monomer escapes the deodorization step in the polymerization process, it will subsequently be hydrolyzed to glycerin in the hydrolysis step. This is followed by drying, 100°C under vacuum, and then filtration of the sodium chloride produced [28, 29].

The polyglycerol backbone is then esterified to varying extents, either by direct reaction with a fatty acid or by interesterification with a triglyceride fat. Again, the number of acid groups esterified to a polyglycerol molecule varies around some central value, so an octaglycerol octaoleate really should be understood as an (approximately octa)-glycerol (approximately octa)-oleate ester. By good control of feedstocks and reaction conditions, manufacturers manage to keep the properties of their various products relatively constant from batch to batch [15].

PGFAs are preferred emulsifiers in baked products due to a variety of properties including enhanced volume and shelf life of yeast dough. They can also act as whipping emulsifiers with improved texture, water binding, and taste in fine baked goods [30]. The most popular food grade PGFAs are polyglycerol monostearates. When mixed with water, these emulsifiers form highly stable α-gels exhibit α-tending properties. These esters are of special interest, for example in sponge cake technology, where they lead to an optimal stabilization of the batter and a more uniform foam structure (Figure 2). α-Gels also improve batter rheology and crumb structure (more regular and softer) and enhance starch complexation leading to increased cake shelf life. In addition, PGFAs exhibit higher heat stability than the equivalent glycerides [31, 32].

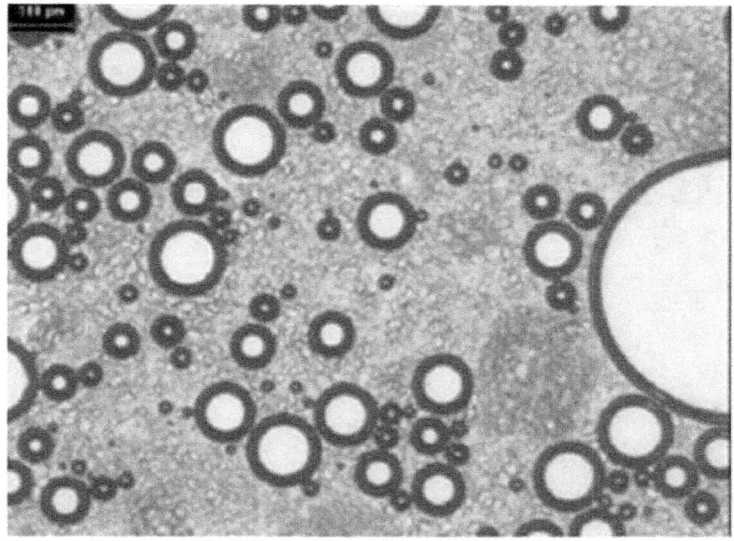

(a)

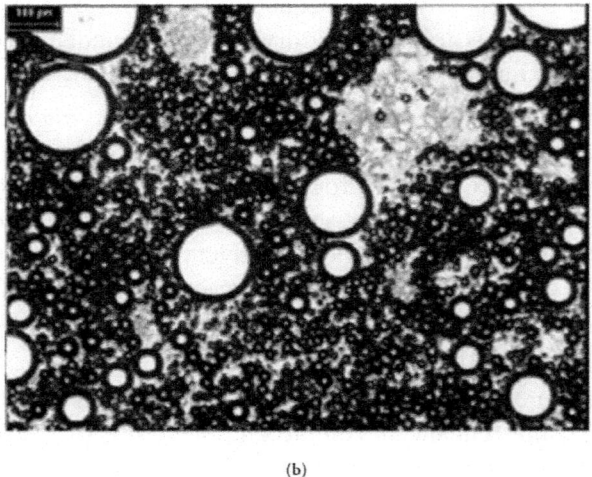

(b)

Figure 2: Cake batter with no emulsifier (a) and with 3 g/kg polyglycerol monostearate (b) [23].

POLYGLYCEROL POLYRICINOLEATE

Polyglycerol polyricinoleate, PGPR, identified with the code E-476, is manufactured from polymerized glycerol (above described) and polymerized ricinoleic acid. A typical structure of PGPR is shown in Figure 3.

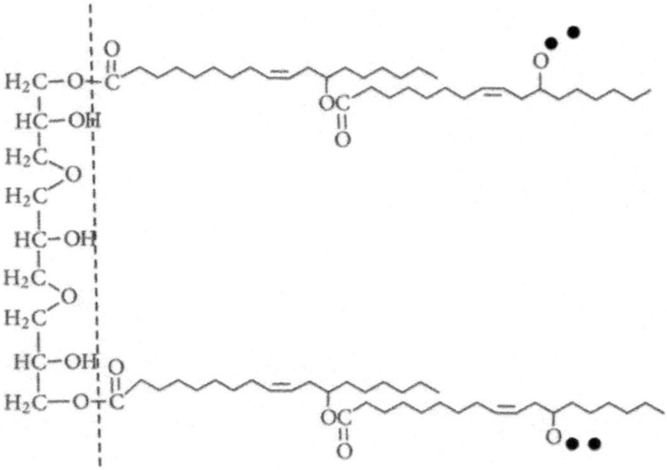

Figure 3: Chemical structure of PGPR. Black dots denote polyricinoleic acid chains.

Ricinoleic acid (12-hydroxy-9-cis-octadecenoic acid) is an unsaturated omega-9 fatty acid (see Figure 4) that naturally occurs in mature castor plant (Ricinus communis L., Euphorbiaceae as shown in Figure 5) seeds.

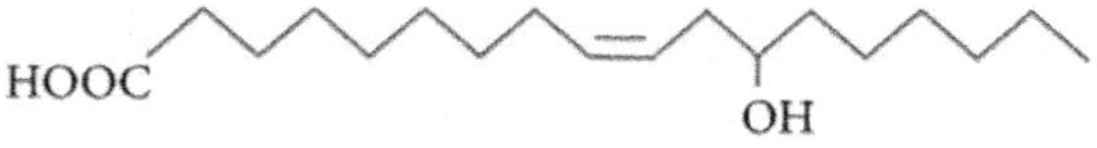

Figure 4: Chemical structure of ricinoleic acid.

(a)   (b)

Figure 5: Photographs of the castor bean plant, Ricinus communis.

Polymerized ricinoleic acid, also called ricinoleic acid estolides, is obtained by different methods which are discussed below. The chemical structure is shown in Figure 6.

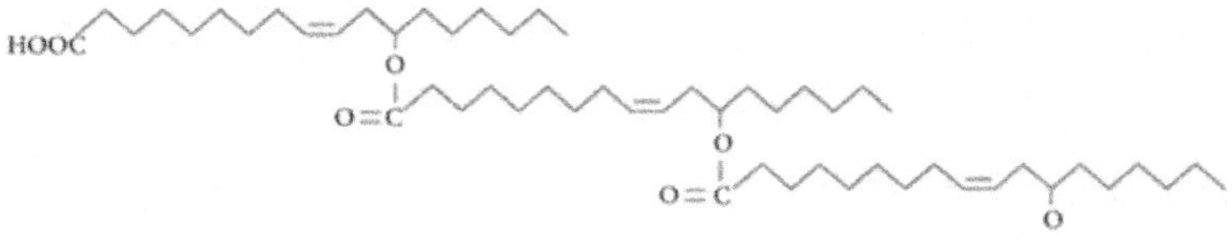

Figure 6: Chemical structure of polyricinoleic acid.

PGPR is widely known as an excellent water-in-oil emulsifier in the food industry, because it forms very stable emulsions even when the water content is very high, such as 80% [33–39]. Therefore, PGPR is used as emulsifier in tin-greasing emulsions for the baking trade [40]. Another important application of PGPR is its use as a water-in-oil emulsifier for the production of low-fat spreads. In this case, PGPR can be used alone or blended with monoglycerides to obtain an optimal quality/cost ratio (see Figure 7) [41, 42].

(a)

(b)

Figure 7: Example of a low-fat spread based on 20% fat and containing milk powder, gelatin, and sodium alginate. (a) 1% distilled monoglyceride (left) and 0.6% distilled monoglyceride + 0.4% PGPR (right). (b) 0.6% distilled monoglyceride (left) and 0.4% distilled monoglyceride + 0.2% PGPR (right) [23].

However, the main application of PGPR is in the chocolate industry. It is especially important for chocolate coatings to flow properly during the enrobing process. Cocoa butter is an expensive raw material, and chocolate manufactures prepare low-fat products to use them only for these applications. Adequate flow properties can be achieved by the addition of PGPR (see Figure 8), which improves the flow characteristics of molten chocolate by reducing the "Casson yield value" (which represents the viscosity of chocolate at low shear rate). By contrast, lecithin tends to be used to decrease the "plastic viscosity" (viscosity at high shear rate) of chocolate [23].

(a)

(b)

Figure 8: Influence of PGPR addition (0.2%) to a molten chocolate (27% cocoa butter) (http://www.sweetcom.de/cgi-bin/inhalt.pl?heft=01_05&nr=45).

Since lecithin and PGPR have complementary rheological properties, they are often used in combination for an optimal control of chocolate rheology [23, 43]. This allows a more even coating of confectionary pieces while reducing the consumption of expensive cocoa butter in the recipe. Lowering yield value also improves the release of entrapped air in chocolate, leading to a smoother and more efficient molding and depositing. This is achieved without compromising quality and taste and with cost savings.

An additional property of PGPR in chocolate is its ability to limit fat bloom (see Figure 9). Fat bloom is not always caused by a simple set of circumstances, such as the chocolate becoming wet. Fat bloom is more complicated, and oftentimes it may be more difficult to discover the actual source of the problem. Fat bloom typically appears as lighter color spots on the chocolate. As the name implies, the bloom is composed of fat, in this case the naturally occurring fat that comes from the cacao bean-cocoa butter.

(a)

(b)

Figure 9: Effect of fat bloom in chocolate (http://www.chocolatisimo.es/fat-bloom-tu-chocolate-se-vuelve-blanquecino/).

When discussing the reasons for fat bloom, it is important to note that when cocoa butter hardens, it forms crystals. Some of the crystals are stable, but other crystals are not and will actually change form over time. During chocolate manufacturing, a process called tempering is used to ensure that only stable crystals form, while the chocolate hardens. Fat bloom is caused by the interaction of the various types of crystals or the tempering process (or lack thereof).

If chocolate is not tempered, the unstable forms of cocoa butter crystal will form, most notability the β' and αforms. After the cocoa butter hardens, these unstable forms will slowly change their forms to the stable βform. The β crystals are slightly smaller than the β' or α forms, so that when this transition occurs, the chocolate contracts. The new stable β crystals then form, projecting above the surface of the chocolate, visible as bloom. If the chocolate is stored in a room where the temperature fluctuates near the melting temperature of the stable β crystals, two additional types of fat bloom may form. In the first, some of the β crystals melt. When they recrystallize, they recrystallize slowly, since the ambient temperature is close to that of the chocolate. This allows the crystals to grow much larger than the original small, compact crystals. In addition to projecting above the surface of the chocolate, these larger crystals may displace cocoa butter, forcing it to the surface.

The second type of bloom is created when the crystals have softened instead of melted. It is during this period that cocoa butter that has slightly melted migrates toward the surface. When it breaks the surface, it pools ever so slightly, and when it cools the cocoa butter appears as spots.

Many people are surprised to learn that fat bloom also occurs in cocoa powder. Cocoa powder contains between 12–20% cocoa butter. Since some cocoa butter is present, it must be tempered during manufacturing, just as chocolate is. Cocoa powder that has been improperly tempered or undergone temperature fluctuations may cause bleaching of the cocoa powder and may cause clumping, as the cocoa butter helps the particles of the cocoa powder adhere to each other. As with chocolate, when bloom occurs it does not affect the edibility of the cocoa powder but may have an aesthetic impact.

Studies on fat bloom indicate that the bloom consists of large, single cocoa butter crystals or collections of crystals of the stable β form of cocoa butter. Other forms of cocoa butter crystals are not present in fat bloom.

Fat blooming actually occurs in a third process. This case affects not so much the chocolate industry directly but the ancillary confectionary industry. When chocolate is used to coat nuts or fillings that contain oils or fats (such as nut butters) that are incompatible with chocolate, the oils may actually seep into or through the chocolate over time. This is called fat migration. As the oils displace the cocoa butter, cocoa butter may sweep onto the surface of the piece of confectionary and recrystallize as bloom. When this occurs, the manufacturing process needs to be examined, or the confectionary needs to be reformulated [44–47].

PGPR is also claimed to increase chocolate's tolerance to the thickening effect caused by small quantities of water, sometimes introduced during enrobing operations [23].

The safety of PGPR consumption has been widely studied. PGPR has been used continuously in greasing emulsions since 1952, following short-term rat feeding trials undertaken in 1951. It was also first used in chocolate couverture in the UK in 1952. During 1952 and 1953, 4.5 tons were used in 1500 tons of chocolate couverture. However, new requirements for biological testing led to the withdrawal of the product for this purpose, and it was not used again in chocolate production until 1958 onwards [38].

The safety testing programme on PGPR in the 1950s included acute toxicity studies in several species, 30 and 45 wk rat feeding trials, a rat reproduction study over three generations, and a number of indirect metabolism studies to show that PGPR is digested and utilized like a normal dietary fat. In the 1960s, the programme was extended to include 2 yr rat and 80 wk mouse feeding studies, a 90-day feeding study in a nonrodent (chicken), studies of PGPR-induced liver and kidney enlargement in rats, mice and chickens, metabolism in the rat using radiolabelled materials, and human studies on the digestibility and absorption of PGPR which included liver and kidney function tests. The results of these studies have been published in several papers [40, 48–52]. PGPR was found to be 98% digested by rats and utilized as a source of energy superior to starch and nearly equivalent to groundnut oil. There was no interference with normal fat metabolism in rats or in the utilization of fat-soluble vitamins. Despite the intimate relationship with fat metabolism, no evidence was found of any adverse effects on such vital processes as growth, reproduction, and maintenance of tissue homeostasis. PGPR was not carcinogenic in either 2-year rat or 80-week mouse feeding studies.

In 1969, the 13th report of the Joint FAO/WHO Expert Committee on Food Additives (JECFA) set a temporary human acceptable daily intake (ADI) for PGPR of 3.75 mg/kg body weight with a request for more biological studies. This ADI, without the temporary prefix, was raised to 7.5 mg/kg body weight in the 17th report of JECFA in 1974 [53]. In 1979, the Scientific Committee for Food (SCF) of the European Community also set an ADI for PGPR of 7.5 mg/kg body weight [54].

On the other hand, the maximum PGPR levels of use in foodstuffs ready for consumption in Europe are listed in Table 2 [55]. The European Commission established in 2008 the purity criteria for PGPR. These are listed in Table 3 [56].

Table 2: Maximum levels of PGPR in foodstuff. European Directive 95/2/EC.

E number	Name	Foodstuff	Maximum level
E-476	Polyglycerol polyricinoleate	Spreadable fats as defined in Annexes A, B, and C of Regulation (EC) Number 2991/94 having a fat content of 41% or less	4 g/kg
		Similar spreadable products with a fat content of less than 10% fat	4 g/kg
		Dressings	4 g/kg
		Cocoa-based confectionery, including chocolate	5 g/kg

Table 3: Purity criteria on food additives other than colours and sweeteners. European Directive 2008/84/EC.

	E 476 polyglycerol polyricinoleate
Synonyms	Glycerol esters of condensed castor oil fatty acids Polyglycerol esters of polycondensed fatty acids from castor oil Polyglycerol esters of interesterified ricinoleic acid PGPR
Definition	Polyglycerol polyricinoleate is prepared by the esterification of polyglycerol with condensed castor oil fatty acids
Description	Clear, highly viscous liquid
Identification	
(A) Solubility	Insoluble in water and in ethanol Soluble in ether, hydrocarbons, and halogenated hydrocarbons
(B) Positive tests for glycerol, polyglycerol, and for ricinoleic acid	
(C) Refractive index $[n]^{65}$	Between 1.4630 and 1.4665
Purity	
Polyglycerols	The polyglycerol moiety shall be composed of not less than 75% of di-, tri-, and tetraglycerols and shall contain not more than 10% of polyglycerols equal to or higher than heptaglycerol
Hydroxyl value	Not less than 80 and not more than 100
Acid value	Not more than 6
Arsenic	Not more than 3 mg/kg
Lead	Not more than 5 mg/kg
Mercury	Not more than 1 mg/kg
Cadmium	Not more than 1 mg/kg
Heavy metals (as Pb)	Not more than 10 mg/kg

PRODUCTION METHODS OF POLYGLYCEROL POLYRICINOLEATE

Chemical Procedures

It is well known that chemical production of PGPR [40] is carried out in four stages:(1)preparation of the castor oil fatty acids,(2)condensation of the castor oil fatty acids,(3)preparation of polyglycerol, and(4)partial esterification of the condensed castor oil fatty acids with polyglycerol.

(1) Preparation of the Fatty Acids

The castor oil fatty acids are prepared by hydrolyzing castor oil with water and steam at elevated pressure without any added catalyst after which the resulting fatty acids are freed from glycerol by water washing. Castor oil contains, as its main fatty acid component, ricinoleic acid (80–90%), and it is this fatty acid which is important in the condensation reaction. Other present fatty acids are oleic acid (3–8%), linoleic acid (3–7%), and stearic acid (0–2%).

(2) Condensation of the Ricinoleic Acid to Produce Polyricinoleic Acid, Also Called Ricinoleic Acid Estolide

Fatty acid condensation is brought about by heating the castor oil fatty acids at elevated temperatures under vacuum and in an inert atmosphere to prevent oxidation. Samples are taken at regular intervals and tested for their free fatty acid content until an acid value of 35–40 mg KOH/g is achieved. This acid value is equivalent to an average of about four-five fatty acid residues per molecule of the condensed product.

The operation conditions can be found in several papers. So, reaction temperatures between 157°C and 230°C [57, 58] or between 205°C and 210°C [49] have been described. Other authors suggest not a range but a single temperature: 220°C [57] or 200°C [57]. Most researchers affirm that reduced pressure facilitates the extent of the reaction: 300–400 mmHg [58], 45 mmHg [57], or 20 mmHg [49]. Either carbon dioxide [40, 59] or nitrogen [51, 57, 58] can be used to ensure an inert atmosphere.

During the condensation phase, ricinoleic acid may react in a number of ways as shown in Figure 10. Simple linear esterification is the desired reaction, but cyclic esterification, which is a chain terminating process, is theoretically possible. However, no evidence was found for the presence of this type of cyclic material in the polyricinoleic acid. Dehydration is also possible but occurs to only a small extent.

(a)

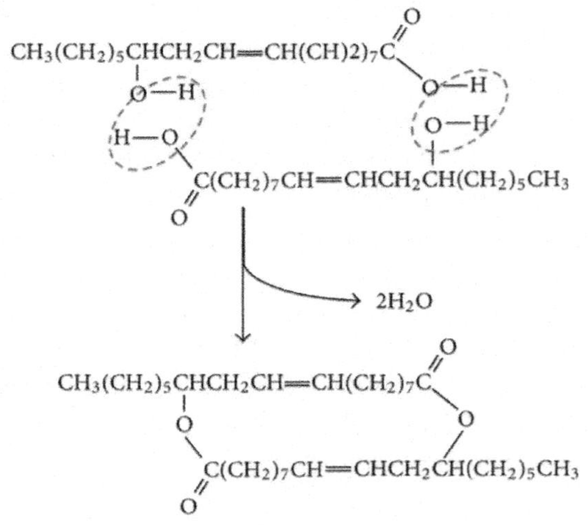

$CH_3(CH_2)_5CHCH_2CH=CH(CH)2)_7C$

(image of linear and cyclic esterification and dehydration reactions)

2H$_2$O

(b)

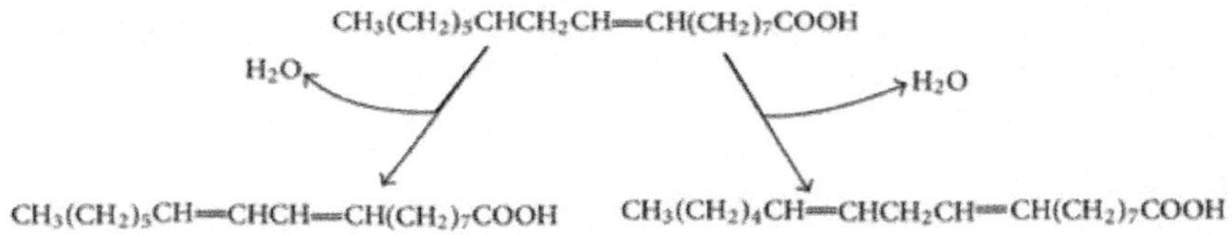

(c)

Figure 10: Possible reactions of ricinoleic acid during condensation. (a) Linear esterification, (b) cyclic esterification, and (c) dehydration (adapted from [40]).

(3) Preparation of Polyglycerol

The preparation of the polyglycerol is achieved by using one of the methods described in Section 3.

(4) Esterification of Polyricinoleic Acid with Polyglycerol

The final stage of the preparation involves heating an appropriate amount of polyglycerol with the polyricinoleic acid. The reaction takes place immediately following the preparation of the latter and in the same vessel, while the charge is still hot. The esterification conditions are the same as those for fatty acid condensation. The process is continued until a sample withdrawn from the reaction mixture is found to have a suitable acid value (acid value $\leq$ 6 mg KOH/g in Europe) and refractive index. The major components have the general structure shown in Figure 11(a), where the average value of n is about 3. , R_1 R_2, and R_3 each may be hydrogen or a linear condensation product of ricinoleic acid with itself, as in Figure 11(b) with n being on average between 5 and 8.

Some of the specifications listed in Table 3 with the exception of refractive index are weight average analysis and do not indicate specific structural characteristics of PGPR. Taken on the whole, however, these weight average-related specifications do indicate correct chemical structure, but with limited accuracy. Refractive index, however, is directly indicative of final chemical structure; for example, if the oligomer distribution of PGPR is not correct or distributed differently on the polyglycerol backbone, the refractive index measurement will not comply with the specifications as described in Table 3

One problem with this conventional process is that the step of polymerization of ricinoleic acid is complicated by the fact that there is a requirement to follow the refractive index (see Table 3) of the mixture, while polymerization of ricinoleic acid and esterificaction of polyglycerol and polyricinoleic acid is under way and to stop the reaction when the key value is indicated by the analysis. The refractive index test is not easily established in a manufacturing facility because the instrument is delicate, requires precise calibration, and requires a circulating temperature bath set to a particular temperature (65°C). It is therefore difficult to run the refractive index test at the kettle, and this often requires that the instrument is used in a laboratory setting, usually in a quality control laboratory, which is both time consuming and inconvenient. Moreover, the refractive index requires a high degree of technical training and precision.

The four-step conventional process also reduces efficiency of production and adds cost to the product. Therefore, recently, a more economical and simplified chemical method for manufacturing PGPR was developed [60]. Further, in the conventional process, there is a tendency to produce compositions of darker color. This is most likely due to the added processing steps of preparing two separate ingredients, each having its own cycle of heating and cooling, along with the additional handling associated with the manufacture of each ingredient. Figure 12 shows different PGPRs with different colors.

The new process proposed in this patent produce noticeably lower color end product, such a clear yellow liquid rather than amber liquid. In this procedure [60] polyglycerol is mixed with ricinoleic, acid and they are allowed to react at 200°C, with stirring and N_2sweep. When the acid value of the reaction mixture falls below 6 mg KOH/g, it was found that such a PGPR product also meets the other EC specifications, including the refractive index.

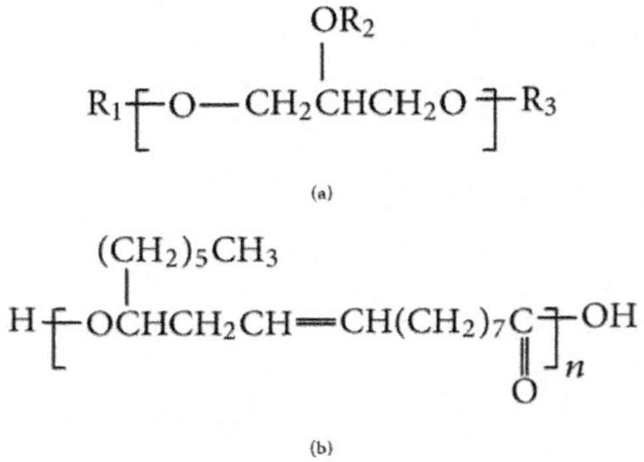

$$\text{R}_1 \underset{\Big[}{\Big[} \text{O} - \text{CH}_2\overset{\overset{\displaystyle \text{OR}_2}{|}}{\text{CHCH}_2}\text{O} \underset{\Big]}{\Big]} \text{R}_3$$

(a)

$$\text{H} \underset{\Big[}{\Big[} \text{OCHCH}_2\overset{\overset{\displaystyle (\text{CH}_2)_5\text{CH}_3}{|}}{\text{CH}}=\text{CH}(\text{CH}_2)_7\underset{\overset{\displaystyle \|}{\text{O}}}{\text{C}} \underset{\Big]_n}{\Big]} \text{OH}$$

(b)

Figure 11: Major components of the polyglycerol polyricinoleate (adapted from [40]).

Figure 12: Different samples of PGPR (http://novicell.ipapercms.dk/PalsgaardAS/Brochurer/Chokolade/StandardPGPRforchocolateandcompoundsPalsgaardPG-PR4120/).

Enzymatic Synthesis

The chemical procedures above describe have the disadvantage of requiring very long reaction times, involving high energy costs. This fact, together with the high operating temperature, can adversely affect the quality of the final product because of problems related with coloration and odours, making it unsuitable for the food industry [57]. As an alternative, a biocatalytic synthesis of PGPR using enzymes has been recently proposed by the author research group [61–66]. Lipases are able to act in mild reaction conditions and produce a final product more suitable for use as a food additive.

Lipases (E.C. 3.1.1.3) are enzymes that are primarily responsible for the hydrolysis of acylglycerides. However, a number of other low- and high-molecular weight esters, thiol esters, amides, polyol/polyacid esters, and so forth are accepted as substrates by this unique group of enzymes. The wide berth for employment in a variety of reactions, endowed by this broad substrate specificity, is further enlarged by the fact that lipases are capable of catalyzing the reverse reaction of synthesis just as efficiently. In fact, some lipases are better suited for synthesis than for hydrolysis applications [67].

The two main categories in which lipase-catalyzed reactions may be classified are as follows [67].

(1)Hydrolysis:

$$RCOOR' + H_2O \longleftrightarrow RCOOH + R'OH. \tag{1}$$

(2)Synthesis: reactions under this category can be further separated as follows.

(a)Esterification:

$$RCOOH + R'OH \longleftrightarrow RCOOR' + H_2O. \tag{2}$$

(b)Interesterification:

$$RCOOR' + R''COOR^* \longleftrightarrow RCOOR^* + R''COOR'. \tag{3}$$

(c)Alcoholysis:

$$RCOOR' + R''OH \longleftrightarrow RCOOR'' + R'OH. \tag{4}$$

(d)Acidolysis:

$$RCOOR' + R''COOH \longleftrightarrow R''COOR' + RCOOH. \tag{5}$$

The last three reactions are often grouped together into a single term: transesterification.

The ability of lipases to catalyze the reaction of synthesis is used in the manufacture of several products: pharmaceuticals, cosmetics, leather, detergents, foods, perfumery, medical diagnostics, and other organic synthetic materials [68]. On the other hand, lipases have been employed to catalyze reactions involving hydroxyl fatty acids (like ricinoleic acid) to narrowly shape the product distribution via their region- and stereoselectivities [69]. Esterification mixtures generally contain only the substrates and enzyme, and water is the only by-product of the reaction [67]. Many reported lipase-catalyzed syntheses are carried out in organic solvents. However, residues of organic solvents in products are undesired, and many solvents which could be used are even toxic and are not allowed for processing procedures to make products for food applications. As well as that, removal of organic-solvent traces in products requires extra expense and increases manufacturing costs. Solvent-free processes are thus desired [70–73] due to their advantages [72] and because they fulfill the twelve principles of the Green Chemistry, as defined by Anastas and Warner [74].

The potential of this relatively easy to perform bioconversion for industrial purposes seems to be enormous, but there are only few examples of successful production processes in practice. There are some important reasons for this. The use of dried enzyme powders, although often reported in laboratory scale experiments, is generally unsuitable for large scale processing in nonaqueous media. Development of the enzymes into active and stable biocatalysts, usually by appropriate immobilization techniques on support materials suitable for large scale production processes on a multitons basis, is very important. At the moment there are only a few off-the-shelf purpose made biocatalysts commercially available which are suitable for industrial production [73].

Taking into account all these considerations, we developed a novel method for PGPR production, using immobilized lipase and in a solvent-free system [61–66]. This process is environmentally friendly and avoids side reaction, so that the product has a higher purity and quality. The enzymatic procedure consists of two steps (similar to chemical procedure). First, the ricinoleic acid is polymerized to obtain the polyricinoleic acid, PR, also known as ricinoleic acid estolide [62–64]. Then, it is esterified with polyglycerol to obtain polyglycerol polyricinoleate, PGPR [61, 65, 66]. Figure 13 shows the reactions involved in the biosynthesis.

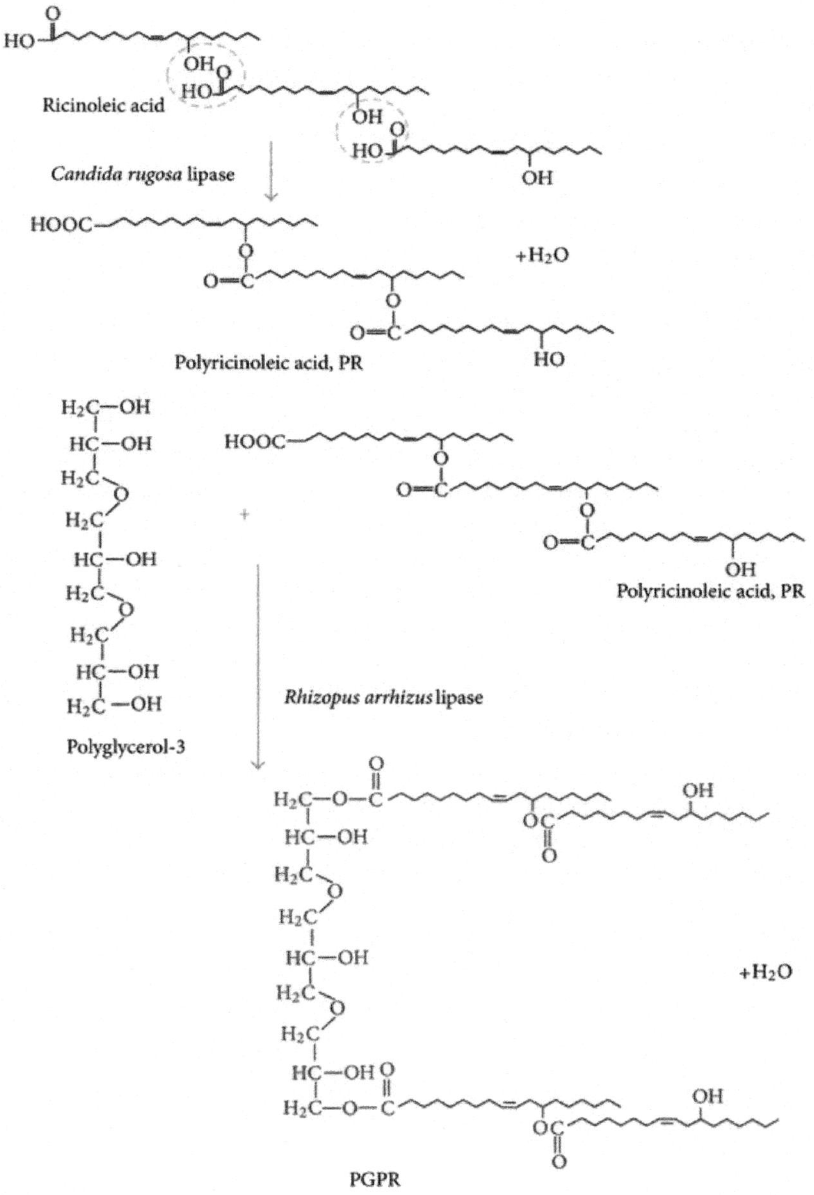

Figure 13: Biocatalytic synthesis of PGPR. First reaction catalyzed by Candida rugosalipase and second reaction catalyzed by Rhizopus arrhizus lipase.

Polymerization of Ricinoleic Acid

The studies about the first reaction step are based on previous works [69, 75–80] and a result of this conscientious bibliographical search; Candida rugosa lipase has been selected to catalyze the autocondensation reaction of ricinoleic acid to obtain the polyricinoleic acid.

Immobilization of Candida rugosa LipaseFirst of all, efforts have been devoted to obtaining an immobilized derivative with a high immobilized protein percentage and enzymatic activity for the present application [63]. It has been described that adsorbed lipase on a ceramic carrier SM-10 [79] is suitable for producing ricinoleic acid estolide. However, the difficulty of acquiring the support led the authors to test different immobilization matrices in an attempt to obtain an immobilized derivative, which could be successfully used to catalyze the production of ricinoleic acid estolide [63]. Eight inorganic supports (two types of BioLite, Celite R-643, Chromosorb W, nonporous glass beads of two particle sizes, and porous glass beads of different pore sizes) and two organic carriers (cationic and anionic exchange resins, Dowex 50×8 and Lewatit MonoPlus MP 64, resp.) have been used. Twelve different immobilized derivatives have been obtained, six of them by physical adsorption and the other six by covalent coupling via the amino groups of the enzyme. Immobilization on glutaraldehyde-activated aminopropyl glass beads was selected because it has been widely used by the authors with different enzymes [81, 82] and has been shown to be very versatile.

The results obtained are shown in Table 4 where percentages of immobilized protein and protein contents are summarized. It is important to note that these values are based on the protein content provided by Lowry's method [83], which showed that the commercial lipase contained only 15% protein.

Table 4: Coupling parameters for the immobilization of Candida rugosa lipase in different supports [63]

Support	Immobilization method	Immobilized protein (%)	Enzyme loading (mg E/g support)
Biolita L2.7	Adsorption	—	—
	Covalent binding	0.64	0.48
Biolita P3.5		—	—
Celite R-643	Adsorption	27.60	4.14
Chromosorb W		—	—
Nonporous glass 425–600 μm		0.89	0.67
Nonporous glass 91–107 μm		—	—
PG 75-400	Covalent binding	41.73	31.30
PG 700-400		28.65	21.49
PG 1000-400		22.83	17.12
Cationic exchange resin Dowex 50 × 8		—	—
Anionic exchange resin			
Lewatit MonoPlus	Adsorption	47.20	7.08
MP 64			

The best results have been obtained when porous glass was used as immobilization matrix and covalent binding as coupling method. In these cases enzyme loading increased, as the pore size became smaller because of the greater internal surface available for immobilization. The percentage of immobilized lipase obtained by physical adsorption on Lewatit MonoPlus MP 64 was higher than those obtained with porous glass because five times less enzyme was offered for immobilization, so that enzyme loading (mg E/g support) was noticeably lower. Celite R-643 was also shown to be suitable for Candida rugosa lipase immobilization.

The five above mentioned immobilized derivatives were used to catalyze the polymerization reaction of ricinoleic acid. The extent of the reaction was monitored by acid value measures [84]. Figure 14 shows the results obtained.

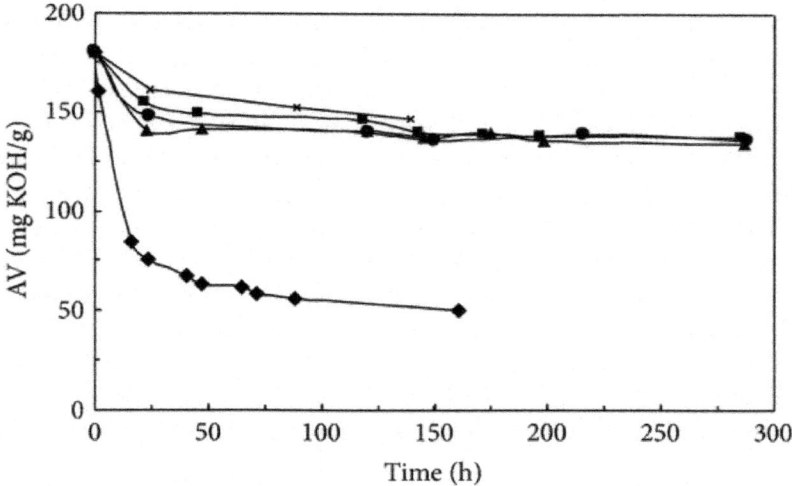

Figure 14: Change in acid value as a function of time for estolide synthesis catalysed by five different immobilized derivatives [63]. (▲) Celite R-643, (■) PG 75-400, and (×) PG 700-400, (•) PG 1000-400, (◆) Lewatit MonoPlus MP 64.

It can be seen that there was a large difference between the activity of the derivative obtained on the anion exchange resin and the activity of other derivatives. In the case of the anion exchange resin, the acid value dropped from 180 to 50 in 150 h, while the best result of the other derivatives was a fall to 136 in 285 h (porous glass 75–400). Therefore, the immobilized lipase obtained by physical adsorption onto Lewatit MonoPlus MP 64 was chosen for PR production. Moreover, other studies showed that the activation of the support with soybean lecithin has a beneficial effect on the enzymatic activity and that the optimum pH for enzyme immobilization is 7 [63].

Optimization of the Reaction Conditions

The optimization of some reaction conditions is especially important in an experimental system like the described one. It is known that temperature is a crucial parameter in every enzyme catalyzed reactor, but in this case, due to the special characteristics of the reaction medium (solvent free), temperature greatly influences viscosity, mass transport phenomena, and, as a consequence, the esterification rate [85]. While high temperature favors the medium fluidity, enzyme has to be prevented from thermal deactivation [62–64,85, 86]. The optimum temperature for polyricinoleic acid synthesis catalyzed with immobilized Candida rugosa lipase was 40°C. Lower reaction rates have been detected below this value, and a slightly unfavourable effect could be observed at high temperature [63].

Another decisive parameter in this process is the water content. Water plays multiple roles in lipase-catalyzed esterifications performed in nonconventional media. It is widely known that water is absolutely necessary for the catalytic function of enzymes because it participates, directly or indirectly, in all noncovalent interactions that maintain the conformation of the catalytic site of enzymes [62–64, 87–89]. However, it has been found that the amount of water necessary for enzyme activity might be very small, and, in the case of lipase, just a few layers around the enzyme surface are needed [90]. On the other hand, in esterification/hydrolysis reactions it is well known that the water content affects the equilibrium conversion of the reactions. Particularly, in the case of estolides production, the water formed by the reaction must be removed from the reaction mixture if polyricinoleic acid with a high degree of condensation is to be obtained [62–64, 79]. In the light of the above considerations, a study on the optimal initial amount of water in the reactor was deemed necessary. With this purpose, the authors carried out four experiments using the immobilized derivative as obtained (soaked), adding different amounts of water and drying the derivative under vacuum at room temperature before use [64]. The time course of these experiments is shown in Figure 15 where the acid value is represented against operation time.

With these experiments it was demonstrated that an optimum in the initial water content exists, although this optimum seems to be quite wide. The same results were obtained when derivative was used as obtained and when small amounts of water were added. However, drying the derivative or adding higher amounts of water led to a lower initial rate (specially the high water content) and a higher final acid value [64].

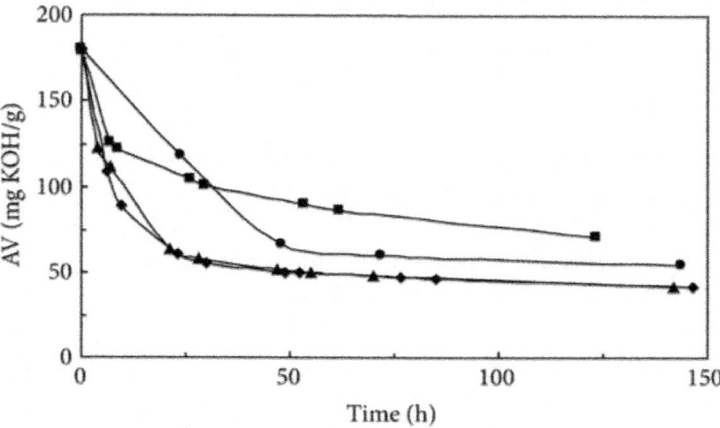

Figure 15: Influence of the initial water content on the evolution of acid value with time for the PR synthesis in the open reactor [64]. (▲) wet resin, (•) dry resin, (■) addition of 0.1 mL water/g immobilized derivative, and (◆) addition of 1 mL water/g immobilized derivative.

Production of PR in a Vacuum ReactorAll the experiments described above were carried out simultaneously, in an air open tank reactor, within a month of each other. However, when the results were tested for reproducibility, great discrepancy between them was observed, where a variation of 30% in the AV value was obtained for experiments carried out in different seasons. Measurements of the water content of the reaction medium revealed that, at 40°C, water continuously evaporated and that after 48 h (approximately) the reactor water content was independent of the initial water content and mainly dependent on environmental relative humidity. The air conditioner/heat pump equipment installed in the laboratory stabilizes the relative humidity at 70% in summer (air conditioner) and at 20% in winter (heat pump). These values correspond with equilibrium water contents of 3600 and 1000 ppm, respectively, which are the cause of the discrepancies in the results obtained in the open-air reactor. Obviously, this poor reproducibility of the results is unacceptable if the process is to be applied on an industrial scale [64]. For this reason, the production of PR should be carried out in a closed system with controlled atmosphere, a suitable level of stirring and low pressure. Moreover, to enhance dehydration, a current of dry nitrogen was passed through the reaction mixture for a period of 7 hours, and after that, maintenance of the vacuum (160 mm Hg). In these conditions, the water content in the reactor can be adequately controlled (3000 ppm), and an estolide with a high degree of polymerisation (AV≈40 mg KOH/g) can be obtained [64].

Esterification of Polyricinoleic Acid with Polyglycerol

Once we are able to produce polyricinoleic acid with the appropriate acid value, it is used as substrate of the second reaction, that is, the esterification of PR with polyglycerol.

Selection of LipasesLipase from Candida rugosa was used to carry out the autocondensation of ricinoleic acid to obtain the estolide, which is the first step in PGPR synthesis. Obviously, it would be very convenient if the same lipase could serve as catalyst for the two reaction steps. However, several studies revealed that Candida rugosa lipase was unsuitable for PGPR synthesis, and therefore others lipases were assayed for this purpose [65]. A further twenty lipases from different sources were used, and the corresponding experiments of PGPR synthesis were performed. Table 5 shows the lipases tested, their specific activities (as declared by the manufacturer), and the amounts of protein used in each experiment. It is important to note that many of the lipases were part of two kits [91] and the amount available was limited. In such cases, the total available protein was added to the reactor. The evolution of the acid value with time for the enzymatic production of PGPR with the above mentioned lipases is plotted in Figures from 16(a) to 16(d). In a first selection, eight lipases were rejected because they were not able to reach acid values lower than 15 mg KOH/g in seven days; they are lipases from Aspergillussp., Candida antarctica, Candida cylindracea, Candida lipolytica, Penicillium roqueforti, porcine pancreas,Rhizopus niveus, and wheat germ [65]. The lipase from wheat germ exhibited a particular behavior. When it was tested to produce PGPR, the acid value of the reaction mixture increased, which indicates that polyricinoleic acid is being hydrolyzed, and therefore, under the experimental conditions, the hydrolytic activity of this lipase is greater than its synthetic activity. None of the twelve remaining lipases were able to produce a PGPR with an acid value lower that 6 mg KOH/g, which is the requirement of the European Commission Directive [56], although we considered that, after applying appropriate optimization procedures, one or more of these enzymes might be able to efficiently catalyze the enzymatic synthesis of PGPR [65]. The twelve chosen lipases were all from microbial sources, being some 1,3 specific and others "random" lipases. It was thought that any acid value decay in the reaction mixture might be due to two possible reactions: (i) the synthesis of estolides with a higher polymerization degree and (ii) the esterification of polyricinoleic acid with polyglycerol-3 (the desired process). It has been described in the literature that the enzymatic synthesis of estolides can only be successfully catalyzed by lipases that lack 1,3 positional selectivity [68, 69, 92], so that lipases from Chromobacterium viscosum and from Pseudomonas (which are "random" lipases and show the best results, Figure 16(d)) should be capable, in theory at least, of catalyzing the first step of the enzymatic synthesis of PGPR. However, it was experimentally demonstrated that, under the assayed experimental conditions, these lipases are not capable of catalyzing the production of estolides with an acid value lower than 50 mg KOH/g (data not shown), so that the noticeable decrease of the acid value observed in the above described experiments can be attributed mainly to the esterification reaction between polyricinoleic acid and polyglycerol. In case of the reactions catalyzed by the remaining lipases tested, there is no doubt about the cause of the decrease of acid value, because they are 1,3-specific and cannot act on hydroxy fatty acids [69].

On the other hand, it may be surprise that Mucor javanicus and Rhizopus sp. lipases (1, 3 specific) performed so well. If polyglycerol-3 is a linear molecule, only two of the five hydroxyl groups available as acyl acceptor groups are primary, and the acid value reached when these lipases are used indicates that more than two hydroxyl groups have been esterified. This fact can be explained if condensation of glycerol takes place between secondary-primary or secondary-secondary hydroxyl groups. In that case more than two primary hydroxyl groups may remain available as acyl acceptor groups. As can be seen in Figure 16, satisfactory results were obtained when the twelve mentioned lipases were used to catalyze the production of PGPR, and some graphs are indistinguishable. Table 6 shows the acid values reached after 7 days of reaction, which permits a better comparison of the obtained results. It can be observed that the lowest acid values were reached when lipases from Pseudomonas (3 enzymes) and Chromobacterium viscosum were used. However, some of the lipases used in the present work are very expensive, which is an aspect that should be carefully considered if the long-term purpose is to develop an industrial procedure for PGPR production. Therefore, in order to finally choose one or more of these lipases, we took into account not only kinetic aspects (reaction rates and final acid value of the reaction mixture) but also the cost of the procedure. In order to evaluate this economic aspect of the enzymatic biosynthesis of PGPR and because lipase is the most expensive material involved in the reaction, the cost of biocatalysts that cause a decrease of one unit of the acid value was calculated, and the results are showed in the last column of Table 6. It can be observed that the cheapest procedures were those catalyzed by lipases from the fungi Rhizopus oryzae, Rhizopus arrhizus,Mucor javanicus, Rhizomucor miehei, and Rhizopus niveus, with which the decrease of one unit in acid value costs less than 1 €. These results, together with those shown in Figure 16, led us to select lipases fromRhizopus oryzae, Rhizopus arrhizus, and Mucor javanicus to catalyze PGPR synthesis [65]. Although the above selected lipases are not very expensive to develop an industrial procedure for PGPR synthesis, it is desirable to use immobilized enzymes because of its well-known advantages: continuous operation of reactors and/or the reusability of the immobilized enzymes, both of which diminish operational costs. Therefore, the three chosen lipases were immobilized by physical adsorption onto an anion exchange resin (Lewatit MonoPlus MP 64). As described previously, the authors have optimized the immobilization process of Candida rugosa lipase, and as a preliminary attempt the same technique was used in this work in order to compare the behavior of these three lipases. In further studies the immobilization process should be optimized. Thus, three immobilized derivatives were prepared, and the results are shown in Table 7, where the protein content of the commercial lipases, the immobilization yields, and the enzyme loadings of all the immobilized derivatives are summarized, all data being based on the protein concentration values provided by Lowry's method [83]. It should be mentioned here that the protein content of the three commercial preparations was quite low, although that of the lipase from Rhizopus arrhizus was slightly higher than the others. However, the percentage of immobilized protein obtained with this lipase was approximately half that obtained with the other two lipases, and so the enzyme loading factor of this immobilized derivative was the lowest (8.59 mg E/g support). The highest immobilization yield was achieved when the lipase from Rhizopus oryzae was adsorbed; in this case an immobilized derivative with adequate enzyme content was obtained despite of the low Lowry protein content of the commercial enzyme. The immobilized derivative of lipase from Mucor javanicus had the higher enzyme loading, 14.11 mg E/g support. The above results did not differ sufficiently to permit us to decide at this stage which of the three lipases should be selected. Therefore the immobilized derivatives were tested for activity, using them to catalyze the synthesis of PGPR. Figure 17 shows the variation of the acid value of the reaction mixtures with time. As can be seen, all the immobilized derivatives showed their ability to catalyze the esterification between polyricinoleic acid and polyglycerol. The use of the lipase from Mucor javanicus should not be totally discarded because reasonably good results were obtained when it was used as biocatalyst, and a PGPR with an acid value of 13 mg KOH/g was reached at the end of the experiment. The highest reaction rates were achieved when lipases from Rhizopus arrhizus and Rhizopus oryzae were used, and in these cases, PGPRs with lower acid values were produced. Comparing these results with those obtained with the soluble enzymes, it can be observed that the acid values reached with the immobilized derivatives (10.42 mg KOH/g with lipase from Rhizopus arrhizus and 9.22 mg KOH/g with lipase fromRhizopus oryzae) were similar to those obtained with the soluble lipases (11.04 mg KOH/g and 13.94 mg KOH/g, respectively, Table 6), even though the amounts of soluble enzymes added to the reactors (500 mg in both cases) were higher than those used in the experiments with immobilized enzymes (42.95 mg lipase fromRhizopus arrhizus and 64.4 mg lipase from Rhizopus oryzae). These results suggest that immobilisation had a beneficial effect on the activity and stability of both lipases [65]. Lipase from Rhizopus arrhizus was selected for PGPR production because lower amount of enzyme is required to achieve the same final acid value [66].

Table 5: Lipases tested to catalyze the enzymatic production of PGPR [64].

	Enzyme	Source	Activity (U/mg solid) (as declared by the manufacturer)	Added amount (mg)
	1	Aspergillus sp.	$0.20^{(1)}$	100
	2	Candida antarctica	$2.9^{(2)}$	50
	3	Candida cylindracea	$3.85^{(2)}$	1000
	4	Mucor miehei	$1.4^{(2)}$	100
Basic kit	5	Pseudomonas cepacia	$46.2^{(2)}$	100
	6	Pseudomonas fluorescens	$36^{(2)}$	50
	7	Rhizopus arrhizus	$9.18^{(3)}$	500
	8	Rhizopus niveus	$1.7^{(4)}$	1000
	9	Porcine pancreas	$20.6^{(5)}$	1000
	10	Aspergillus oryzae	$48^{(2)}$	100
	11	Candida lipolytica	$0.0011^{(2)}$	1000
	12	Mucor javanicus	$11.6^{(5)}$	500
	13	Penicillium roqueforti	$0.65^{(5)}$	500
Extension kit	14	Pseudomonas fluorescens	$309^{(2)}$	50
	15	Rhizomucor miehei recombinant from Aspergillus oryzae	$0.51^{(2)}$	50
	16	Wheat germ	$0.1^{(1)}$	500
	17	Chromobacterium viscosum	$2711^{(2)}$	25
	18	Pseudomonas sp.	$2324^{(2)}$	10
	19	Pseudomonas sp. (Type B)	$256^{(6)}$	50
	20	Rhizopus oryzae	$58.4^{(7)}$	500

[1] 1 unit corresponds to the amount of enzyme which liberates 1 μmol acetic acid per minute at pH 7.4 and 40°C, using triacetin as substrate.

[2] 1 unit corresponds to the amount of enzyme which liberates 1 μmol oleic acid per minute at pH 8.0 and 40°C, using triolein as substrate.

[3] 1 unit corresponds to the amount of enzyme which liberates 1 μmol butyric acid per minute at pH 8.0 and 40°C, using tributyrin as substrate.

[4] 1 unit corresponds to the amount of enzyme which liberates 1 μmol fatty acid from a triglyceride per minute at pH 7.7 and 37°C, using olive oil as substrate.

[5] As [4] but at pH 8.0.

[6] 1 unit corresponds to the amount of enzyme which liberates 1 μmol oleic acid per minute at pH 8.0 and 37°C, using cholesteryl oleat as substrate.

[7] As [4] but at pH 7.2.

Table 6: Selection of lipases based on kinetic and economic aspects [65].

Enzyme[1]	Source	Final AV (mg KOH/g) after 7 days	ΔAV[2]	Enzyme cost[3] (€)	€/unit AV[4]
4	*Mucor miehei*	8.0	34.0	46.8	1.4
5	*Pseudomonas cepacia*	8.1	33.9	35.7	1.1
6	*P. fluorescens* (36 U/mg solid)	7.2	34.8	35.3	1.0
7	*Rhizopus arrhizus*	11.0	31.0	22.9	0.7
10	*Aspergillus oryzae*	11.4	30.6	85.2	2.8
12	*Mucor javanicus*	8.7	33.3	22.0	0.7
14	*P. fluorescens* (309 U/mg solid)	9.6	32.4	110.5	3.4
15	*Rhizomucor miehei*	9.3	32.7	27.4	0.8
17	*Chromobacterium viscosum*	7.1	34.9	43.4	1.2
18	*Pseudomonas* sp. (2324 U/mg solid)	7.7	34.3	44.6	1.3
19	*Pseudomonas* sp. Type B (256 U/mg solid)	7.6	34.4	43.0	1.2
20	*Rhizopus oryzae*	13.9	28.1	1.8	0.06

[1]Enzyme identification numbers are the same that those used in Table 5.
[2]Calculated as the difference between initial AV (42.0 mg KOH/g) and final AV (3rd column).
[3]Estimated from commercial price lists.
[4]Calculated as column 5 and divided by column 4.

Table 7: Coupling parameters for the immobilization of lipases onto Lewatit MonoPlus MP 64 [65].

Lipase source	Protein content of the commercial lipase (%)	Immobilisation yield (%)	Enzyme loading (mg E/g support)
Mucor Javanicus	22.07	63.90	14.11
Rhizopus arrhizus	26.35	32.59	8.59
Rhizopus oryzae	19.80	65.03	12.88

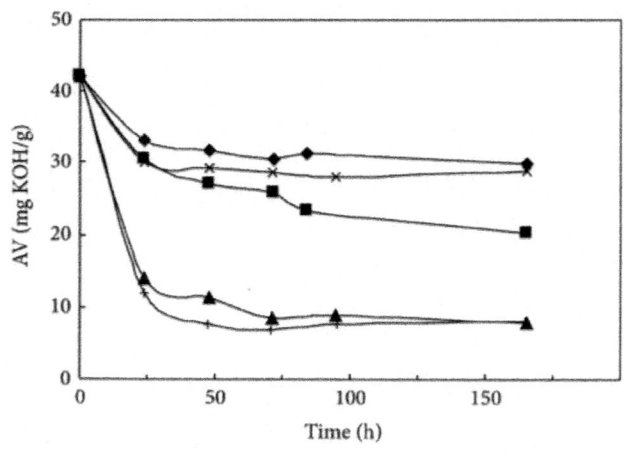

(a)

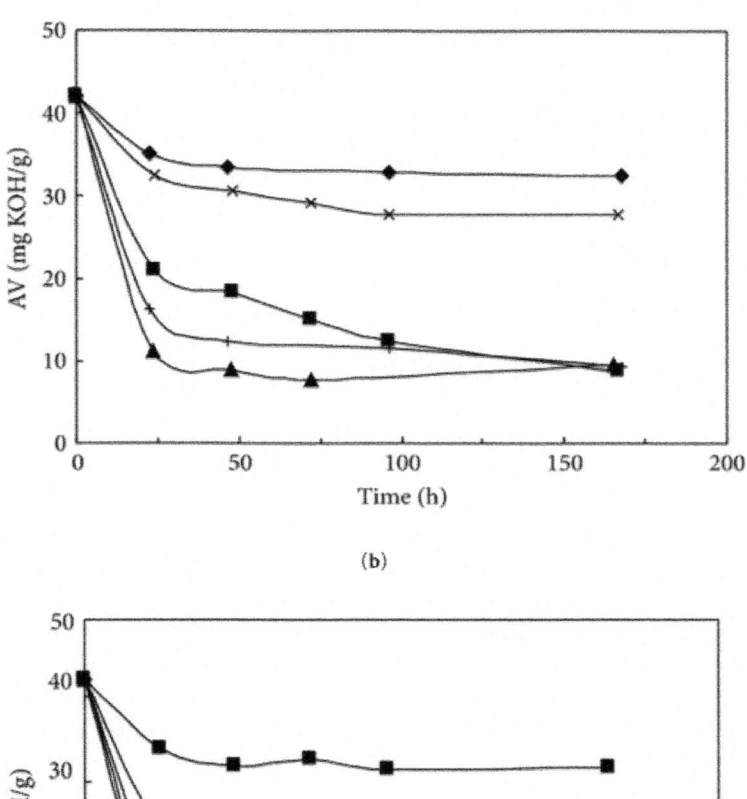

(b)

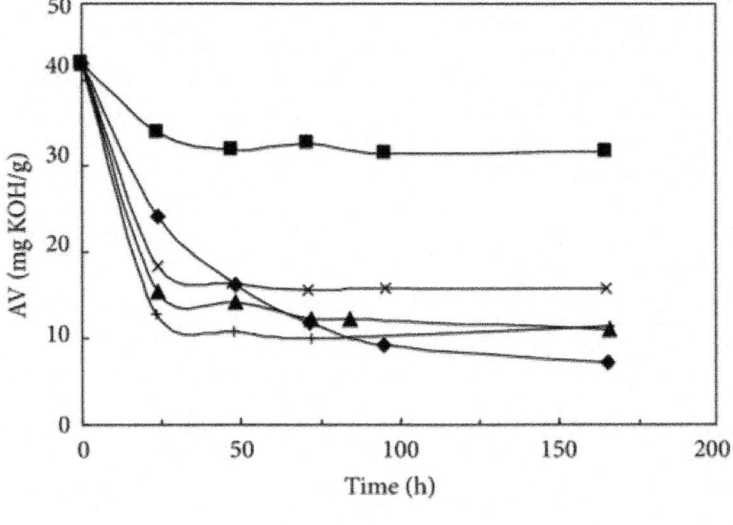

(c)

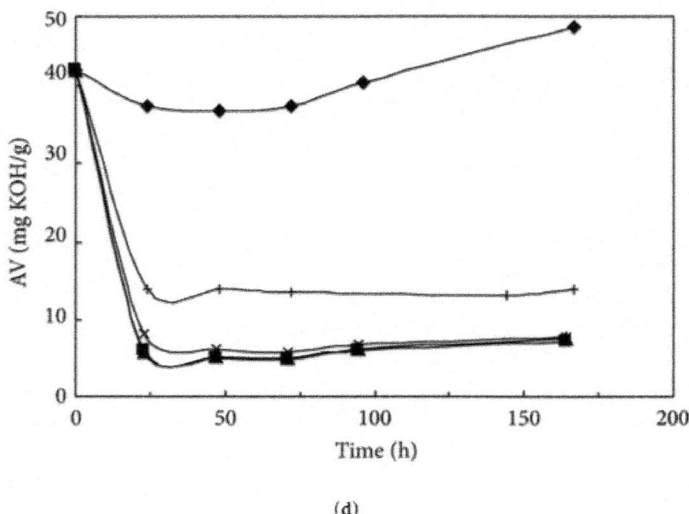

(d)

Figure 16: Evolution of acid value with time for the PGPR synthesis catalysed by free lipases from different sources [65]. (a) ◆ Aspergillus sp.; ■ Candida antarctica; × Candida cylindracea; ▲ Mucor miehei; + Pseudomonas cepacia. (b) ◆ Candida lipolytica; ■ Mucor javanicus; × Penicillium roquefortii; ▲ Pseudomonas fluorescens (300 U/mg solid); + Rhizomucor miehei. (c) ◆ Pseudomonas fluorescens (40 U/mg solid); ■ Porcine pancreas; × Rhizopus niveus; ▲ Rhizopus arrhizus; + Aspergillus oryzae. (d) ◆ Wheat germ; ■ Chromobacterium viscosum; × Pseudomonas sp. (1200 U/mg solid); ▲ Pseudomonas sp. Type B (≥60 U/mg solid); + Rhizopus oryzae.

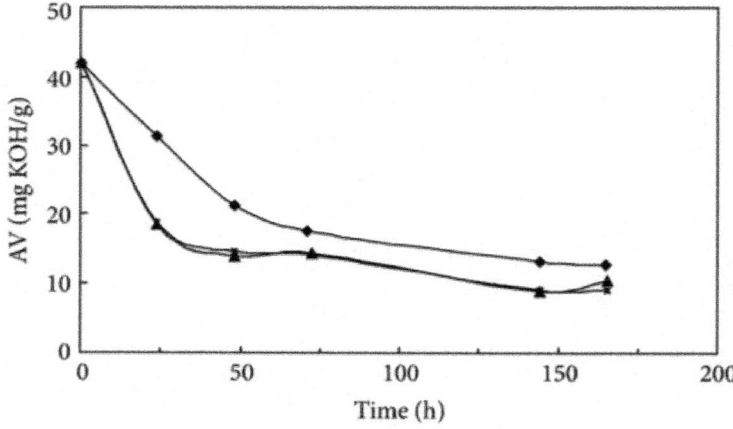

Figure 17: Evolution of acid value with time for the PGPR synthesis catalysed by immobilized lipases from different sources [65]. (◆) Mucor javanicus, (▲) Rhizopus arrhizus, and (×) Rhizopus oryzae.

Optimization of the Reaction ConditionsAfter a detailed literature search, no references were found, which might indicate, even approximately, the experimental conditions in which the reaction should be carried out. Therefore, it was necessary to start the study of the enzymatic synthesis of PGPR by establishing the optimal experimental conditions [66]. The polyricinoleic acid (PR), used as substrate, has an acid number lower than 50 mg KOH/g. At this acid number value, the average length of the PR chains is four, which is considered as optimum for food use by most authors [62, 79]. Among the commercially available polyglycerols, polyglycerol-3 from Solvay (PR-3) was considered the most appropriate because it produces a high performance PGPR [23]. As described previously, the European Commission Directive 2008/84/EC [56] establishes an acid value (AV) lower than 6 mg KOH/g for PGPR and a hydroxyl value (HV) of between 80 and 100 mg KOH/g. HV is a measurement of the free hydroxyl groups, and any reduction is concomitant with a decrease in AV because one of each is consumed when an ester linkage is formed [93]. When the PR/PG-3 mass ratio used is lower than 12, both requirements cannot be fulfilled because, even if all the acid groups react, and AV is close to zero, too many hydroxyl groups remain unreacted (HV > 100 mg KOH/g). On the other hand when the PR/PG-3 ratio is too high, the final product will contain too many acid groups or too few hydroxyl groups. Therefore, the substrate mass ratio (PR/PG-3) in all the experiments was maintained constant at a value of 15, which means that three of the five hydroxyl groups of the polyglycerol could be esterified.

Other optimal reaction conditions were determined: 40°C and no additional water added to the reactor at the beginning of the reaction but only that soaked in the immobilized derivative [66].

Production of PGPR in a Vacuum Reactor

In all the experiments described until now the final acid value was far from this objective, which means that the equilibrium has to be shifted towards the esterification pathway. This can be done by using a more anhydrous medium. On the other hand, the crucial importance of the amount of water in the reaction medium is illustrated by the poor reproducibility of the processes taking place in open reactors, since this parameter is heavily influenced by seasonality and weather conditions. This is particularly important in our case, when the final purpose is the production of the additive on an industrial scale, which requires rigorous standardization. For this reason, a high performance reactor was tested for PGPR production. The reactor is thermostated, it can work in a wide range of pressures, and is also able to mix to the reaction medium in accordance with its high viscosity. In this reactor the amount of water can be manipulated through the pressure and the entry of dry nitrogen, making it independent of environmental conditions [66]. Figure 18 shows the results of PGPR synthesis using immobilized lipase in the high performance reactor. This reactor provides a controlled atmosphere that facilitates the adjusting of the water content in the reactor medium. In the tested conditions, the water content in the reaction medium was established at around 2000 ppm (Karl-Fisher) after 10 h (a totally anhydrous medium would lead to enzyme inactivation). In these conditions, the objective of the European Commission Directive was attained after 100 h (AV = 5.9 mg KOH/g) but even lower values can be reached at longer times (after 125 h the AV was 4.9 mg KOH/g). Although Figure 18 only depicts one curve corresponding to the vacuum reactor, several experiments were carried out in identical conditions and the same results were obtained [66].

Figure 18 also reveals the enormous difference between PGPR biosynthesis in open-air and vacuum reactors. Both esterification processes were made with the same amount of substrate, immobilized derivative, and initial amount of water. In the atmospheric reactor the target was not reached even after one week, whereas in the vacuum reactor PGPR was ready after 100 h.

The PGPR produced by biocatalytic synthesis fulfills all the requirements of the European Commission Directive, as shown in Table 8.

Finally, it is important to note that the organoleptic properties of the enzymatic PGPR are better than those showed by other commercial PGPRs. In Figure 19 a photograph of the enzymatic PGPRs is shown, where the absence of color can be observed [94]. This fact, together with the properties displayed in Table 8, makes the enzymatic PGPR a valuable alternative to the chemical synthesized one.

Nowadays the general applicability of this enzymatic method to obtain different fatty acid esters is under study.

Table 8: Enzymatic PGPR characterization values compared with those established by the European Directive 2008/84/EC.

	European Commission Directive [56]	Enzymatic PGPR
Acid value (mg KOH/g)	≤6	4.9
Hydroxyl value (mg KOH/g)	80–100	82.77
Refractive index $[n]^{65}$	1.4630–1.4665	1.4655

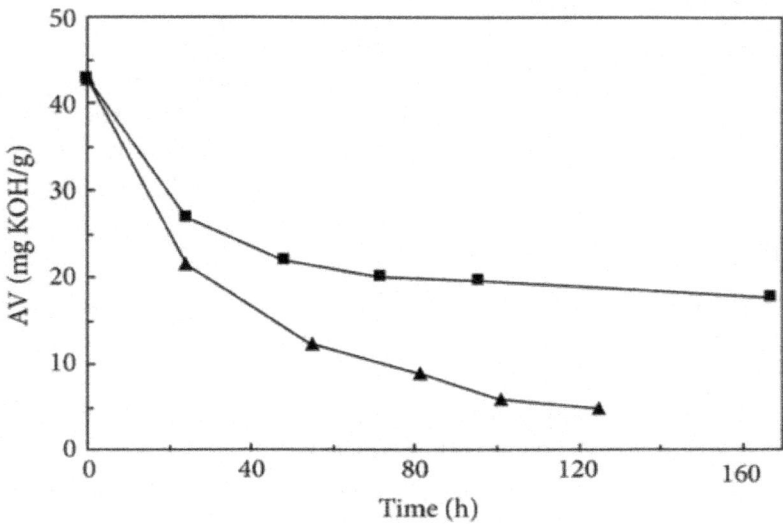

Figure 18: Influence of the reactor device on the production of PGPR using immobi-lizedRhizopus arrhizus lipase [66]. (■) open-air reactor and (▲) vacuum reactor.

Figure 19: Photograph of enzymatic PGPR [94].

ACKNOWLEDGMENTS

The author wishes to express her deep gratitude to the members of the research group "Análisis y Simulación de Procesos Químicos, Bioquímicosy de Membrana", Universidad de Murcia, España, without whose knowledge and assistance this article would not have been successful. This work was partially support by a grant CTQ2011-24091, MICINN, Spain.

REFERENCES

1.   R. MacRae, R. K. Robinson, and M. J. Sadler, Eds., Encyclopaedia of Food Science, Food Technology, and Nutrition, vol. 8, Academic Press, New York, NY, USA, 1993.

2.   Y. H. Hui, Ed., Encyclopedia of Food Science and Technology, vol. 4, John Wiley & Sons, New York, NY, USA, 1992.

3.   P. J. Fellows, Food Processing Technology: Principles and Practices, Woodhead Publishing Limited, Cambridge, UK, 3rd edition, 2009.

4.   A. L. Branen, P. M. Davidson, S. Salminen, and J. Thorngate, Eds., Food Additives, Taylor & Francis, Boca Raton, Fla, USA, 2001.

5.   T. E. Furia, Ed., Handbook of Food Additives, vol. 2, CRC Press, New York, NY, USA, 2nd edition, 1980.

6.   J. A. Maga and A. T. Tu, Eds., Food Additive Toxicology, Taylor & Francis, London, UK, 1995.

7.   J. S. Smith, Ed., Food Additive User's Handbook, Springer, New York, NY, USA, 1991.

8.   Codex alimentarius, 2012, http://www.codexalimentarius.org.

9.   H. Moonen and H. Bas ", "Mono- and diglycerides," in Emulsifiers in Food Technology, R. J. Whitehurst, Ed., John Wiley & Sons, New York, NY, USA, 2008.

10. NPCS Board of Consultants & Engineers, The Complete Book on Emulsifiers With Uses, Formulae and Processes, 2007.

11. G. L. Hasenhuettl and R. W. Hartel, Eds., Emulsifiers and Their Applications, Springer, New York, NY, USA, 2008.

12. C. Stauffer, Emulsifiers, Eagan Press, 1999.

13. N. Garti, "Food emulsifiers. structure-reactivity relationships, design, and applications," in Physical Properties of Lipids, A. G. Marangoni and S. S. Narine, Eds., Taylor & Francis, London, UK, 2002.

14. C. C. Cai, "Emulsifiers used in food applications, focusing on the meat processing industry," Palsgaard Technical Paper, 2011.

15. C. E. Stauffer, "Emulsifiers for the food industry," in Bailey'a Induatrial Oil and Fat Producta, F. Shahidi, Ed., John Wiley & Sons, New York, NY, USA, 2005.

16. M. Gómez, S. Del Real, C. M. Rosell, F. Ronda, C. A. Blanco, and P. A. Caballero, "Functionality of different emulsifiers on the performance of breadmaking and wheat bread quality," European Food Research and Technology, vol. 219, no. 2, pp. 145–150, 2004.

17. L. Stampfli and B. Nersten, "Emulsifiers in bread making," Food Chemistry, vol. 52, pp. 353–360, 1995.

18. Z. Kohajdová, J. Karovičová, and Š. Schmidt, "Significance of emulsifiers and hydrocolloids in bakery industry," Acta Chimica Slovaca, vol. 2, pp. 46–461, 2009.

19. R. T. McIntyre, "Polyglycerol esters," Journal of the American Oil Chemists› Society, vol. 56, pp. A835–A840, 1979.

20. M. F. Stewart and E. J. Hughes, "Polyglycerol esters as food additives," Process Biochemistry Journal, vol. 7, pp. 27–28, 1972.

21. N. Garti, G. F. Remon, and B. Zaidman, "Polyglycerol esters of vegetable oils," Tenside, Surfactants, Detergents, vol. 23, no. 6, pp. 320–324, 1986.

22. N. Garti, A. Aserin, and B. Zaidman, "Polyglycerol esters: optimization and techno-economic evaluation," Journal of the American Oil Chemists› Society, vol. 58, no. 9, pp. 878–883, 1981.

23. "Polyglycerols in food applications," Application data sheet. Solvay Chemicals International.

24. J. Holstborg, B. V. Pedersen, N. Krog, and S. K. Olesen, "Physical properties of diglycerol esters in relation to rheology and stability of protein-stabilised emulsions," Colloids and Surfaces B, vol. 12, no. 3–6, pp. 383–390, 1999.

25. S. K. Olesen and N. Krog, "Phase behaviour of new food emulsifiers and their application," in Oils and Fats in Food Applications: Proceedings of the Food Applications Session of the 22nd Congress of the International Society of Fats Research (ISF), Kuala Lumpur, Malaysia, 7–12 September 1997, K. G. Berger, Ed., 1997.

26. A. Sein, J. A. Verheij, and W. G. M. Agterof, "Rheological characterization, crystallization, and gelation behavior of monoglyceride gels," Journal of Colloid and Interface Science, vol. 249, no. 2, pp. 412–422, 2002.

27. N. Krog, "Crystallization properties and lyotropic phase behavior of food emulsifiers," inCrystallization Processes in Fats and Lipid Systems, N. Garti and K. Sato, Eds., Taylor & Francis, London, UK, 2001.

28. P. Seiden and J. B. Martin, "Process for preparing polyglycerol," US 3,968,169, 1976.

29. "Polyglycerols—general overview," Product data sheet. Solvay Chemical International.

30. V. Norn, "Polyglycerol esters," in Emulsifiers in Food Technology, R. J. Whitehurst, Ed., 2008.

31. T. Ushikusa, T. Maruyama, I. Nhya, and M. Okada, "Pyrolysis behaviors and thermostability of polyglycerols and polyglycerol fatty acid esters," Journal of the Japan Oil Chemists› Society, vol. 39, no. 5, pp. 314–320, 1990.

32. "Polyglycerols for ester production," Application data sheet. Solvay Chemical International.

33. F. van de Velde, F. Weinbreck, M. W. Edelman, E. Van Der Linden, and R. H. Tromp, "Visualisation of biopolymer mixtures using confocal scanning laser microscopy (CSLM) and covalent labelling techniques," Colloids and Surfaces B, vol. 31, no. 1–4, pp. 159–168, 2003.

34. I. Kobayashi, X. Lou, S. Mukataka, and M. Nakajima, "Preparation of monodisperse water-in-oil-in-water emulsions using microfluidization and straight-through microchannel emulsification," Journal of the American Oil Chemists› Society, vol. 82, no. 1, pp. 65–71, 2005.

35. J. Su, J. Flanagan, Y. Hemar, and H. Singh, "Synergistic effects of polyglycerol ester of polyricinoleic acid and sodium caseinate on the stabilisation of water-oil-water emulsions," Food Hydrocolloids, vol. 20, no. 2-3, pp. 261–268, 2006.

36. J. Surh, G. T. Vladisavljević, S. Mun, and D. J. McClements, "Preparation and characterization of water/oil and water/oil/water emulsions containing biopolymer-gelled water droplets," Journal of Agricultural and Food Chemistry, vol. 55, no. 1, pp. 175–184, 2007.

37. A. Benichou, A. Aserin, and N. Garti, "Polyols, high pressure, and refractive indices equalization for improved stability of W/O emulsions for food applications," Journal of Dispersion Science and Technology, vol. 22, no. 2-3, pp. 269–280, 2001.

38. S. Mun, Y. Choi, S. J. Rho, C. G. Kang, C. H. Park, and Y. R. Kim, "Preparation and characterization of water/oil/water emulsions stabilized by polyglycerol polyricinoleate and whey protein isolate," Journal of Food Science, vol. 75, no. 2, pp. E116–E125, 2010.

39. D. Saᵤlam, P. Venema, R. de Vries, L. M. C. Sagis, and E. van der Linden, "Preparation of high protein micro-particles using two-step

emulsification," Food Hydrocolloids, vol. 25, no. 5, pp. 1139–1148, 2011.

40. R. Wilson, B. J. van Schie, and D. Howes, "Overview of the preparation, use and biological studies on polyglycerol polyricinoleate (PGPR)," Food and Chemical Toxicology, vol. 36, no. 9-10, pp. 711–718, 1998.

41. E. Flack, "Margarines and spreads," in Food Emulsifiers and Their Applications, G. L. Hasenhuettl and R. W. Hartel, Eds., Springer, New York, NY, USA, 1997.

42. S. M. Clegg, A. K. Moore, and S. A. Jones, "Low-fat margarine spreads as affected by aqueous phase hydrocolloids," Journal of Food Science, vol. 61, no. 5, pp. 1073–1079, 1996.

43. B. Schantz and H. Rohm, "Influence of lecithin-PGPR blends on the rheological properties of chocolate," Lebensm-wiss u-Technol, vol. 38, no. 1, pp. 41–45, 2005.

44. H. F. Banford, K. J. Gardiner, G. R. Howat, and A. F. Thomson, "The use of polyglycerol polyricinoleate in chocolate," Confectionery Production, vol. 36, pp. 359–365, 1970.

45. P. Lonchampt and R. W. Hartel, "Fat bloom in chocolate and compound coatings," European Journal of Lipid Science and Technology, vol. 106, pp. 241–274, 2004.

46. R. Peschar, M. M. Pop, D. J. A. de Ridder, J. B. van Mechelen, R. A. J. Driessen, and H. Schenk, "Crystal structures of 1,3-distearoyl-2-oleoylglycerol and cocoa butter in the β(V) phase reveal the driving force behind the occurrence of fat bloom on chocolate," Journal of Physical Chemistry B, vol. 108, no. 40, pp. 15450–15453, 2004.

47. H. Schenk and R. Peschar, "Understanding the structure of chocolate," Radiation Physics and Chemistry, vol. 71, pp. 829–835, 2004.

48. R. Wilson and M. Smith, "A three-generation reproduction study on polyglycerol polyricinoleate (PGPR) in wistar rats," Food and Chemical Toxicology, vol. 36, no. 9-10, pp. 739–741, 1998.

49. R. Wilson and M. Smith, "Human studies on polyglycerol polyricinoleate (PGPR)," Food and Chemical Toxicology, vol. 36, no. 9-10, pp. 743–745, 1998.

50. R. Wilson, B. H. Doell, W. Groger, J. Hope, and J. B. M. Gellatly, "The physiology of liver enlargement," in Metabolic Aspects of Food Safety, F. J. C. Roe, Ed., 1970.

51. D. Howes, R. Wilson, and C. T. James, "The fate of ingested glyceran esters of condensed castor oil fatty acids [polyglycerol polyricinoleate (PGPR)] in the rat," Food and Chemical Toxicology, vol. 36, no. 9-10,

pp. 719–738, 1998.

52. J. Philp McL, "Evaluation of the safety of foods," Proceedings of the Nutrition Society, vol. 40, pp. 47–56, 1981.

53. JECFA, "17th Report on the Joint FAO/WHO Expert Committee on Food Additives," 1973,http://www.fao.org/.

54. "Report from the Commission on Dietary Food Additive Intake in the European Union," Commission of the European Communities, Brussels, Belgium, 2001.

55. Directive No 95/2/EC of the European Parliament and of the Council of 20 February 1995 on food additives other than colours and sweeteners.

56. Commission Directive 2008/84/EC of the European Parliament and of the Council of 27 August 2008 laying down specific purity criteria on food additives other than colours and sweeteners.

57. P. Denecke, G. Börner, and V. V. Allmen, "Method of preparing polyglycerol polyricinoleic fatty acid esters," GB2073232A, 1981.

58. S. N. Modak and J. G. Kane, "Studies in estolides. I. Kinetics of estolide formation and decomposition,"Journal of the American Oil Chemists› Society, vol. 42, pp. 428–232, 1965.

59. K. T. Achaya, "Chemical derivatives of castor oil," Journal of the American Oil Chemists› Society, vol. 48, no. 11, pp. 758–763, 1971.

60. R. Tenore, "Process for the direct manufacture of polyglycerol polyricinoleate," WO 2007/027447 A1, 2007.

61. A. Manresa, A. Bódalo, J. L. Gómez et al., "Method for obtaining polyglycerol polyricinoleate," WO2088/031908 A1, 2008.

62. A. Bódalo, J. Bastida, M. F. Máximo, M. C. Montiel, and M. D. Murcia, "Enzymatic biosynthesis of ricinoleic acid estolides," Biochemical Engineering Journal, vol. 26, pp. 155–158, 2005.

63. A. Bódalo, J. Bastida, M. F. Máximo, M. C. Montiel, M. Gómez, and M. D. Murcia, "Production of ricinoleic acid estolide with free and immobilised lipase from Candida rugosa," Biochemical Engineering Journal, vol. 39, pp. 450–456, 2008.

64. A. Bódalo, J. Bastida, M. F. Máximo, M. C. Montiel, M. D. Murcia, and S. Ortega, "Influence of the operating conditions on lipase-catalysed synthesis of ricinoleic acid estolides in solvent-free systems,"Biochemical Engineering Journal, vol. 44, no. 2-3, pp. 214–219, 2009.

65. A. Bódalo, J. Bastida, M. F. Máximo, M. C. Montiel, M. Gómez, and S. Ortega, "Screening and selection of lipases for the enzymatic production of polyglycerol polyricinoleate," Biochemical Engineering Journal, vol.

46, pp. 217–222, 2009.

66. J. L. Gómez, J. Bastida, M. F. Máximo, M. C. Montiel, M. D. Murcia, and S. Ortega, "Solvent-free polyglycerol polyricinoleate synthesis mediated by lipase from Rhizopus arrhizus," Biochemical Engineering Journal, vol. 54, pp. 111–116, 2011.

67. N. N. Gandhi, "Applications of lipase," Journal of the American Oil Chemists› Society, vol. 74, pp. 621–634, 1997.

68. D. G. Hayes and R. Kleiman, "Lipase-catalyzed synthesis and properties of estolides and their esters,"Journal of the American Oil Chemists› Society, vol. 72, no. 11, pp. 1309–1316, 1995.

69. D. G. Hayes, "The catalytic activity of lipases toward hydroxyl fatty acids. A review," Journal of the American Oil Chemists› Society, vol. 73, pp. 543–549, 1996.

70. F. Ergan, M. Trani, and G. André, "Production of glycerides from glycerol and fatty acid by immobilized lipases in non-aqueous media," Biotechnology and Bioengineering, vol. 35, no. 2, pp. 195–200, 1990.

71. D. Charlemagne and M. D. Legoy, "Enzymatic synthesis of polyglycerol-fatty acid esters in solvent-free system," Journal of the American Oil Chemists› Society, vol. 72, pp. 61–65, 1995.

72. G. W. V. Cave, C. L. Raston, and J. L. Scott, "Recent advances in solventless organic reactions: towards benign synthesis with remarkable versatility," Chemical Communications, no. 21, pp. 2159–2169, 2001.

73. G. Hills, "Industrial use of lipases to product fatty acid esters," European Journal of Lipid Science and Technology, vol. 105, pp. 601–607, 2003.

74. P. T. Anastas and J. C. Warner, Green Chemistry: Theory and Practice, Oxford University Press, New York, NY, USA, 1998.

75. C. Yamaguchi, M. Akita, S. Asaoka, and F. Osada, "Enzymatic manufacture of castor oil fatty acid estolides," JP 8916591, 1988.

76. C. Yamaguchi, A. Tooyama, S. Asaoka, and F. Osada, "Manufacture of glycerin-free estolides from castor oil with lipase," JP 9013387, 1989.

77. C. Yamaguchi, et al., "Production of estolide," JP 5211878, 1993.

78. Y. Yoshida, et al., "Production of estolide from ricinoleic acid," JP 5304966, 1993.

79. Y. Yoshida, M. Kawase, C. Yamaguchi, and T. Yamane, "Enzymatic synthesis of estolides by a bioreactor," Journal of the American Oil Chemists› Society, vol. 74, no. 3, pp. 261–267, 1997.

80. Y. Yoshida, M. Kawase, C. Yamaguchi, and T. Yamane, "Synthesis of estolides with immobilized lipase,"Yukagaku, vol. 44, pp. 328–333,

1995.

81. A. Bódalo, E. Gómez, J. L. Gómez, J. Bastida, M. F. Máximo, and F. Díaz, "A comparison of different methods of β-galactosidase immobilization," Process Biochemistry Journal, vol. 26, pp. 349–353, 1991.

82. J. L. Gómez, A. Bódalo, E. Gómez, J. Bastida, A. M. Hidalgo, and M. Gómez, "Immobilisation of peroxidase on glass beads: an improved alternative for phenol removal," Enzyme and Microbial Technology, vol. 39, pp. 2016–2022, 2006.

83. E. F. Hartree, "Protein determination and improved modification of the Lowry's method which gives a linear photometric response," Analytical Biochemistry, vol. 42, pp. 422–427, 1973.

84. ASTM D974-06, "Standard test method for acid and base number by color indicator titration".

85. J. C. Santos, G. F. M. Nunes, A. B. R. Moreira, V. H. Perez, and H. F. de Castro, "Characterization of Candida rugosa lipase immobilized on poly(N-methylolacrylamide) and its application in butyl butyrate synthesis," Chemical Engineering and Technology, vol. 30, no. 9, pp. 1255–1261, 2007.

86. L. Guo, Z. Zhang, Y. Zhu, J. Li, and Z. Xie, "Synthesis of polysiloxane-polyester copolymer by lipase-catalyzed polycondensation," Journal of Applied Polymer Science, vol. 108, no. 3, pp. 1901–1907, 2008.

87. H. T. Dang, O. Obiri, and D. G. Hayes, "Feed batch addition of saccharide during saccharide-fatty acid esterification catalyzed by immobilized lipase: time course, water activity, and kinetic model," Journal of the American Oil Chemists› Society, vol. 82, no. 7, pp. 487–493, 2005.

88. M. Goldberg, D. Thomas, and M. D. Legoy, "The control of lipase-catalysed transesterification and esterification reaction rates. Effects of substrate polarity, water activity and water molecules on enzyme activity," European Journal of Biochemistry, vol. 190, no. 3, pp. 603–609, 1990.

89. A. R. M. Yahya, W. A. Anderson, and M. Moo-Young, "Ester synthesis in lipase catalysed reactions,"Enzyme and Microbial Technology, vol. 23, pp. 438–450, 1998.

90. A. M. Klibanov, "Enzymatic catalysis in anhydrous organic solvents," Trends in Biochemical Sciences, vol. 14, no. 4, pp. 141–144, 1989.

91. "Sigma Aldrich catalog," En, http://www.sigmaaldrich.com/.

92. Y. Yesiloglu, "Utilization of bentonite as a support material for immobilisation of Candida rugosalipase," Process Biochemistry Journal, vol. 40, pp. 2155–2159, 2005.

93. Committee on Food Chemicals Codex, Food and Nutrition Board, and Institute of Medicine, Food Chemicals Codex, 5th edition, 2004.

94. S. Ortega Requena, Síntesis biocatalítica de polirricinoleato de poliglicerol [Ph.D. thesis], University of Murcia, Murcia, Spain, 2012.

Chapter 12

MECHANISM OF BACTERIAL INACTIVATION BY (+)-LIMONENE AND ITS POTENTIAL USE IN FOOD PRESERVATION COMBINED PROCESSES

Laura Espina[1], Tilahun K. Gelaw[2], Sı́lvia de Lamo-Castellvı́ [2], Rafael Pagán[1], Diego Garcı́a-Gonzalo[1]

[1] Departamento de Produccio´n Animal y Ciencia de los Alimentos, Facultad de Veterinaria, Universidad de Zaragoza, Zaragoza, Spain,

[2] Departament d'Enginyeria Quı́mica, Universitat Rovira i Virgili, Avinguda Paı̈ssos Catalans, Tarragona, Spain

ABSTRACT

This work explores the bactericidal effect of (+)-limonene, the major constituent of citrus fruits' essential oils, against *E. coli*. The degree of *E. coli* BJ4 inactivation achieved by (+)-limonene was influenced by the pH of the treatment medium, being more bactericidal at pH 4.0 than at pH 7.0. Deletion of *rpoS* and exposure to a sub-lethal heat or an acid shock did not modify *E. coli* BJ4 resistance to (+)-limonene. However, exposure to a sub-lethal cold shock decreased its resistance to (+)-limonene. Although no sub-lethal injury was detected in the cell envelopes after exposure to (+)-limonene by the selective-plating technique, the uptake of propidium iodide by inactivated *E. coli* BJ4 cells pointed out these structures as important targets in the mechanism of action. Attenuated Total Reflectance Infrared Microspectroscopy (ATR-IRMS) allowed identification of altered *E. coli* BJ4 structures after (+)-limonene treatments as a function of the treatment pH: β-sheet proteins at pH 4.0 and phosphodiester bonds at pH 7.0. The increased sensitivity to (+)-limonene observed at pH 4.0 in an *E. coli* MC4100 *lptD4213*mutant with an increased outer membrane permeability along with the identification of altered β-sheet proteins by ATR-IRMS indicated the importance of this structure in the mechanism of action of (+)-limonene. The study of mechanism of inactivation by (+)-limonene led to the design of a synergistic combined process with heat for the inactivation of the pathogen *E. coli*O157:H7 in fruit juices. These results show the potential of (+)-limonene in food preservation, either acting alone or in combination with lethal heat treatments.

INTRODUCTION

The compound (+)-limonene is the major constituent of citrus fruits' essential oils (EOs) [1],[2]. Because of its citrus-like flavor, (+)-limonene is employed as a flavoring agent in perfumes, creams, soaps, household cleaning products and in some food products such as fruit beverages and ice creams [3]. In addition, (+)-limonene has been found to possess antifungal [4], [5], bacteriostatic [6], [7] and bactericidal [8] properties. Therefore, its use as a food preservative has also been proposed [9].

Antimicrobial compounds have been successfully combined with other preservation technologies in order to achieve a synergistic effect in the inactivation of the target pathogens, following the hurdle theory proposed by Leistner and Gorris [10]. For example, exposure of*Escherichia coli* or *Cronobacter sakazakii* to citral combined with high hydrostatic pressure (HHP), pulsed electric fields (PEF) or heat treatments, respectively, increased the inactivation degree achieved for each hurdle acting alone [11], [12], [13]. Similarly, plenty of other compounds present in EOs were found to be effective in combination with heat in the inactivation of *E. coli* and *Listeria monocytogenes* [14]. Combinations of (+)-limonene with heat or non-thermal technologies could likewise yield a similar synergistic effect in the inactivation of the target pathogens while preserving the organoleptic properties of the fresh food product.

The use of (+)-limonene in the design of food preservation processes requires a proper understanding of its mechanism of inactivation and of the influence of environmental factors that might affect it. (+)-Limonene belongs to the cyclic monoterpene hydrocarbon family, which are considered to accumulate in the microbial plasma membrane and thus cause a loss of membrane integrity and dissipation of the proton motive force [15]. Previous studies on the inactivation of *E. coli* by other terpenes and terpenoids (such as carvacrol or citral) have demonstrated the occurrence of sub-lethal injury in the outer and cytoplasmic membranes [13],[14], pointing out the membrane disruption as a mechanism of inactivation by these compounds. However, the precise targets of terpenes and terpenoids are not yet completely understood.

Description of cellular target of antimicrobial compounds could be assisted by the use of Fourier transform-infrared (FT-IR) spectroscopy [16]. FT-IR spectroscopy is a physico–chemical analytical technique based on measurement of vibration of a molecule excited by IR radiation at a specific wavelength range. Specially, attenuated total reflectance infrared microspectroscopy (ATR-IRMS) provides bands from all the cellular components of microorganisms (e.g. cell membrane and wall components, proteins and nucleic acids), giving spectral signatures or "fingerprints" that permit the classification of a microorganism at the strain and serovar level [17].

Regulation of gene expression by alternative sigma factors, which are proteins that act as transcription initiation factors through specific binding of RNA polymerase to gene promoters is key in bacterial resistance to food preservation technologies [18]. In many Gram-negative and Gram-positive genera, sigma factors, σ^S (encoded by *rpoS* gene) and σ^B (encoded by *sigB* gene), respectively, are responsible for the transcription of specific stationary-phase genes[19]. Besides, these sigma factors could also be responsible for cell protection under environmental stresses such as acid, cold, heat or osmotic shocks [19], [20], [21]. Since previous work showed that the expression of RpoS contributed to the higher resistance of *E. coli* to the terpene aldehyde citral [13], a similar regulation could be expected for other chemical compounds such as (+)-limonene.

The aims of this work were: (a) to study the inactivation of *Escherichia coli* BJ4 by (+)-limonene, describing the effect of the pH of the treatment medium, deletion of sigma factor σ^S and sub-lethal shocks; (b) to study the occurrence of lethal and sub-lethal injuries caused by (+)-limonene in bacterial envelopes of *E. coli* BJ4 and MC4100; (c) to identify the *E. coli* BJ4 structures affected by (+)-limonene through ATR-IRMS spectroscopy, and (d) to determine the synergistic lethal effect obtained when combining (+)-limonene with heat and PEF treatments to inactivate *E. coli* O157:H7.

To accomplish these objectives we used different *E. coli* strains. In the first part, dedicated to describing the mechanism of inactivation by (+)-limonene the strains *E. coli* BJ4 and its Δ*rpoS* mutant [22] were used to study the influence of this alternative sigma factor in the bacterial resistance to (+)-limonene, and *E. coli* K-12 MC4100 and its Δ*lptD4213* mutant [23] to study the role of the outer membrane in this resistance. In the second part, dedicated to demonstrating that knowledge of the mechanism of inactivation by (+)-limonene may have an applied interest to develop food preservation combined processes. Thus, we evaluated the efficacy of a combined process using (+)-limonene to inactivate the foodborne bacterium *E. coli* O157:H7 in fruit juices in which this pathogen uses to cause food safety problems.

MATERIALS AND METHODS

Micro-organisms and growth conditions

The strains used were *Escherichia coli* BJ4 and its Δ*rpoS* null mutant BJ4L1 [22], *E. coli* K-12 MC4100 and its Δ*lptD4213* mutant [23] and *E. coli* O157:H7 VTEC - (Phage type 34) [24]. The cultures were maintained in cryovials at -80 °C prior to use. Broth subcultures were prepared by inoculating one single colony from a plate, a test tube containing 5 mL of sterile tryptic soy broth (Biolife, Milan, Italy) with 0.6% yeast extract added (Biolife) (TSBYE). After inoculation, the tubes were incubated overnight at 37 °C. With these subcultures, 250 mL Erlenmeyer flasks containing 50 mL of TSBYE were inoculated to a final concentration of 10^4 CFU/mL. These flasks were incubated with agitation (130 rpm; Selecta, mod. Rotabit, Barcelona, Spain) at 37° C until the stationary growth phase was reached.

Bacterial treatment with (+)-limonene

(+)-Limonene (97% purum) was purchased from Sigma-Aldrich (Sigma-Aldrich Chemie, Steinheim, Germany). This compound is practically immiscible in water, so a vigorous shaking method was used to prepare suspensions. (+)-Limonene was added at a final concentration of 200 μL/L to tubes containing 10 mL of citrate-phosphate buffer of pH 4.0 (23.85 g/L) and 7.0 (27.09 g/L). Before treatments, bacterial cultures were centrifuged at 6,000•g for 5 min and resuspended in the same buffer that of each treatment. Microorganisms were added at a final concentration of $3 \cdot 10^7$ CFU/mL and maintained under constant agitation (130 rpm) at 20 °C. Samples were taken at regular intervals, and survivors were enumerated. According to previous studies [13], [14], treatment time and temperature; and initial concentrations of (+)-limonene and bacteria were chosen to detect 5 $\log_{10}$ cycles of cell inactivation (i.e. from $3 \cdot 10^7$ to $3 \cdot 10^2$ CFU/mL). Minimal inhibitory concentration (MIC) of (+)-limonene determined using the tube dilution method [25] for *E. coli* BJ4 and O157:H7 was 5 μL/mL (data not shown).

   Previous experiments showed that native *E. coli* was insensitive to incubation in citrate–phosphate buffer at pH 7.0 or pH 4.0 for 24 h at 20 °C.

Sub-lethal heat, cold and acid shock treatments

One 1-mL aliquot of bacterial suspensions was centrifuged at 10,000•g for 5 min and resuspended in 1 mL of TSBYE at 45°C or 0°C (sub-lethal heat and cold shocks, respectively) or in TSBYE acidified to pH 4.5 with HCl at 20 °C (sub-lethal acid shock). Sub-lethal heat shock was performed by immersing the bacterial suspensions in a thermostatic water bath (Bunsen, mod. BTG, Madrid, Spain) and holding at 45 °C for 2 h. Suspensions were kept on ice for 4.5 h (sub-lethal cold shock) or at 20°C (sub-lethal acid shock) for 1 h. Microbial resistance to (+)-limonene was assessed as explained above. These conditions were chosen from previous published work [26], [27].

Cell permeabilization by (+)-limonene

Permanent cell permeabilization of *E. coli* BJ4 was evaluated after the treatment with 200 μL/L of (+)-limonene (initial cell concentration: $3 \cdot 10^7$ CFU/mL) for different treatment times (10 min, 25 min, 1 h, 6 h, 24 h) at pH 4.0 and 7.0 at 20° C. Cells were centrifuged, supernatant was removed, and propidium iodide (PI) (Sigma – Aldrich, Madrid, Spain) was added to a final concentration of 0.08 mmol/L [28]. Cell suspensions were incubated for 15 min at 20° C, centrifuged at 10,000·g for 5 min, and washed three times until no extracellular PI remained in the buffer. Cell permeabilization was analyzed using a fluorescence microscope (Nikon, Mod. L-Kc, Nippon Kogaku KK, Japan).

Attenuated total reflectance infrared microspectroscopy (ATR-IRMS) with multivariate analysis

An aliquot of cell suspensions was centrifuged at 6,000·g for 10 min at 4° C. Pellets were washed three times with 1 mL of 0.9% NaCl and centrifuged at 6,000·g for 10 min. Pellets were placed onto grids of hydrophobic membrane (HGM; ISO-GRID, Neogen Corporation, Lansing, MI, USA) and dried out under laminar flow at room temperature for 1 h. Samples were analyzed by IR equipment (Illuminate IR, Smiths detection, The Genesis Centre Science Park South Birchwood Warrington, United Kingdom) interfaced with mercury-cadmium-telluride photoconductive detector and equipped with a microscope with a motorized x-y stage, 20x and 50x objectives, and slide-on attenuated total reflection (ATR) diamond objective (Smiths detection, United Kingdom). The hydrophobic membranes were placed on the stage of the microscope and a specific position of the microbial pellet was selected with the assistance of the microscope and live camera (Leica OM 2,500, Module FT-IR, Renishaw plc, New Mills, Wotton-under-Edge, Gloucestershire, United Kingdom). The microscope was software-controlled using Wire 3.2 version software (Renishaw plc, United Kingdom). Spectra were collected from 4,000 to 800 cm^{-1} with a resolution of 4 cm^{-1}. The spectrum of each sample was obtained by taking the average of 128 scans. Spectra were displayed in terms of absorbance obtained by rationing the single beam spectrum against that of the air background. The spectrometer was completely software controlled by synchronize IR basic version 1.1 software (SensIR Technologies, Smiths detection, United Kingdom). Pirouette® multivariate analysis software (version 4.0, InfoMetrix, Inc., Woodville, WA) was used to analyze the raw spectra of bacterial cells. The IR spectral data were mean-centred, transformed to their second derivative using a 15-point Savitzky-Golay polynomial filter, and vector-length normalized; sample residuals and Mahalanobis distance were used to determine outliers [29], [30]. Soft independent modeling of class analogy (SIMCA) was used to build a predictive model based on the construction of separate principal component analysis (PCA) models for each class. SIMCA class models were interpreted based on class projections, misclassifications and discriminating power. Class projections were visible through three-dimensional graphics of clustered samples. Probability clouds (95%) are built around the clusters based on PCA scores, allowing SIMCA to be used as a predictive modeling system. Total misclassifications were analyzed and interpreted for the input data. Variable importance, also known as discriminating power, was used to define the variables (wavenumbers) that have a predominant effect on bacterial classification, minimizing the difference between samples within a cluster, and maximizing differences between samples from different clusters. SIMCA analysis assesses itself by predicting each sample included in the training set comparing that prediction to its assigned class; this assessment is referred to as misclassifications. Zero misclassifications typify a model in which all samples were correctly predicted to the pre-assigned class. Compared samples were *E. coli* BJ4 (initial concentration: $3 \cdot 10^7$ CFU/mL) after being incubated for 24 h at 20° C in absence or presence of 200 µL/L of (+)-limonene in citrate-phosphate buffer of pH 4.0 or 7.0.

Duration of lag phase in untreated and (+)-limonene-treated cells

E. coli BJ4 at an initial concentration of $3 \cdot 10^7$ CFU/mL were treated for 20 min with 200 µL/L of (+)-limonene in citrate – phosphate buffer of pH 4.0 so that 1 $\log_{10}$ cycle of inactivation was reached. At this moment, cells were centrifuged and resuspended in TSBYE without (+)-limonene. Non-treated cells were also centrifuged, resuspended in TSBYE without (+)-limonene and adjusted at the same final concentration of live cells ($3 \cdot 10^6$ CFU/mL). Optical absorbance was measured at 590 nm during growth for 14 h of both samples at 37 °C with a spectrophotometer (GENios, Tecan, Austria).

Microbial inactivation by the combination of a lethal heat treatment and (+)-limonene

Tubes containing 5 mL of apple juice or orange juice in absence or presence of (+)-limonene added to a final concentration of 200 µL/L were placed in a shaking thermostatic bath at 54 °C (Bunsen, mod. BTG, Madrid, Spain). Before treatments, bacterial suspensions of *E. coli*O157:H7 were centrifuged at 10,000·g for 5 min and resuspended in apple or orange juice. Once the treatment temperature was reached, the microbial suspension was added to a final concentration of $3·10^7$ CFU/mL. Samples were taken after 10 min and survivors were enumerated. These treatment conditions were chosen to make these data comparable with previously published data obtained under the same conditions [1], [12], [14], [31].

Microbial inactivation by the combination of pulsed electric fields and (+)-limonene

PEF treatments were carried out in an equipment that delivered an exponential-decay pulse previously described by García et al. [32], provided with a parallel-electrode treatment chamber with a distance of 0.25 cm between electrodes and an area of 2.01 cm². Before treatments, bacterial suspensions of *E. coli* O157:H7 were centrifuged at 10,000·g for 5 min and resuspended in shelf-stable apple juice (pH 3.6) or orange juice (pH 3.8) (Don Simón, Murcia, Spain). Bacterial cultures were added to tubes containing 5 mL of each of these media with or without 200 µL/L of (+)-limonene, and 0.5 mL of these suspensions were placed into the treatment chamber with a sterile syringe. Exponential waveform pulses at an electrical field strength of 30 kV/cm and a pulse repetition rate of 1 Hz were used in this study. Experiments started at room temperature and after the application of 25 pulses the temperature of the samples was below 35° C. After treatment, samples were taken and survivors were evaluated.

Survival counts

After treatments, samples were adequately diluted in 0.1% w/v Peptone Water (Biolife), containing 1% v/v Tween 80 (Biolife) as a neutralizer. 0.1 ml aliquots from the neutralized samples were pour-plated onto Tryptic Soy Agar with 0.6% Yeast Extract added (TSAYE) as non-selective medium. Plates were incubated at 37°C for 24 h. Previous experiments showed that longer incubation times did not influence the survival counts. In order to determine bacterial cell injury, treated samples were also plated onto selective media: TSAYE with 4% (*E. coli* BJ4 and MC4100) or 3% (*E. coli* BJ4L1, O157:H7 and MC4100 Δ*lptD4213*) sodium chloride (Sigma-Aldrich, Madrid, Spain) added (TSAYE-SC) to evaluate cytoplasmic membrane damage; and onto TSAYE with 0.35% (*E. coli* O157:H7) or 0.2% (*E. coli* BJ4 and BJ4L1) bile salts (Biolife) added (TSAYE-BS) to evaluate outer membrane damage. These levels of sodium chloride and bile salts were determined as the maximum noninhibitory concentrations for native cells. Plates containing selective media were incubated for 48 h at 37°C. Previous experiments showed that longer incubation times did not influence survival counts.

After incubation of plates, colonies were counted with an improved image analyzer automatic counter (Protos; Analytical Measuring Systems, Cambridge, United Kingdom) as described by Condón et al. [33]. The extent of sub-lethal injury was expressed as the difference between the $\log_{10}$ count (CFU) on non-selective medium (TSAYE) and the $\log_{10}$ count on selective media (TSAYE-SC and TSAYE-BS) after treatments. According to this representation, "2 $\log_{10}$ cycles of injured cells" means a 2-$\log_{10}$ difference in the count on selective and non-selective media or that 99% of survivors were sub-lethally injured.

Statistical analysis

Experiments were carried out in triplicate on different working days. Inactivation was expressed in terms of the extent of reduction in $\log_{10}$ counts after every treatment. The error bars in the figures indicate the mean ± standard error from the data obtained from at least three independent experiments. Analyses of variance (p=0.05) were performed using SPSS software (SPSS, Chicago, IL, USA).

To characterize the growth kinetics, the absorbance values were fit using nonlinear regression with the Gompertz model [34], which in this case can be described by the equation: $A(t) = C \times \exp(-\exp(-B \times (t - M)))$ where $A(t)$ is the absorbance value in time t, C is the absorbance value in the stationary phase, B is the relative growth rate in point M, and M is the time in which the cells reach their maximum growth rate. The lag phase duration was calculated as $M - (1/B)$.

RESULTS

Inactivation of *Escherichia coli* BJ4 by (+)-limonene: effect of treatment medium pH, deletion of *rpoS* and sub-lethal shocks

The antimicrobial activity of 200 µL/L of (+)-limonene on the survival of $3 \cdot 10^7$ CFU/ml of *E. coli* BJ4 and its *rpoS* mutant BJ4L1 was tested at pH 4.0 and 7.0 for 10 min, 6 h and 24 h (Figure 1). Both *E. coli* strains were less resistant at pH 4.0 than at pH 7.0: after 24 h of treatment less than 2 $\log_{10}$ cycles of the initial populations were inactivated at pH 7.0 (Figure 1B), while a treatment of 6 h at pH 4.0 was able to inactivate more than 3 $\log_{10}$ cycles of both strains, and about 5 $\log_{10}$ cycles of inactivation were achieved after 24 h of storage (Figure 1A).

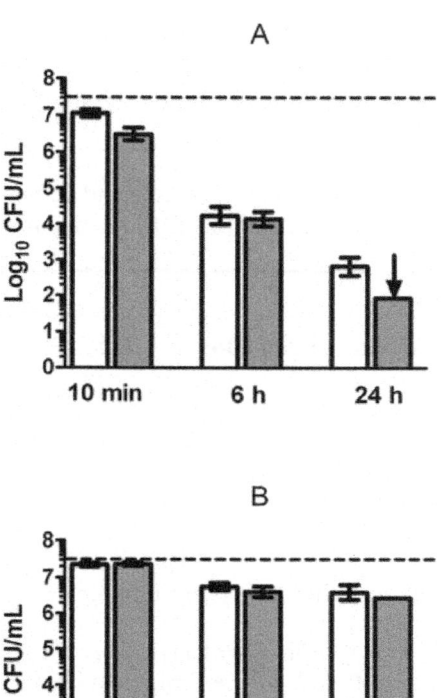

Figure 1: Study of the effect of time, pH and *rpoS* deletion on *Escherichia coli* BJ4 inactivation by (+)-limonene. Log_{10} of survival counts of *Escherichia coli* BJ4 (wild type: □) and BJ4L1 (Δ*rpoS*: ▪) after 10 min, 6 h and 24 h with 200 µL/L of (+)-limonene in citrate-phosphate buffer of pH 4.0 (A) or 7.0 (B) at 20° C. Cells were recovered in TSAYE. Discontinuous line indicates initial cell concentration ($3 \cdot 10^7$ CFU/mL). Arrow indicates survival counts under detection limit. Error bars indicate standard error. doi:10.1371/journal. pone.0056769.g001

Regarding the comparison between the wild and the mutant strain, both wild type and *rpoS* mutants showed a similar (+)-limonene resistance for all the treatments assayed ($p > 0.05$).

Development of cross-resistance to (+)-limonene as a consequence of sub-lethal shocks was studied. On the one hand, exposure to a previous sub-lethal heat or acid shock did not affect the final inactivation reached by (+)-limonene in *E. coli*, since no statistically significant differences were found when compared to the control treatment ($p > 0.05$). On the other hand, exposure to a sub-lethal cold shock significantly decreased ($p < 0.05$) the resistance of both strains to (+)-limonene (Table 1).

Table 1: Influence of previous sub-lethal heat, cold and acid treatments on*Escherichia coli* BJ4 resistance to (+)-limonene. doi:10.1371/journal.pone.0056769.t001

	Escherichia coli **BJ4**	*Escherichia coli* **BJ4L1**
Control	$4.24^a \pm 0.24$	$4.16^a \pm 0.21$
Heat-shock	$4.00^a \pm 0.18$	$3.72^a \pm 0.33$
Cold-shock	$3.00^b \pm 0.41$	$<2.18^c$
Acid-shock	$4.89^a \pm 0.21$	$4.55^a \pm 0.23$

$^{a,\ b,\ c}$: same letters indicate non-significant differences among mean values; $p > 0.05$.
Log_{10} of survival counts (CFU/mL) of *Escherichia coli* BJ4 (wild type) and BJ4L1 ($\Delta rpoS$) (initial concentration: $3 \cdot 10^7$ CFU/mL) by a treatment with 200 µL/L of (+)-limonene in citrate-phosphate buffer of pH 4.0. Table includes data from non-stressed cells (control) and cells exposed to different sub-lethal shocks before (+)-limonene treatments (mean ± standard error). Sub-lethal heat shock: 45 °C/2 h; sub-lethal cold shock: ice/4.5 h; sub-lethal acid shock: pH 4.5/1 h. doi:10.1371/journal.pone.0056769.t001

Bacterial counts were not modified ($p > 0.05$) by incubation in citrate–phosphate buffer at pH 7.0 or pH 4.0 without (+)-limonene for 24 h at 20 °C (data not shown).

Occurrence of sub-lethal damage after (+)-limonene treatments in *E. coli*BJ4

The enumeration of the survivors on the selective medium TSAYE-SC (with sodium chloride) and TSAYE-BS (with bile salts) (Figure 2) revealed that storage for 6 h with 200 µL/L of (+)-limonene caused sub-lethal damages neither to the cytoplasmic nor to the outer membrane of*E. coli* BJ4, since the levels of inactivation in these media were similar to those detected in TSAYE for each pH ($p > 0.05$, Fig. 1). The same conclusion was drawn from the survival counts after 10 min and 24 h of treatment (data not shown).

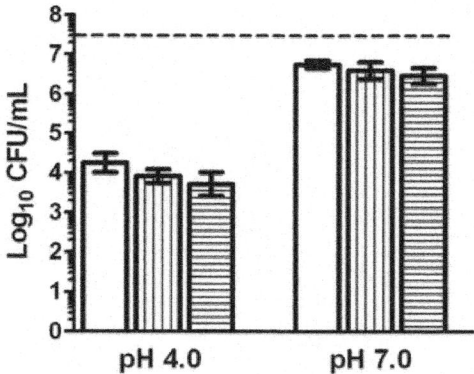

Figure 2: Study of sub-lethal injury caused by (+)-limonene on *Escherichia coli*BJ4 determined using selective plating technique. Log_{10} of survival counts of *Escherichia coli* BJ4 after 6 h with 200 μL/L of (+)-limonene in citrate-phosphate buffer of pH 4.0 or 7.0 at 20° C. Cells were recovered in TSAYE (□), TSAYE-SC (sodium chloride: vertical stripes) or TSAYE-BS (bile salts: horizontal stripes). Discontinuous line indicates initial cell concentration ($3 \cdot 10^7$ CFU/mL). Error bars indicate standard error. doi:10.1371/journal.pone.0056769.g002

Bacterial counts in selective or non-selective media were not modified ($p > 0.05$) by incubation in citrate–phosphate buffer at pH 7.0 or pH 4.0 without (+)-limonene for 24 h at 20 °C (data not shown).

Moreover, we evaluated the growth of treated and non-treated cells after exposure/non-exposure to (+)-limonene. Since exponential phase started after 2 h post-inoculation in both populations (2.19 ± 0.19 h) no lag phase delay was detected in treated cells ($p > 0.05$) (data not shown).

Membrane permeabilization of *E. coli* BJ4 by (+)-limonene

Permanent membrane permeabilization of *E. coli* BJ4 was demonstrated by the uptake of the fluorescent probe PI. As can be seen in Figure 3, a direct correlation ($R^2 = 0.96$) was found between the percentage of inactivated cells and the percentage of permeabilized cells after adding 200 μL/L of (+)-limonene for different treatment times. Furthermore, as seen with cell inactivation, the percentage of permeabilized cells after 10 min of exposure to (+)-limonene was influenced by pH: after 1 h, *E. coli* showed maximum cell permeabilization (>90%) at pH 4.0, corresponding to more than 2 log_{10} cycles of cell inactivation, while at pH 7.0 only about 50% of cells were permeabilized and inactivated. After incubation in the presence of (+)-limonene for 24 h, maximum cell permeabilization (>90%) was observed at both pHs.

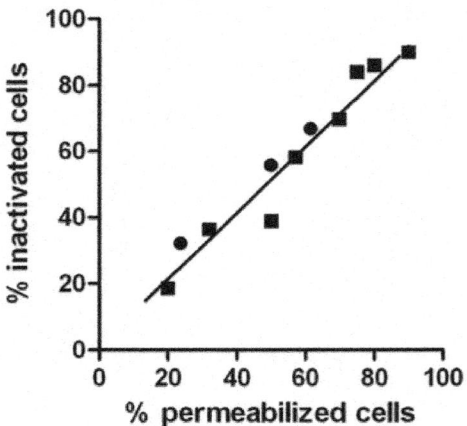

Figure 3: Correlation between permeabilized and inactivated *Escherichia coli* BJ4 cells. Correlation between the percentage of inactivated *Escherichia coli* BJ4 and the percentage of cells stained with propidium iodide (permeabilized cells) after different treatment times (10 min, 25 min, 1 h, 6 h, 24 h) with 200 μL/L of (+)-limonene (initial cell concentration: $3 \cdot 10^7$ CFU/mL) at pH 4.0 (•) or 7.0 (■). doi:10.1371/journal.pone.0056769.g003

No membrane permeabilization ($p > 0.05$) was detected after incubation in citrate–phosphate buffer at pH 7.0 or pH 4.0 without (+)-limonene for 24 h at 20 °C (data not shown).

Attenuated total reflectance infrared microspectroscopy of *E. coli* BJ4 cells after (+)-limonene treatments

Typical spectra of *E. coli* BJ4 with the presence or absence of (+)-limonene at pH 4.0 and 7.0 are shown in Figures 4A and 4D, respectively. Class projections illustrate the ability of SIMCA to differentiate IR data based on the first 3 principal components. Since the range of 4000 to 2000 cm^{-1} was not significant to describe the biochemical differences among our samplesFigures 4B, 4C, 4E and 4F only includes data obtained from the range 1,900–800 cm^{-1}. Our classification models obtained from derivatized infrared spectra (1900–800 cm^{-1}) of *E. coli*BJ4 cells (Figures 4B and 4E) allowed for the tight clustering and clear differentiation of *E. coli*BJ4 samples according to the presence or absence of (+)-limonene for each pH. Discriminating power of SIMCA is a measure of variable importance in infrared frequency and contributes to the development of the classification model [30]. Figures 4C and 4F show the wavenumbers that had a predominant effect on discrimination of (+)-limonene-treated and untreated cells at pH 4.0 and 7.0, respectively. As can be seen, the discriminating power of non-treated and (+)-limonene treated samples at pH 4.0 (Figure 4C) showed two spectral bands at 1,624 and 1,395 cm^{-1}, corresponding to changes in the amide I absorption band of β-sheet proteins [35], [36]; and in the symmetric stretching of COO$^-$ groups in amino acids and/or fatty acids [37], [38],[39]. At pH 7.0 (Figure 4F), comparison of (+)-limonene treated and non-treated cells showed that the major discriminating bands were those located at 1,083, 1,250 and 992 cm^{-1}, corresponding to the symmetric and asymmetric stretching of P=O groups in phosphodiester bonds and ring vibrations of carbohydrates [16], [37], [40]. ATR-IRMS spectra of (+)-limonene treated *E. coli* O157:H7 allowed us obtaining similar conclusions (data not shown).

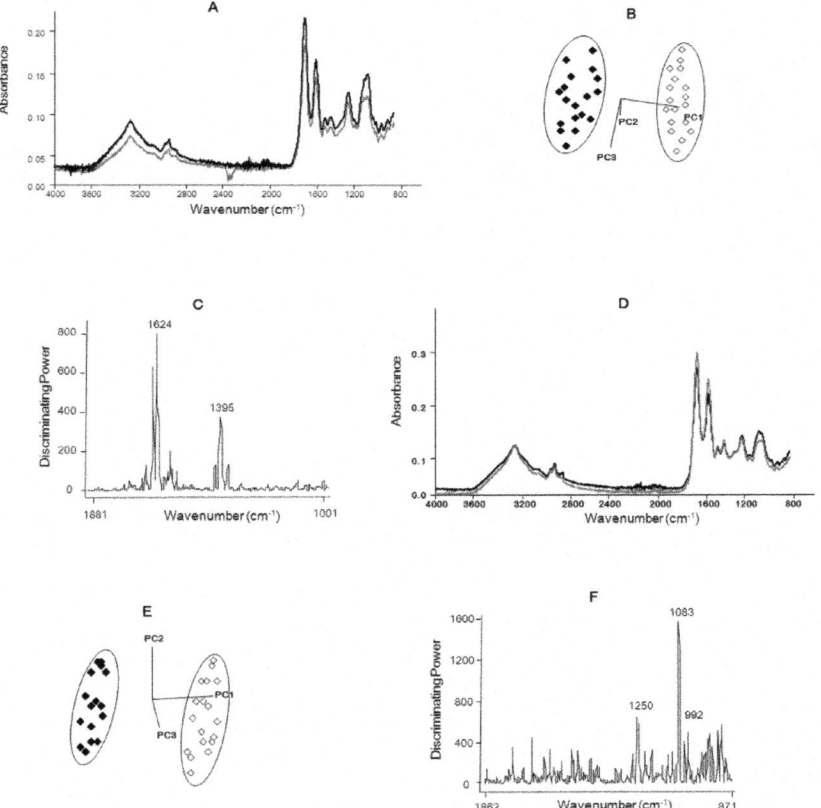

Figure 4: ATR-IRMS spectra of (+)-limonene treated *Escherichia coli* BJ4 cells. Typical raw spectra (A, D) and soft independent modeling class analogy (SIMCA) of class projections (B, E) and discriminating power (C, F) of non-treated and (+)-limonene treated (200 μL/L) *Escherichia coli* BJ4 (initial concentration: 3·10^7 CFU/mL) at pH 4.0 (A, B, C) or 7.0 (D, E, F) of transformed attenuated total reflectance infrared micro spectroscopy (ATR-IRMS) spectra. Black and gray lines and symbols represent

non-treated (+)-limonene treated cells, respectively. Spectra were obtained from three independent samples. doi:10.1371/journal.pone.0056769.g004

Role of *E. coli* MC4100 outer membrane in (+)-limonene resistance

For this study we used an *E. coli* MC4100 Δ*lptD4213* strain. This mutation disrupts the outer membrane permeability barrier, making *E. coli* sensitive to antimicrobial compounds that are not normally effective against Gram-negative bacteria [41].

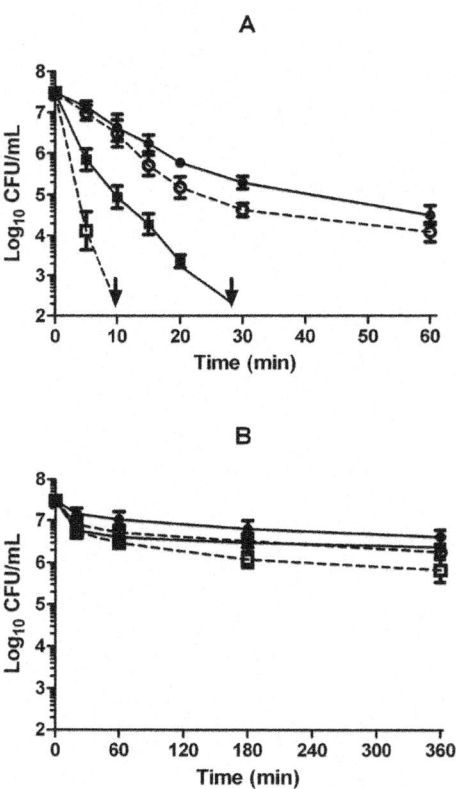

Figure 5: Effect of increased outer membrane permeability on *Escherichia coli* MC4100 resistance to (+)-limonene. Survival curves of *Escherichia coli* MC4100 (•) and its defective mutant Δ*lptD4213* (▪) after exposure to 200 μL/L of (+)-limonene in citrate-phosphate buffer of pH 4.0 (A) and pH 7.0 (B). Cells were recovered in TSAYE (•, ▪) and TSAYE-SC (sodium chloride: ○, □). Arrows indicate survival counts under detection limit. Error bars indicate standard error. doi:10.1371/journal.pone.0056769. g005

The *lptD4213* mutant was less resistant to (+)-limonene at pH 4.0 than its wild type strain (Figure 5A). For example, after 30 min more than 5 $\log_{10}$ cycles (>99.999%) of the initial *lptD4213* population were dead, whereas less than 2.5 $\log_{10}$ cycles (99.7%) of the wild strain population were inactivated. Surviving counts in Figure 5A indicate that (+)-limonene did not induce sub-lethal injuries in the cytoplasmic membrane of the wild type strain MC4100. However, a high proportion (>2.5 $\log_{10}$ cycles or 99.7% of survivors) of *lptD4213* cells had sub-lethal damages in their cytoplasmic membrane after (+)-limonene treatment at pH 4.0. On the contrary, the *lptD4213* mutants treated by (+)-limonene at pH 7.0 showed the same resistance as wild type cells and sub-lethal injuries in the cytoplasmic membrane were not detected (Figure 5B).

Bacterial counts in selective or non-selective media were not modified (*p*>0.05) by incubation in citrate–phosphate buffer at pH 7.0 or pH 4.0 without (+)-limonene for 60 min at 20 °C (data not shown).

Combined preservation processes: lethal heat treatment of *E. coli* O157:H7 in presence of (+)-limonene

To study a combined process of (+)-limonene with a lethal heat treatment, a pathogenic *E. coli* serotype, *E. coli* O157:H7, and acid fruit juices as treatment medium were chosen. Preliminary results showed that *E. coli* O157:H7 (+)-limonene resistance at 20°C was similar (data not shown).

Figure 6A shows the inactivation of *E. coli* O157:H7 by a lethal heat treatment (54°C for 10 min) alone or in combination with 200 μL/L of (+)-limonene in apple or orange juice. A lethal heat treatment alone inactivated 0.5 $\log_{10}$ cycles of the initial population of *E. coli* O157:H7; and caused sub-lethal damages in the cytoplasmic membrane in about 2 and 0.5 $\log_{10}$ cycles of survivors (as seen by the difference in $\log_{10}$ counts between recovery in TSAYE and TSAYE with sodium chloride) when cells were treated in apple and orange juice, respectively. Moreover, 4.5 and 2 $\log_{10}$ cycles of survivors showed sub-lethal damages in their outer membrane (as seen by the difference in $\log_{10}$ counts between recovery in TSAYE and TSAYE with bile salts) after lethal heat treatments in apple or orange juice, respectively.

The combined process of lethal heat and (+)-limonene in both juices caused the inactivation of more than 4 extra $\log_{10}$ cycles as compared with application of separate treatments. Hence, this combination in juices resulted in a synergistic effect on the final inactivation. A synergistic effect between (+)-limonene and lethal heat treatments under the same treatment conditions was also observed in citrate-phosphate buffer at pH 7.0. As observed at pH 4.0, simultaneous application of both treatments at pH 7.0 allowed the inactivation of more than 4 extra $\log_{10}$ cycles (data not shown).

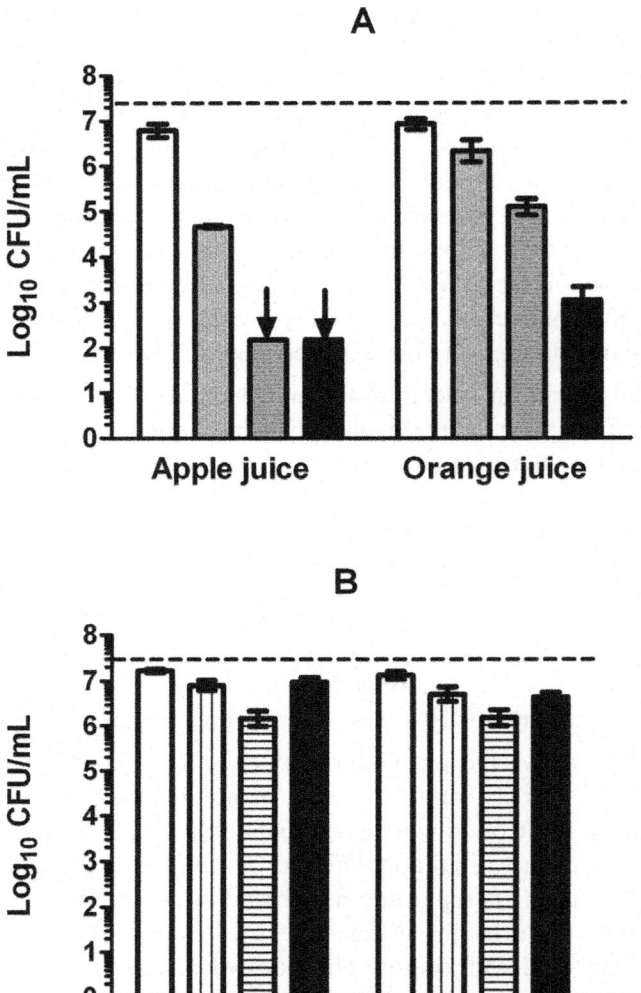

Figure 6: Inactivation of *Escherichia coli* O157:H7 by combined processes of heat and Pulsed Electric Treatments with (+)-limonene. Log_{10} of survival counts of *Escherichia coli* O157:H7 after lethal heat treatments (54° C for 10 min) (A) or Pulsed Electric Fields treatments (30 kV/cm and 25 pulses) (B) in absence (□) or presence of 200 μL/L of (+)-limonene (▪) and recovery onto TSAYE. Cells after heat and PEF treatments without (+)-limonene were also recovered onto TSAYE-SC (sodium chloride: vertical stripes) or TSAYE-BS (bile salts: horizontal stripes). Discontinuous line indicates initial cell concentration ($3 \cdot 10^7$ CFU/mL). Arrows indicate survival counts under detection limit. Error bars indicate standard error. doi:10.1371/journal.pone.0056769.g006

Bacterial counts in selective or non-selective media were not modified ($p > 0.05$) by incubation in apple or orange juice without (+)-limonene for 60 min at 20 °C (data not shown).

Combined preservation processes: Pulsed Electric Fields treatment of *E. coli* O157:H7 combined with (+)-limonene

Figure 6B shows the inactivation of *E. coli* O157:H7 by a mild PEF treatment (25 pulses at 30 kV/cm) alone or in combination with 200 µL/L of (+)-limonene in apple or orange juice.

On the one hand, a separate treatment of 200 µL/L (+)-limonene at 20°C against $3 \cdot 10^7$ CFU/mL of *E. coli* O157:H7 when suspended in these juices for 10 min inactivated less than 0.5 $\log_{10}$ cycles of the initial population (data not shown). On the other hand, a separate PEF treatment in absence of (+)-limonene inactivated less than 0.5 $\log_{10}$ cycles of the initial *E. coli* O157:H7 population, and caused sub-lethal injury in the outer membrane in less than 1 $\log_{10}$ cycle of surviving cells.

The final level of inactivation resulting from the combined process (PEF with (+)-limonene) was additive, i.e. was equal to the sum of the levels of inactivation of both treatments applied separately, not observing any extra inactivation because of the simultaneous application of a lethal heat treatment in presence of (+)-limonene.

Bacterial counts in selective or non-selective media were not modified ($p > 0.05$) by incubation in apple or orange juice without (+)-limonene for 60 min at 20 °C (data not shown).

DISCUSSION

Previous research on the antibacterial activity of (+)-limonene has been mostly focused on its bacteriostatic activity [7], [42], but little is known about its activity as a bactericidal agent in food preservation. In this respect, an important aspect to consider is the pH of the treatment medium (or the food matrix), since the final inactivation achieved by (+)-limonene was considerably higher in acid conditions (Figure 1). It is generally considered that the bacterial resistance to essential oils (EO) and their components decreases with lowering pH values because of the increase in EO hydrophobicity at low pH. As a consequence, there is an easier EO dissolution in the lipids of the cell membrane [43].

Our research was divided into two well-differentiated parts. The first part is focused on the study of mechanism of bacterial inactivation by (+)-limonene for which two wild-type and mutant strains (BJ4 and its *rpoS* mutant, and MC4100 and its *lptD4213* mutant) were used. The second part of our study is dedicated to a practical application in fruit juices of the knowledge obtained in the first part in order to demonstrate the key role of the outer membrane in microbial protection against (+)-limonene. For this objective, *E. coli* O157:H7 was used owing to its importance in food safety of fruit juices.

The expression of RpoS has been reported to cause physiological and morphological modifications that increase microbial resistance to various stresses [44]. Since deletion of *rpoS* did not decrease *E. coli* BJ4 resistance to (+)-limonene (Figure 1), probably the expression of σ^S-controlled genes under stationary-phase conditions, such as *dps* (a stress response DNA-binding protein) or *uspB* (universal stress protein B) [19], [20], [21] did not play a role in cell resistance to (+)-limonene. This finding would suggest different mechanisms of inactivation and microbial resistance for (+)-limonene in relation to other antimicrobial compounds such as citral [13] and food preservation technologies [44].

The application of a sub-lethal heat, cold or acid shock has been demonstrated to induce cross resistance to multiple stresses (see review [45]) in *E. coli*. In this study, we have shown that a previous sub-lethal heat or acid shock did not influence subsequent *E. coli* BJ4 resistance to (+)-limonene (Table 1). Interestingly, *rpoS* deletion did not modify (+)-limonene resistance of sub-lethally heat- and acid-shocked cells. However, a previous sub-lethal cold-shock decreased the resistance of both wild type and *rpoS* mutant cells to (+)-limonene (Table 1). Exposure to cold temperatures leads to a decrease in the membrane fluidity which triggers an increase in the ratio of unsaturated fatty acids [46], [47], that could be responsible of the decreased resistance to (+)-limonene (Table 1).

The occurrence of sub-lethal injuries after food preservation treatments can be evaluated using different techniques, such as different survival counts obtained between plating treated cells in non-selective and selective media [48], and delay in lag phase before starting growth in treated with regards non-treated cells [49], [50]. At the conditions assayed in this study no sub-lethal damage in the cell envelopes was detected after exposure of *E. coli* BJ4 to (+)-limonene by the selective media plating technique (Figure 2). Furthermore, the same duration of lag phase was observed for treated and untreated *E. coli* BJ4 cells, suggesting that neither the cell envelopes nor other cell structures were sub-lethally injured. Therefore, the action of (+)-limonene could be catalogued under the "quantal" effect, a response which can be expressed in binary terms: it is either present or absent ("all or nothing") [51] in which bacteria are either killed or intact after the treatment. Occurrence of sub-lethal damage in *E. coli* BJ4 cell envelopes by other EO compounds, such as citral or carvacrol [13], [14], would suggest a different mechanism of inactivation between these compounds and (+)-limonene.

Food preservation technologies, such as heat, pulsed electric fields (PEF), high hydrostatic pressure (HHP) and essential oils (EOs) normally target cell envelopes [52], [53], [54]. Thus, we evaluated membrane permeabilization in *E. coli* BJ4 using propidium iodide. A direct correlation between the percentage of cell inactivation and the percentage of membrane permeabilization was obtained (Figure 3). The simultaneous occurrence of both phenomena identifies the cell envelopes as an important target in the mechanism of *E. coli* BJ4 inactivation by (+)-limonene.

To further study the damages caused by (+)-limonene, we included an analysis by ATR-IRMS. We used this technique that evaluates the biochemical composition of the bacterial cell constituents [16], such as water, proteins, nucleic acids, fatty acids and polysaccharides, to describe the changes caused by (+)-limonene. ATR-IRMS results allowed selecting two major discriminating bands at both pH 4.0 (Figure 4C) and pH 7.0 (Figure 4F) as the main responsible for the differences between untreated and (+)-limonene-treated *E. coli* BJ4 cells. At pH 4.0, the $1,624$ cm^{-1} band corresponding to the amide I absorption band of β-sheet proteins [35], [36]; and the band at $1,395$ cm^{-1} reflecting the symmetric stretching of COO$^-$ groups in amino acids and/or fatty acids [37], [38], [39]. Since β-barrel membrane proteins occur in the outer membranes of Gram-negative bacteria [55], the main contribution to the discrimination between untreated and (+)-limonene-treated cells at pH 4.0 could come from affected outer membrane proteins that form membrane-spanning β-barrels. However, at pH 7.0, the discriminating bands for (+)-limonene treatments were found at $1,083$, $1,250$ and 992 cm^{-1}, corresponding to the symmetric ($1,083$ cm^{-1}) and asymmetric ($1,250$ cm^{-1}) stretching of P=O groups in phosphodiester bonds and ring vibrations of carbohydrates (992 cm^{-1}) [16], [37], [40]. Phosphodiester bonds are present in phospholipids of the cytoplasmic membrane and of the inner leaflet of the outer membrane [56], while carbohydrates are found in the lipopolysaccharide (LPS) fraction of the cell wall [57]. In consequence, (+)-limonene would target phospholipids and LPS cell fraction at pH 7.0, and the protein fraction at pH 4.0 in *E. coli* BJ4.

It should be noted that these conclusions related to the mechanism of bacterial inactivation by (+)-limonene were drawn from experiments using the Gram-negative strains *E. coli* BJ4 and its Δ*rpoS* mutant. Further experiments using different microorganisms are needed to extrapolate these conclusions to other Gram-negative strains.

Once we confirmed the role of cell envelopes in the (+)-limonene antimicrobial activity, we used the mutant strain Δ*lptD4213* (formerly known as Δ*imp4213*) to evaluate the role of outer membrane in the mechanism of inactivation by (+)-limonene (Figure 5). LptD is an essential β-barrel protein of the outer membrane [58] which is implicated in lipopolysaccharide (LPS) assembly [59]. Depletion of this protein results in increased outer membrane permeability to lipophilic compounds, such as novobiocin or rifampin [24]. In effect, at pH 4.0 *lptD4213*mutants showed a decreased (+)-limonene resistance, and occurrence of sub-lethal damage in the cytoplasmic membrane was demonstrated after (+)-limonene treatments. This finding, together with ATR-IRMS observations, could indicate that, at pH 4.0, (+)-limonene should damage the outer membrane in order to gain access to the periplasmic space and cytoplasmic membrane and inactivate the bacterial cell. Once outer membrane permeability to (+)-limonene is increased, there would be an enhanced interaction of (+)-limonene molecules at pH 4.0 with the components in the cytoplasmic membrane. However, the bactericidal action of (+)-limonene at pH 7.0 was not enhanced by higher outer membrane permeability in *lptD4213*mutants, indicating that facilitation of (+)-limonene access to the periplasmic space and cytoplasmic membrane would not be required at pH 7.0. Furthermore, results shown by ATR-IRMS would indicate that LPS damage was related to mechanism of inactivation by (+)-limonene at pH 7.0. Therefore, mechanism of *E. coli* BJ4 and MC4100 inactivation by (+)-limonene was different as a function of the pH of the treatment medium. In spite of differences between *E. coli* BJ4 and MC4100, (+)-limonene resistance shown by both strains was similar (Figures 1 and 5). However, further research using other microorganisms is needed in order to increase the knowledge on the mechanism of bacterial inactivation by (+)-limonene and to use this compound in practical applications.

From the study of the mechanism of inactivation by (+)-limonene in *E. coli* BJ4 and MC4100 we could expect that application of a food preservation technology causing sub-lethal damages in outer membrane would increase the lethal effect induced by (+)-limonene, leading to an advantageous combined process. In order to prove this hypothesis and to provide a practical application of this knowledge we studied the effect of (+)-limonene in a combined process with heat or PEF in *E. coli* O157:H7 because of the presence of an outer membrane in this pathogenic serotype and its importance in food safety of fruit juices [60], [61]. We determined that combinations of a lethal heat treatment that damaged outer membrane with (+)-limonene also achieved a synergistic effect to inactivate *E. coli* O157:H7 in juice. Thus, a facilitated access of (+)-limonene to the periplasmic space and cytoplasmic membrane would cause the inactivation of these sub-lethally damaged cells (Fig. 6A). On the contrary, since PEF did not cause sub-lethal damage to the outer membrane of *E. coli* O157:H7 the combination of (+)-limonene and PEF did not yield any extra inactivation when compared to the inactivation by PEF alone at the assayed conditions (Figure 6B). Since *E. coli* O157:H7 is a virulent strain whose genome has a significant number of differences from other *E. coli* strains, such as the presence of more than 1,300 new genes [62], [63], transfer of the knowledge on mechanism of microbial inactivation by (+)-limonene from *E. coli* BJ4 and MC4100 to *E. coli* O157:H7 would require further studies on the influence of the factors investigated in this research.

Although preliminary results indicate that (+)-limonene concentrations used in this study were accepted by consumers, sensory analysis of apple juice with (+)-limonene should be performed to evaluate commercial viability. Previous work with citral and PEF in *E. coli* BJ4 reached a similar conclusion [13], as well as combined processes between PEF and different antimicrobials against *E. coli* O157:H7 in apple and orange juices [14].

CONCLUSION

The study of the mechanism of bacterial inactivation by (+)-limonene showed that the lethality of this compound was higher at pH 4.0 than at neutral pH. Contrary to other food preservation treatments, deletion of *rpoS* did not modify *E. coli* BJ4 resistance to (+)-limonene. Furthermore, a previous sub-lethal heat or acid shock did not change *E. coli* BJ4 resistance to (+)-limonene, independently of *rpoS* deletion. However, a previous sub-lethal cold shock decreased the resistance of wild-type *E. coli* BJ4 and even more the resistance of *rpoS* mutant to (+)-limonene. Assessment of *E. coli* BJ4 permeabilization with propidium iodide showed that this phenomenon occurred simultaneously with bacterial inactivation, identifying the cell envelopes as important (+)-limonene targets. In contrast to other essential oils compounds, (+)-limonene did not cause sub-lethal injuries in any *E. coli* BJ4 structure, cataloguing its lethal action under the "quantal" effect ("all or nothing"). Different resistance pattern of *lptD4213*mutants and ATR-IRMS results showed the importance of outer membrane in the mechanism of inactivation by (+)-limonene at pH 4.0. At pH 7.0, increased outer membrane permeability did not lead to a decreased (+)-limonene resistance and ATR-IRMS spectra demonstrated the importance of LPS in the mechanism of *E. coli* BJ4 inactivation at this pH. Considering the orange-like flavor of (+)-limonene and its consideration as a GRAS (Generally Recognized As Safe) substance [64], [65], we propose the simultaneous application of (+)-limonene with other preservation technologies that damage outer membrane, such as heat treatments, in order to design combined food preservation processes with a synergistic lethal effect, as demonstrated for *E. coli* O157:H7 in this study. Although bacterial resistance of the studied *E. coli* strains was similar, further research is needed in order to increase the knowledge on the mechanism of inactivation by (+)-limonene in other bacteria and to use this compound in practical applications.

AUTHOR CONTRIBUTIONS

Conceived and designed the experiments: LE RP DGG. Performed the experiments: LE. Analyzed the data: LE TKG SLC RP DGG. Wrote the paper: LE RP DGG.

REFERENCES

1.   Espina L, Somolinos M, Lorán S, Conchello P, García D, et al. (2011) Chemical composition of commercial citrus fruit essential oils and evaluation of their antimicrobial activity acting alone or in combined processes. Food Control 22: 896–902. doi: 10.1016/j.foodcont.2010.11.021

2.   Fisher K, Phillips C (2008) Potential antimicrobial uses of essential oils in food: is citrus the answer? Trends Food Sci Technol 19: 156–164. doi: 10.1016/j.tifs.2007.11.006

3.   Bakkali F, Averbeck S, Averbeck D, Idaomar M (2008) Biological effects of essential oils – A review. Food Chem Toxicol 46: 446–475. doi: 10.1016/j.fct.2007.09.106

4.   Chee HY, Kim H, Lee MH (2009) In vitro antifungal activity of limonene against*Trichophyton rubrum*. Mycobiology 37: 243–246.

5.   Dambolena JS, López AG, Cánepa MC, Theumer MG, Zygadlo JA, et al. (2008) Inhibitory effect of cyclic terpenes (limonene, menthol, menthone and thymol) on*Fusarium verticillioides* MRC 826 growth and fumonisin B1 biosynthesis. Toxicon 51: 37–44. doi: 10.1016/j.toxicon.2007.07.005

6.   Jaroenkit P, Matan N, Nisoa M (2011) In vitro and in vivo activity of citronella oil for the control of spoilage bacteria of semi dried round scad (*Decapterus maruadsi*). Int J Med Arom Plants 1: 234–239.

7.   Vuuren SFv, Viljoen AM (2007) Antimicrobial activity of limonene enantiomers and 1, 8-cineole alone and in combination. Flavour Fragr J 22: 540–544. doi: 10.1002/ffj.1843

8.   Zukerman I (1951) Effect of oxidized d-limonene on micro-organisms. Nature 168: 517–517. doi: 10.1038/168517a0

9.   Dorman HJD, Deans SG (2000) Antimicrobial agents from plants: antibacterial activity of plant volatile oils. J Appl Microbiol 88: 308–316. doi: 10.1046/j.1365-2672.2000.00969.x

10.   .Leistner L, Gorris LGM (1995) Food preservation by hurdle technology.

Trends Food Sci Technol 6: 41–46. doi: 10.1016/s0924-2244(00)88941-4

11. Arroyo C, Somolinos M, Cebrián G, Condón S, Pagán R (2010) Pulsed electric fields cause sublethal injuries in the outer membrane of *Enterobacter sakazakii* facilitating the antimicrobial activity of citral. Lett Appl Microbiol 51: 525–531. doi: 10.1111/j.1472-765x.2010.02931.x

12. Espina L, Somolinos M, Pagán R, García-Gonzalo D (2010) Effect of citral on the thermal inactivation of *Escherichia coli* O157:H7 in citrate phosphate buffer and apple juice. J Food Prot 73: 2189–2196.

13. Somolinos M, García D, Condón S, Mackey B, Pagán R (2010) Inactivation of*Escherichia coli* by citral. J Appl Microbiol 108: 1928–1939.

14. Ait-Ouazzou A, Cherrat L, Espina L, Lorán S, Rota C, et al. (2011) The antimicrobial activity of hydrophobic essential oil constituents acting alone or in combined processes of food preservation. Innov Food Sci Emerg 12: 320–329. doi: 10.1016/j.ifset.2011.04.004

15. Sikkema J, de Bont J, Poolman B (1994) Interactions of cyclic hydrocarbons with biological membranes. J Biol Chem 269: 8022–8028.

16. Alvarez-Ordóñez A, Mouwen DJM, López M, Prieto M (2011) Fourier transform infrared spectroscopy as a tool to characterize molecular composition and stress response in foodborne pathogenic bacteria. J Microbiol Meth 84: 369–378. doi: 10.1016/j.mimet.2011.01.009

17. Grasso EM, Yousef AE, de Lamo Castellvi S, Rodriguez-Saona LE (2009) Rapid detection and differentiation of *Alicyclobacillus* species in fruit juice using hydrophobic grid membranes and attenuated total reflectance infrared microspectroscopy. J Agric Food Chem 57: 10670–10674. doi: 10.1021/jf902371j

18. Abee T, Wouters JA (1999) Microbial stress response in minimal processing. Int J Food Microbiol 50: 65–91. doi: 10.1016/s0168-1605(99)00078-1

19. Kazmierczak MJ, Wiedmann M, Boor KJ (2005) Alternative sigma factors and their roles in bacterial virulence. Microbiol Mol Biol Rev 69: 527–543. doi: 10.1128/mmbr.69.4.527-543.2005

20. Weber H, Polen T, Heuveling J, Wendisch VF, Hengge R (2005) Genome-wide analysis of the general stress response network in *Escherichia coli*: σ^S-dependent genes, promoters, and sigma factor selectivity. J Bacteriol 187: 1591–1603. doi: 10.1128/jb.187.5.1591-1603.2005

21. Ait-Ouazzou A, Mañas P, Condón S, Pagán R, García-Gonzalo D (2012) Role of general stress-response alternative sigma factors σ^S (RpoS)

and σ^B (SigB) in bacterial heat resistance as a function of treatment medium pH. Int J Food Microbiol 153: 358–364. doi: 10.1016/j. ijfoodmicro.2011.11.027

22. Krogfelt KA, Hjulgaard M, Sørensen K, Cohen PS, Givskov M (2000) *rpoS* gene function is a disadvantage for *Escherichia coli* BJ4 during competitive colonization of the mouse large intestine. Infect Immun 68: 2518–2524. doi: 10.1128/iai.68.5.2518-2524.2000

23. Sampson BA, Misra R, Benson SA (1989) Identification and characterization of a new gene of *Escherichia coli* K-12 involved in outer membrane permeability. Genetics 122: 491–501.

24. Chapman PA, Siddons CA, Wright DJ, Norman P, Fox J, et al. (1993) Cattle as a possible source of verocytotoxin-producing *Escherichia coli* O157 infections in man. Epidemiol Infect 111: 439–448. doi: 10.1017/ s0950268800057162

25. Rota C, Carraminana JJ, Burillo J, Herrera A (2004) In vitro antimicrobial activity of essential oils from aromatic plants against selected foodborne pathogens. J Food Prot 67: 1252–1256.

26. Somolinos M, Espina L, Pagán R, Garcia D (2010) *sigB* absence decreased *Listeria monocytogenes* EGD-e heat resistance but not its Pulsed Electric Fields resistance. Int J Food Microbiol 141: 32–38. doi: 10.1016/j.ijfoodmicro.2010.04.023

27. Somolinos M, García D, Mañas P, Condón S, Pagán R (2008) Effect of environmental factors and cell physiological state on Pulsed Electric Fields resistance and repair capacity of various strains of *Escherichia coli*. Int J Food Microbiol 124: 260–267. doi: 10.1016/j.ijfoodmicro.2008.03.021

28. Pagán R, Mackey B (2000) Relationship between membrane damage and cell death in pressure-treated *Escherichia coli* cells: differences between exponential- and stationary-phase cells and variation among strains. Appl Environ Microbiol 66: 2829–2834. doi: 10.1128/aem.66.7.2829-2834.2000

29. Hruschka WR (2001) Data analysis: wavelength selection methods. In: Williams P, Norris K, editors. Near-Infrared technology in the agricultural and food industries. Minnesota: AACC International. pp. 39–58.

30. Dunn WJ, Wold S (1995) SIMCA pattern recognition and classification In: van de Waterbeemd H, editor. Chemometric methods in molecular design.New York: Wiley-VCH Publishers.pp. 179–193.

31. Espina L, Somolinos M, Ouazzou AA, Condón S, García-Gonzalo D, et al. (2012) Inactivation of *Escherichia coli* O157:H7 in fruit juices by combined treatments of citrus fruit essential oils and heat. Int J Food

Microbiol 159: 9–16. doi: 10.1016/j.ijfoodmicro.2012.07.020

32. García D, Gómez N, Raso J, Pagán R (2005) Bacterial resistance after pulsed electric fields depending on the treatment medium pH. Innov Food Sci Emerg Technol 6: 388–395. doi: 10.1016/j.ifset.2005.04.003

33. Condón S, Palop A, Raso J, Sala FJ (1996) Influence of the incubation temperature after heat treatment upon the estimated heat resistance values of spores of *Bacillus subtilis*. Lett Appl Microbiol 22: 149–152. doi: 10.1111/j.1472-765x.1996.tb01130.x

34. Gibson AM, Bratchell N, Roberts TA (1988) Predicting microbial growth: growth responses of salmonellae in a laboratory medium as affected by pH, sodium chloride and storage temperature. Int J Food Microbiol 6: 155–178. doi: 10.1016/0168-1605(88)90051-7

35. Barth A (2007) Infrared spectroscopy of proteins. BBA-Bioenergetics 1767: 1073–1101. doi: 10.1016/j.bbabio.2007.06.004

36. Kong J, Yu S (2007) Fourier transform infrared spectroscopic analysis of protein secondary structures. Acta Biochim Biophys Sin 39: 549–559.

37. Belfer S, Gilron J, Daltrophe N, Oren Y (2005) Comparative study of biofouling of NF modified membrane at SHAFDAN. Desalination 184: 13–21. doi: 10.1016/j.desal.2005.04.035

38. Legal JM, Manfait M, Theophanides T (1991) Applications of FTIR spectroscopy in structural studies of cells and bacteria. J Mol Struct 242: 397–407. doi: 10.1016/0022-2860(91)87150-g

39. Parikh SJ, Chorover J (2006) ATR-FTIR spectroscopy reveals bond formation during bacterial adhesion to iron oxide. Langmuir 22: 8492–8500. doi: 10.1021/la061359p

40. Yu C, Irudayaraj J (2005) Spectroscopic characterization of microorganisms by Fourier transform infrared microspectroscopy. Biopolymers 77: 368–377. doi: 10.1002/bip.20247

41. Ruiz N, Kahne D, Silhavy TJ (2006) Advances in understanding bacterial outer-membrane biogenesis. Nat Rev Microbiol 4: 57–66. doi: 10.1038/nrmicro1322

42. Mourey A, Canillac N (2002) Anti-*Listeria monocytogenes* activity of essential oils components of conifers. Food Control 13: 289–292. doi: 10.1016/s0956-7135(02)00026-9

43. Juven BJ, Kanner J, Schved F, Weisslowicz H (1994) Factors that interact with the antibacterial action of thyme essential oil and its active constituents. J Appl Microbiol 76: 626–631. doi: 10.1111/j.1365-2672.1994.tb01661.x

44. Hengge-Aronis R (2002) Signal transduction and regulatory mechanisms involved in control of the σ^S (RpoS) subunit of RNA polymerase. Microbiol Mol Biol Rev 66: 373–395. doi: 10.1128/mmbr.66.3.373-395.2002

45. Chung HJ, Bang W, Drake MA (2006) Stress response of *Escherichia coli*. Compr Rev Food Sci Food Saf 5: 52–64. doi: 10.1111/j.1541-4337.2006.00002.x

46. Fulco AJ (1974) Metabolic alterations of fatty acids. Annu Rev Biochem 43: 215–241. doi: 10.1146/annurev.bi.43.070174.001243

47. Yamanaka K (1999) Cold shock response in *Escherichia coli*. J Mol Microbiol Biotechnol 1: 193–202.

48. Mackey BM (2000) Injured bacteria. In: Lund BM, Baird-Parker TC, Gould GW, editors. The Microbiological Safety and Quality of Food. Gaithersburg: Aspen Publisher, Inc. pp. 315–341.

49. Mackey BM, Derrick CM (1982) The effect of sublethal injury by heating, freezing, drying and gamma-radiation on the duration of the lag phase of *Salmonella typhimurium*. J Appl Microbiol 53: 243–251. doi: 10.1111/j.1365-2672.1982.tb04683.x

50. Shin J-K, Pyun Y-R (1997) Inactivation of *Lactobacillus plantarum* by pulsed-microwave irradiation. J Food Sci 62: 163–166. doi: 10.1111/j.1365-2621.1997.tb04391.x

51. Hucl T, Gallmeier E, Kern SE (2007) Distinguishing rational from irrational applications of pharmacogenetic synergies from the bench to clinical trials. Cell Cycle 6: 1336–1341. doi: 10.4161/cc.6.11.4359

52. Burt S (2004) Essential oils: their antibacterial properties and potential applications in foods--a review. Int J Food Microbiol 94: 223–253. doi: 10.1016/j.ijfoodmicro.2004.03.022

53. Gould GW (1989) Heat induced injury and inactivation. In: Gould GW, editor. Mechanisms of action of food preservation procedures. London: Elsevier Applied Science. pp. 11–42.

54. Mañas P, Pagán R (2005) Microbial inactivation by new technologies of food preservation. J Appl Microbiol 98: 1387–1399. doi: 10.1111/j.1365-2672.2005.02561.x

55. Tamm LK, Hong H, Liang B (2004) Folding and assembly of β-barrel membrane proteins. BBA-Biomembranes 1666: 250–263. doi: 10.1016/j.bbamem.2004.06.011

56. Silhavy TJ, Kahne D, Walker S (2010) The bacterial cell envelope. Cold Spring Harb Perspect Biol 2 doi: 10.1101/cshperspect.a000414

57. Gmeiner J, Schlecht S (1980) Molecular composition of the outer membrane of *Escherichia coli* and the importance of protein-lipopolysaccharide interactions. Arch Microbiol 127: 81–86. doi: 10.1007/bf00428010

58. Braun M, Silhavy TJ (2002) Imp/OstA is required for cell envelope biogenesis in *Escherichia coli*. Mol Microbiol 45: 1289–1302. doi: 10.1046/j.1365-2958.2002.03091.x

59. Wu T, McCandlish AC, Gronenberg LS, Chng S-S, Silhavy TJ, et al. (2006) Identification of a protein complex that assembles lipopolysaccharide in the outer membrane of *Escherichia coli*. Proc Natl Acad Sci U S A 103: 11754–11759. doi: 10.1073/pnas.0604744103

60. Food and Drug Administration (2001) Hazard analysis and critical control point (HACCP): procedures for the safe and sanitary processing and importing of juice: final rule (21 CFR Part 120). Fed. Regist., vol. 66 . Washington, D. C.: U.S. Food and Drug Administration.pp. 6137–6202.

61. Rangel JM, Sparling PH, Crowe C, Griffin PM, Swerdlow DL (2005) Epidemiology of *Escherichia coli* O157:H7 outbreaks, United States, 1982–2002. Emerg Infect Dis 11: 603–609. doi: 10.3201/eid1104.040739

62. Hayashi T, Makino K, Ohnishi M, Kurokawa K, Ishii K, et al. (2001) Complete genome sequence of enterohemorrhagic *Escherichia coli* O157:H7 and genomic comparison with a laboratory strain K-12. DNA Res 8: 11–22. doi: 10.1093/dnares/8.1.11

63. Perna NT, Plunkett G 3rd, Burland V, Mau B, Glasner JD, et al (2001) Genome sequence of enterohaemorrhagic *Escherichia coli* O157:H7. Nature 409: 529–533. doi: 10.1038/35054089

64. Food and Drug Administration (Revised 2011) GRAS – Essential oils, oleoresins (solvent-free), and natural extractives (including distillates). 21CFR182.20.

65. Food and Drug Administration (Revised 2012) GRAS – Synthetic flavoring substances and adjuvants. 21CFR182.60

CITATION

CHAPTER 1

Christian W. Huck (2012). Novel Analytical Tools for Quality Control in Food Science, Latest Research into Quality Control, Dr. Isin Akyar (Ed.), ISBN: 978-953-51-0868-9, InTech, DOI: 10.5772/51915.

CHAPTER 2

G.Z. Qin, S.P. Tian, Y. Xu, Z.L Chan, B.Q. Li, Combination of antagonistic yeasts with two food additives for control of brown rot caused by Monilinia fructicola on sweet cherry fruit, DOI: 10.1111/j.1365-2672.2005.02821.x

CHAPTER 3

M. Iammarino and A. Taranto, "Development and Validation of an Ion Chromatography Method for the Simultaneous Determination of Seven Food Additives in Cheeses," Journal of Analytical Sciences, Methods and Instrumentation, Vol. 3 No. 3A, 2013, pp. 30-37. doi: 10.4236/jasmi.2013.33A005.

CHAPTER 4

S. Roller (1991) 5 the Biotechnological Development of New Food Preservatives, Biotechnology and Genetic Engineering Reviews, 9:1, 183-206, DOI: 10.1080/02648725.1991.10750002

CHAPTER 5

Renato Souza Cruz, Geany Peruch Camilloto and Ana Clarissa dos Santos Pires (2012). Oxygen Scavengers: An Approach on Food Preservation, Structure and Function of Food Engineering, Prof. Ayman Amer Eissa (Ed.), ISBN: 978-953-51-0695-1, InTech, DOI: 10.5772/48453.

CHAPTER 6

Bürge Aşçı, Şule Dinç Zor, and Özlem Aksu Dönmez, "Development and Validation of HPLC Method for the Simultaneous Determination of Five Food Additives and Caffeine in Soft Drinks," International Journal of Analytical Chemistry, vol. 2016, Article ID 2879406, 8 pages, 2016. doi:10.1155/2016/2879406

CHAPTER 7

Eberhard Ritz, Kai Hahn, Markus Ketteler, Martin K. Kuhlmann, Johannes Mann, Phosphate Additives in Food—a Health Risk, DOI: 10.3238/arztebl.2012.0049

CHAPTER 8

Emad I. Hussein, Ghassan J. M. Kanan, Khalid M. Al- Batayneh, Khalaf Alhussaen, Wesam Al Khateeb, Janti Qar, Jacob H. Jacob, Riyadh Muhaidat and Mohamed I. Hegazy, 2012. Evaluation of Food Preservatives, Low Toxicity Chemicals, Liquid Fractions of Plant Extracts and their Combinations as Alternative Options for Controlling Citrus Post-harvest Green and Blue Moulds in vitro. Research Journal of Medicinal Plants, 6: 551-573. DOI: 10.3923/rjmp.2012.551.573

CHAPTER 9

Courage Kosi Setsoafia Saba , 2015. Proliferation of Illegal and Potentially Hazardous Food Additives in Processed and Packaged Foods in Africa: A Case Study and Hazard Identification in Ghana. Journal of Food Resource Science, 4: 73-81. DOI: 10.3923/jfrs.2015.73.81

CHAPTER 10

Hélène Barreteau, Cédric Delattre and Philippe Michaud, Production of Oligosaccharides as Promising New Food Additive Generation, ISSN 1330-9862

CHAPTER 11

Josefa Bastida-Rodríguez, "The Food Additive Polyglycerol Polyricinoleate (E-476): Structure, Applications, and Production Methods," ISRN Chemical Engineering, vol. 2013, Article ID 124767, 21 pages, 2013. doi:10.1155/2013/124767

CHAPTER 12

Espina L, Gelaw TK, de Lamo-Castellví S, Pagán R, García-Gonzalo D (2013) Mechanism of Bacterial Inactivation by (+)-Limonene and Its Potential Use in Food Preservation Combined Processes. PLoS ONE 8(2): e56769. doi:10.1371/journal.pone.0056769

INDEX

Y